Science, Technology and Innovation Studies

Science, technology and innovation (STI) studies are interrelated, as are STI policies and policy studies. This series of books aims to contribute to improved understanding of these interrelations. Their importance has become more widely recognized, as the role of innovation in driving economic development and fostering societal welfare has become almost conventional wisdom. Interdisciplinary in coverage, the series focuses on the links between STI, business, and the broader economy and society. The series includes conceptual and empirical contributions, which aim to extend our theoretical grasp while offering practical relevance. Relevant topics include the economic and social impacts of STI, STI policy design and implementation, technology and innovation management, entrepreneurship (and related policies), foresight studies, and analysis of emerging technologies. The series is addressed to professionals in research and teaching, consultancies and industry, government and international organizations.

Ozcan Saritas • Alexander Sokolov •
Mlungisi Cele
Editors

21st Century Foresight

Shaping the Future for Sustainable Social, Economic and Environmental Development in South Africa

Springer

Editors
Ozcan Saritas
Graduate Programs and Research Department
Rochester Institute of Technology Dubai (RIT Dubai)
Dubai, United Arab Emirates

Alexander Sokolov
National Research University Higher School of Economics
Moscow, Russia

Mlungisi Cele
National Advisory Council on Innovation
Pretoria, South Africa

ISSN 2570-1509 ISSN 2570-1517 (electronic)
Science, Technology and Innovation Studies
ISBN 978-3-031-69556-8 ISBN 978-3-031-69557-5 (eBook)
https://doi.org/10.1007/978-3-031-69557-5

This Springer imprint is published by the registered company Springer Nature Switzerland AG
The registered company address is: Gewerbestrasse 11, 6330 Cham, Switzerland

Contents

About the Editors

Ozcan Saritas is the Director of Future Foresight at the Rochester Institute of Technology (RIT) in Dubai, UAE; and Leading Research Fellow at the HSE University, Moscow. He holds the title of Honorary Professor at the Manchester Institute of Innovation Research (MIoIR), The University of Manchester. Furthermore, Professor Saritas is the Editor-in-chief of "Foresight," the journal futures studies, strategic thinking, and policy. With a doctoral degree from the Foresight and Prospective Studies Program at Manchester Business School, Professor Saritas has gained international recognition in Foresight and Futures Studies, Innovation, and Strategy. His expertise extends across a broad spectrum of qualitative and quantitative approaches, such as horizon scanning, scenario planning, strategic roadmapping, big data analytics, and utilizing collective human-artificial intelligence (CHAI) for Foresight. Professor Saritas collaborates with prominent international organizations, including the United Nations (UNIDO, UNCTAD, UNDP), the Organization for Economic Co-operation and Development (OECD), the World Bank, and the European Commission. He is advising national governments as well as leading corporations in the development of their long-term innovation policies and strategies. E-mail: oxscad1@rit.edu; osaritas@hse.ru.

Alexander Sokolov is Deputy Director of HSE ISSEK and Director of HSE International Foresight Center. His main professional interests are related to Foresight, STI priorities, indicators and policies. Prof. Sokolov is tenure professor at HSE, he teaches Foresight for undergraduate and postgraduate students. He authored over 120 publications in Russia and internationally and managed many Foresight projects, including: Russian S&T Foresight: 2030; Foresight for Russian ICT sector (2012); Innovation Priorities for the Sector of Natural Resources (2008–2010); Russian S&T Delphi Study: 2025 (2007–2008); Russian Critical Technologies (2009) et al. Prof. Sokolov is member of several high-level working groups at OECD and other international organizations, serves for advisory boards at several international conferences and journals. E-mail: sokolov@hse.ru.

Mlungisi Cele studied, worked and developed passion and interest and networks in science, technology and innovation policy, and higher education policy research. He holds a PhD in Higher Education Studies and an MPhil in Science and Technology Studies. He worked as a researcher at the former University of Durban-Westville (now part of the University of KwaZulu Natal) and University of the Western Cape. Academic outputs include edited book entitled innovation policy at the intersection: global and local experiences" (in press). Also presented at local and international academic conferences. Dr. Cele is serving as an Acting Head of the National Advisory Council on Innovation (NACI), which is a statutory advisory body to the South African government on STI issues. Some of his policy contributions include providing leadership to the following initiatives: the review of the 1996 White Paper on Science and Technology, development of the high-level framework for the development of new STI decadal plan, the 2030 National Foresight Exercise on STI, the development of Monitoring and Evaluation Framework for the National System of Innovation (NSI) and review of the National Research and Development Strategy and Ten-Year Innovation Plan, which is underway. In addition, Dr. Cele represents South Africa in various multilateral forums and think tanks. E-mail: Mlungisi.Cele@dst.gov.za.

List of Abbreviations

Terms

4IR	Fourth Industrial Revolution
AI	Artificial Intelligence
BDA	Big Data Analysis
CHAI	Collective Human Artificial Intelligence
ICT	Information and Communication Technologies
iFORA	Intelligent Foresight Analytics System (HSE)
IoT	Internet of Things
NSI	National System of Innovation
R&D	Research and Development
S&T	Science and Technology
SAForSTI	South African Foresight for Science, Technology and Innovation
SDG	Sustainable Development Goal
SFM	Systemic Foresight Methodology
STEEPV	Social, Technological, Economic, Environmental, Political and Value
STI	Science, Technology, and Innovation

Institutions

HSE	National Research University Higher School of Economics (Moscow)
ISSEK	Institute for Statistical Studies and Economics of Knowledge, National Research University Higher School of Economics
NACI	National Advisory Council on Innovation (South Africa)
OECD	Organization for Economic Co-operation and Development
UN	United Nations
UNESCO	United Nations Educational, Scientific and Cultural Organization

Introduction

Ozcan Saritas, Alexander Sokolov, and Mlungisi Cele

Abstract "21st Century Foresight: Shaping the Future of Sustainable Social, Economic & Environmental Development in South Africa" introduces a pioneering methodology for Foresight, blending qualitative and quantitative methods with big data analysis to drive science, technology and innovation (STI) policies. Through a comprehensive national Foresight study in South Africa, this book offers guidelines for practitioners and policymakers, aiming to enhance living conditions and advance Sustainable Development Goals (SDGs). The South African case study demonstrates how social and economic challenges can be tackled inclusively and sustainably. With a focus on emerging and developing economies, the book presents novel methodological innovations applicable to global challenges of the twenty-first century. Divided into four parts, the book delves into foundational aspects of foresight, showcases the South African case study, explores key STI domains and concludes with policy implications. The study presented underscores the importance of context, stakeholder engagement, capacity building, adequate resourcing, infrastructure, partnerships and continuous learning in driving successful foresight exercises. Foresight emerges as a vital tool for shaping desired futures, empowering decision-makers and fostering inclusive growth and social justice.
This book presents a novel Foresight methodology with the combination of qualitative and quantitative methods including horizon scanning and scenarios planning, as well as Big Data Analysis with scientometric and semantic methods. The innovative approaches developed throughout the book are demonstrated through a National

O. Saritas (✉)
National Research University Higher School of Economics, Moscow, Russia

Rochester Institute of Technology Dubai, Dubai, United Arab Emirates
e-mail: sokolov@hse.ru

A. Sokolov
National Research University Higher School of Economics, Moscow, Russia
e-mail: sokolov@hse.ru

M. Cele
National Advisory Council on Innovation, Pretoria, South Africa
e-mail: mlungisi.cele@dst.gov.za

O. Saritas et al. (eds.), *21st Century Foresight*, Science, Technology and Innovation Studies, https://doi.org/10.1007/978-3-031-69557-5_1

Foresight study in South Africa. The book provides guidelines for practitioners by introducing contemporary ways of formulating Science, Technology and Innovation (STI) policies. Hence, the purpose of the book is to present Foresight for STI in a unique background, innovative methodological contribution, and results and policy recommendations to improve the living conditions and make progress toward meeting the Sustainable Development Goals (SDGs). The case of South African shows how social and economic challenges are addressed in an inclusive and sustainable way. The outputs and outcomes generated by the fully-fledged Foresight study is expected to inform and influence the identification and selection of priorities for the new decadal plan for STI in the next decade. While contributing to the academic and practical discourse by presenting novel methodological innovations applied through the case study, the book comprises a case for emerging and developing economies in Africa and in the world in dealing with the grand challenges of the twenty-first century such as water, food, energy, climate change, health and education among the others.

Foresight studies have been undertaken for the purpose of investigating in the future of Science, Technology and Innovation (STI) and how these can serve for the improvement of quality of life and creation of wealth, while achieving the Sustainable Development Goals (SDGs). Building upon the long tradition of Foresight and making use of Big Data and advanced computing power combined with Machine Learning technologies, the proposed book develops new methodological approaches for Foresight. Conventional qualitative approaches are combined with advance quantitative approaches like scientometric and semantic analysis to enhance the evidence base of Foresight, identify new and emerging trends, develop future scenarios and visions and formulate strategies. The proposed book first introduces novel Foresight approaches. Going further, the book demonstrates these approaches through a fully-fledged National Foresight exercise undertaken for South Africa. This first national Foresight case utilizing Big Data and Machine learning technologies combined with expert methods describe how STI priorities and strategies are formulated toward 2030. The case of South African National STI Foresight 2030 aims at contributing to the following STI objectives of the country:

- Advancing the capacity of the national innovation system to contribute to sustainable socio-economic development;
- Enhancing South Africa's capacity for generating knowledge to produce world class research outputs, as well as innovative products and processes thereafter;
- Developing appropriate human capabilities in the STI sector to meet the needs of society;
- Building a world class infrastructure in the STI sector; and,
- Positioning South Africa as a strategic international R&D and innovation partner and destination.

The case of South Africa is of particular interest as the country is the leading economy in Africa and represents one of the BRICS countries. Besides the methodological innovations, the South African case represents how an emerging economy addresses the challenges and opportunities toward meeting SDGs in the next decade.

Therefore, the book is expected to be a valuable source and guidebook both for the countries in Africa, as well as other developing countries in the world. Overall, the book adds significant value to readers in the public and private sector policy and strategy makers, as well as academic community and practitioners of Foresight and STI policy studies. Chapters on the background and methodology of the Foresight study are complemented by chapters on specific STI domains investigated in South Africa. The book is concluded with a synthesis chapter, which summarizes future recommendations and implications for policy and practitioners.

Following this first Introduction chapter, the rest of the book consists of four parts:

1. Introduction to Foresight for Science, Technology and Innovation.
2. Using Foresight for Science, Technology and Innovation: The Case of South Africa.
3. Science, Technology and Innovation Domains for South Africa.
4. Policy Implications.

The First Part on "Introduction to Foresight for Science, Technology and Innovation", which provides a background to the book, establishes a solid foundation for Foresight for Science, Technology and Innovation with methodological advancements and best practices. This provides a background to the Foresight study undertaken in South Africa as well as a guide for future Science, Technology and Innovation (STI) Foresight studies in Africa and other emerging economies in the world.

- Second Chapter, Foresight in the Twenty-First Century and Methodological Advancements, by Ozcan Saritas makes an assessment of Foresight activities over the decades of theoretical practical evolution. He argues the increasing needs for Systems Thinking in an ever complex and dynamic world and sheds light into the future methodological developments in Foresight.
- Third Chapter, Foresight for Science, Technology and Innovation Policy: Relevance and Uses for Emerging Economies, by Alexander Sokolov describes the relevance and uses of Foresight for emerging economies in formulating their STI policies and presents best practices of international and national STI Foresight activities.
- Fourth chapter, Foresight Methodology and Commonly Used Methods in the New Generation Foresight, by Ozcan Saritas, Natalia Veselitskaya and Alexander Sokolov provides an overview of significant qualitative and quantitative Foresight methods with recent methodological advancements. Fifteen Foresight methods are presented, which are vital for designing and implementing Foresight and many of which were used in the South African STI Foresight exercise.

The Second Part of the book, "Using Foresight for Science, Technology and Innovation: The Case of South Africa", describes one of the most recent Foresight exercises for national STI policy making. The South African case is an illustrative case for a number of other economies in Africa and other parts of the developing world as a fully-fledged Foresight exercise with policy impact and methodological advancements.

- Fifth Chapter, South African Foresight for STI (SAForSTI): Background, Rationales, and Objectives, by Ozcan Saritas gives a summary of the SAForSTI Foresight exercise.
- Sixth Chapter, Research Landscape of South Africa: Current Trends in Research Output, Thematic Focus and Scientific partnerships, by Ozcan Saritas and Maxim Kotsemir provides a comprehensive analysis of research landscape to provide necessary capacity for STI transformation in South Africa.
- Seventh Chapter, Using Big Data for Foresight: Scientometric and Semantic Analysis for South Africa, by Ozcan Saritas and Maxim Kotsemir shows the key competences of the country in knowledge generation and emerging topics under each of the domains covered by the SAForSTI exercise.

The Third Part of the book, "Science, Technology and Innovation Domains for South Africa," focuses on the key industries and topics covered by the SAForSTI Foresight exercise. Each chapter gives an overview, results and policy recommendations generated through the Foresight process.

- Eighth Chapter, The Circular Economy, by Petrus Letaba discusses the future of the Circular Economy in South Africa.
- Ninth Chapter on Educating for 2030 by Nisha Sewdass focuses on the future of education to achieve South African National Development plan 2030.
- Tenth Chapter on Sustainable Energy by Mkhulu Mathe considers the future of the Energy sector in South Africa within the framework of the seventh UN Sustainable Development Goal, energy access for all.
- Eleventh Chapter, the Future of Society, by Ozcan Saritas discusses the meta-dimension of inequalities in accessing to technology in South Africa, and how challenges in the enabling environment can be overcome.
- Twelfth Chapter, Health Innovation, by Michael Kahn considers health as essential to social and economic development and focuses on technological solutions for health innovations.
- Thirteenth Chapter on High-Tech Industrialization by Fiona Tregenna and Alexis Habiyaremye examines the emerging trends in South Africa and internationally as well as future challenges and opportunities leading to the actions that are needed to be taken now in the country.
- Fourteenth Chapter, ICTs and Smart Systems, by Judy Backhouse discusses policies and strategies for South Africa to optimally benefit from Information and Communication Technologies (ICTs) in the future.
- Fifteenth Chapter on Food Security by Jennifer Thomson seeks answers for the question of how agriculture, and specifically precision agriculture, can contribute to increasing food security in South Africa.
- Sixteenth Chapter, Water Security, by Ozcan Saritas focuses on reducing water wastage, water pollution and water usage, with particular emphasis on the implications for rural areas.

The Fourth Part of the book Policy Implications concludes the book. Hence,

- Seventeenth Chapter of the book by Mlungisi Cele, the Future of Foresight Exercises for Science, Technology and Innovation in Africa, seeks to draw lessons from the SAForSTI 2030 exercise to inform the future implementation of science, technology and innovation foresight exercises in Africa.

Through the invaluable contributions of the aforementioned authors into this book, the editors conclude that Foresight is an invaluable instrument for the governments of developing countries to construct their own narratives of a desired future, instead of relying on futures foreseen by highly developed countries, which are not necessarily relevant to the contexts of developing countries. Foresight enables public service organizations to frame future policy environments better and present decision-makers with more and better choices for inclusive growth and social justice. Participatory foresight frees countries from over-reliance on foreign or local technical experts and invites citizens to contribute to discussions and decisions concerning their future for a desirable future for all.

Acknowledgments This book was prepared within the framework of the Basic Research Program at the HSE University and supported within the framework of the subsidy granted to the HSE by the Government of the Russian Federation for the implementation of the Global Competitiveness Program.

Part I
Introduction to Foresight for Science, Technology and Innovation

Foresight in the Twenty-First Century and Methodological Advancements

Ozcan Saritas

Abstract The evolution of Foresight has witnessed a transformative journey from individual foresight to the Institutional Foresight known today. Originating in defense-oriented activities post-World War II, foresight has adapted to address diverse challenges, incorporating a shift from forecasting to scenario thinking. The 1980s marked a turn toward innovation systems, emphasizing collaboration across sectors. As global challenges like climate change emerged, foresight exercises became inclusive, involving various stakeholders. With the introduction of Systemic Foresight Methodology, the approach became interconnected, acknowledging the systemic nature of the world. Foresight, encompassing multiple time horizons, evolved to address immediate as well as long-term futures. Beyond the traditional strongholds of the USA and Europe, nations like Japan, Australia, Asia, the Middle East, and Africa have embraced foresight. Toward the Systemic Foresight Methodology (SFM), a conceptual framework emphasizing systems thinking, addresses the complexity and uncertainty of the future. SFM involves iterative phases such as initiation, intelligence, imagination, integration, interpretation, implementation, impact, and interaction. The systemic perspective highlights interconnected actors, emphasizing cooperation and conflicts. The 21st century Foresight envisions new horizons shaped by technologies like Artificial Intelligence, blockchain, and virtual reality. Foresight-on-site advocates incorporating foresight into daily routines, making it a collective endeavor involving diverse stakeholders. The era of "Collective Human Artificial Intelligence" (CHAI) underlines the synergy between human creativity and machine analytics in shaping foresight, posing questions about what is possible, feasible, and desirable for the future.

O. Saritas (✉)
Rochester Institute of Technology Dubai, Dubai, UAE

National Research University Higher School of Economics, Moscow, Russia
e-mail: oxscad1@rit.edu; osaritas@hse.ru

O. Saritas et al. (eds.), *21st Century Foresight*, Science, Technology and Innovation Studies, https://doi.org/10.1007/978-3-031-69557-5_2

1 The Evolution of Foresight

Foresight is a concept that is spreading rapidly to the extent that it can now be seen everywhere globally. If the idea of Foresight seems familiar, it is because it has been around for several decades. Individual Foresight has always existed as a human trait of looking ahead, and earlier systematic foresight applications were seen during the First and Second Industrial Revolutions.

Foresight took many different turns before it became the Institutional Foresight we know today. However, regarding the foresight activities that we speak of today, the history is relatively newer, dating back to the 1950s during the post-World War II era. At the time, the activities were mostly defense-oriented, focusing on technological advancement in defense. Another example is the Delphi forecasting method, which rose to prominence around the same period. Over time, this activity has changed and developed significantly.

The definition of Foresight has evolved over several decades, with several new challenges and opportunities. Oil shocks marked the 1970s, and it was at this time that we understood we could not wholly anticipate the future. Future events are unclear, so scenario thinking became more popular around this period. We began to believe that, indeed, the future is unpredictable. Multiple futures have been incorporated as a new element to our foresight research since the future is genuinely multiple, and there are alternative futures.

As time passed, new issues regarding innovation systems began to surface around the 1980s. The word "systems" means that a single player cannot start innovation and must involve a group effort from the public sector, business, or academic institutions. This was when the system concept was introduced into science, technology, innovation, and policymaking.

Foresight exercises were added to the mix to make it more inclusive and participatory at that point in time. The focus of foresight exercises has shifted from forecasting the most likely future to considering multiple futures that involve multiple actors, i.e., social actors, financial institutions, technology developers, educational institutions, trade, commerce, and many more stakeholders. As we encounter increasing difficulties, such as the effects of climate change, energy, water, and food problems, we have begun to realize that some of these difficulties call for international cooperation and transcend the borders or spheres of influence of any one country.

Different aspects regarding technologies, science, and innovation were investigated to address not only defense issues or economic issues but also broader societal issues, environmental issues, and cultural transformations. Because of how quickly the world is evolving, with the new pandemic in 2020 being a clear example, foresight evolution is never-ending. Regarding policies, our policymaking agenda and our exercise in Foresight have recently included systems thinking (Saritas, 2006). For this reason, the Systemic Foresight Methodology (Saritas, 2013) concept was introduced, and Systemic Foresight exercises (Saritas & Cheah, 2020) have taken place. Systemic Foresight suggests that the world should be viewed as an interconnected system, and we want to know what kinds of synergies result from this. Positive and negative synergies emerge as an interaction of these societal, technological, economic, and political issues, etc. (Miles et al., 2016).

The systemic view shows that all actors are interconnected and have simultaneous possibilities for cooperation and conflicts. To incorporate all facets and stages of the foresight process, we need to view the actors from a systemic perspective and employ a systemic methodology. Since the world is systemic, Foresight should be similarly systemic to bring different activities together, i.e., scanning for horizons, planning scenarios, articulating visions, setting priorities, and then translating those priorities into strategies, moving to actions, and monitoring those actions.

Foresight is about futures, looking at multiple time horizons due to different speeds of changes in the systems constructing our world. Given the pace of change, it is crucial to plan for the immediate future utilizing Foresight. As we think about not only the long-term future, e.g., several decades ahead and so on, foresight exercises should also be concerned with next year, next week, and maybe tomorrow even due to the increasing complexity and uncertainty, which may also arise in rapidly developing situations just as in the case of the Covid-19 pandemic.

Foresight efforts are becoming more prevalent outside of the USA and Europe than historically. Japan has been one of the nations with a solid commitment to Foresight, and over the past 50 years, has produced some incredibly excellent foresight work. Foresight is widespread in Australia, Asia, the Middle East, and Africa. The world community is currently discussing Foresight, being concerned about the future, and creating innovative approaches to anticipate this future and build strong and resilient policies and actions to deal with our rising difficulties.

2 Toward the Systemic Foresight Methodology (SFM)

For transformations in the global context and technological advancements, new and more systemic approaches are required. Observing this necessity, the Systemic Foresight Methodology (SFM) has been developed to respond to the increasing complexity and uncertainty of the future to propose a conceptual framework for designing and implementing foresight activities. The fundamental premises of the SFM are established on the basic notions of systems thinking (Saritas, 2006).

The SFM considers situations by recognizing the interconnections and interdependencies between different events, actors, and processes. The holistic view of the events suggests that social, technological, economic, ecological, political, and legal (STEEPV) systems coexist and are interlinked in cascaded governance systems from global to international, national to regional, and sectoral to thematic levels. Actors within these systems are networked with each other, and their systemic and synchronized action is needed to achieve any successful change process while meeting the expectations of the stakeholders with power, urgency, and legitimacy.

The process of the SFM is also systemic in that the proposed methodology combines divergent and convergent methods with exploratory, normative, and action-oriented thinking. The SFM process can be viewed as a series of phases, including:

(a) Initiation (Scoping)
(b) Intelligence (Scanning)
(c) Imagination (Scenarios)
(d) Integration (Setting priorities)
(e) Interpretation (Strategies)
(f) Implementation (Steps & Actions)
(g) Impact (Evaluation) with a continuous.
(h) Interaction (Stakeholders) across the process.

Briefly, the initiation phase of SFM is concerned with the scope of the foresight exercise. Major decisions are made on the scope and coverage of the activity. This phase examines the goals and specific objectives of the activity, its scope and coverage, intended uses, and users of its outputs and outcomes. Initiation is critical as the boundaries of the foresight exercise are determined in this phase. Steps a to g are iterative and can be repeated as often until the practitioners conclude that their complete function has been fulfilled. The eighth step (h) represents the participatory character of Foresight, which is applied throughout the entire process of Foresight. The overall process investigates how systems (whether human and social systems or industrial and innovation systems) are understood, approached, and intervened for a successful change program. Therefore, this can also be considered as an initial prioritization stage. Decisions on the process and methodology of the exercise, stakeholders to be engaged, key milestones, deliverables, and dissemination activities are also made at this stage.

Following the preparatory Initiation phase, the actual SFM process begins with the Intelligence phase. This may also be considered as the scanning and surveying phase. This phase aims to comprehensively analyze systems and situations the foresight exercise is concerned with. Foresight activities usually deal with complex systems with many interacting elements. Both environmental and horizon scanning activities are conducted to understand trends, drivers of change, weak signals of emerging developments, wild cards/surprises/shocks, and discontinuities (Saritas & Smith, 2011). A STEEPV framework is commonly used to scan various political, economic, social, technological, environmental, and legal issues. This phase typically identifies and prioritizes important problems and uncertainties, which will shape the future.

Imagination is the creative and innovative phase of SFM where future-oriented divergent thinking is applied to explore alternative futures. Unlike future prediction and forecasting, Foresight explores alternative futures and develops normative visions and priorities. The input generated from the Intelligence phase is synthesized as future narratives in the form of scenarios and models of the future. These give a complete and more systemic picture of the future by showing the interplay between different systems and their elements. Qualitative and quantitative methods such as scenario planning, modeling, gaming, system dynamics, and simulation are used to help explore alternative futures and make assessments of their impacts. The use of weak signals and wild cards from the Intelligence phase may help test systems' adaptive capacity under extreme conditions, surprises, and shocks.

Integration is the phase where alternative futures are appraised, normative visions are articulated, and priorities are identified. This process involves thoroughly analyzing the scenarios and models developed in the previous phase. This is a multifaceted process, where the claims of different stakeholders must be considered adequately to ensure that future systems designed are resilient in the long term. The end product of this phase is an agreed model of the future, which will imply the targets to be achieved within the time horizon set for the foresight activity.

The Interpretation phase is primarily considered as the strategy phase. The earlier phases of SFM extensively discussed the question of "where we are" at the intelligence phase. Then, during the Imagination and Integration phases, the "where we want to go" question was answered. Hence, the Interpretation phase fills the gap between where we are and where we want to go, with the question of "how to get there." The gap between the future and the present is filled in this phase by suggesting transitions and transformations needed in the long, medium, and short terms. Thus, this phase represents a more "convergent" phase. The outputs of this phase are usually expected to be in the form of strategic roadmaps.

Foresight exercises are expected to inform policies and actions. Therefore, the Intervention phase of the SFM is concerned with the actions, where formulated priorities, strategies, and follow-up steps are outlined and communicated with the key actors and stakeholders. The aim is to inform present-day decisions concerning immediate change actions to begin structural and behavioral transformations and to influence and shape broader contexts to add value to the viability and development of systems.

Impact is the phase where Foresight is evaluated and embedded to understand the extent to which the activity has achieved its objectives, the impact generated, and the extended activities that should be laid out to move forward. The Foresight process requires substantive investments, often through public funding, and implies considerable time, logistics, and human resources costs. If the impacts of Foresight cannot be made clear throughout the activity, the commitment to dedicating resources will decrease, and as a result, the activity will be discontinued. This phase examines the impacts of Foresight during the process of implementation (e.g., production of baseline reports, articulation of visions, and building new linkages), immediately after the completion of the activity (e.g., new integrated projects and programs), and sometimes later (e.g., innovation impacts and new working communities).

Interaction is one of the main distinguishing features of Foresight from other long-range planning activities, as the activity is inclusive and participative. Unlike the other phases, Interaction should be observed throughout the foresight process. A thorough stakeholder analysis is essential for any foresight exercise to ensure that all legitimate viewpoints are reflected in the process. The SFM recognizes inclusiveness and equity through freedom of association and expression and the role of the democratic society. This phase emphasizes the need for effectiveness and efficiency in meeting the stakeholders' expectations and sustainable use of resources, and therefore aims to develop mechanisms to provide engagement of system actors ranging from policymakers to industry, research to higher education, as well as associations and NGOs among the others.

The ForSTAR model developed by Saritas (2013) brings together the phases of the Systemic Foresight Methodology (SFM) described earlier with the most frequently used methods in each phase (Fig. 1).

It is important to remember that the phases from a to g are iterative and can be repeated as many times as possible until the practitioners conclude that their complete function has been fulfilled. The eighth step (h) represents the participatory character of Foresight, which is applied throughout the entire process of Foresight. The outcome of the process is Innovations as illustrated in the figure. The overall process investigates how systems (whether human and social systems, or industrial and innovation systems) are understood, approached and intervened for a successful change program.

3 The Twenty-First Century Foresight: New Horizons

Artificial Intelligence and other newly emerging and evolving technologies give the first signals of the next generation of Foresight will be characterized by human–technology interaction and creativity. In this respect, it is expected the whole process of Foresight will be affected using advanced analytics, Artificial Intelligence, human–machine systems, artificial and virtual reality technologies, robotics, new production technologies, and social networks, as well as blockchain technologies. Burmaoglu et al. (2020) describe the use of Blockchain for Foresight and Science and Technology Policymaking and introduce the concept of "Ideachain."

We must first comprehend how the functions of the policymakers have evolved. For a large portion of it, the mechanisms for making policies shifted from a top-down approach to a more bottom-up one. We often require evidence to persuade the stakeholders to participate in the policymaking process as well as in the execution of the policies. The results will be far more successful if relevant qualitative and quantitative evidence is used to support a policy and necessary stakeholders are involved. The stakeholder base is broadening now. Stakeholders include entities in a broader ecosystem, like nature and the oceans. Foresight is now becoming an activity as part of an everyday routine. Not only should Foresight be on our agenda as researchers or practitioners, but it should also be on the agenda of all stakeholders in Foresight. Hence, the term Foresight-on-site involves incorporating Foresight into people's routines and agendas to foster this kind of future-focused thinking. It needs to be stressed that Foresight exercise should not be about bringing people, experts, and stakeholders into the foresight process but rather the other way around. Regarding research, we should ensure that foresight thinking permeates scientific labs and daily operations as a scientist concentrates on specific research or technological development. From the very beginning of their study, scientists are the ones

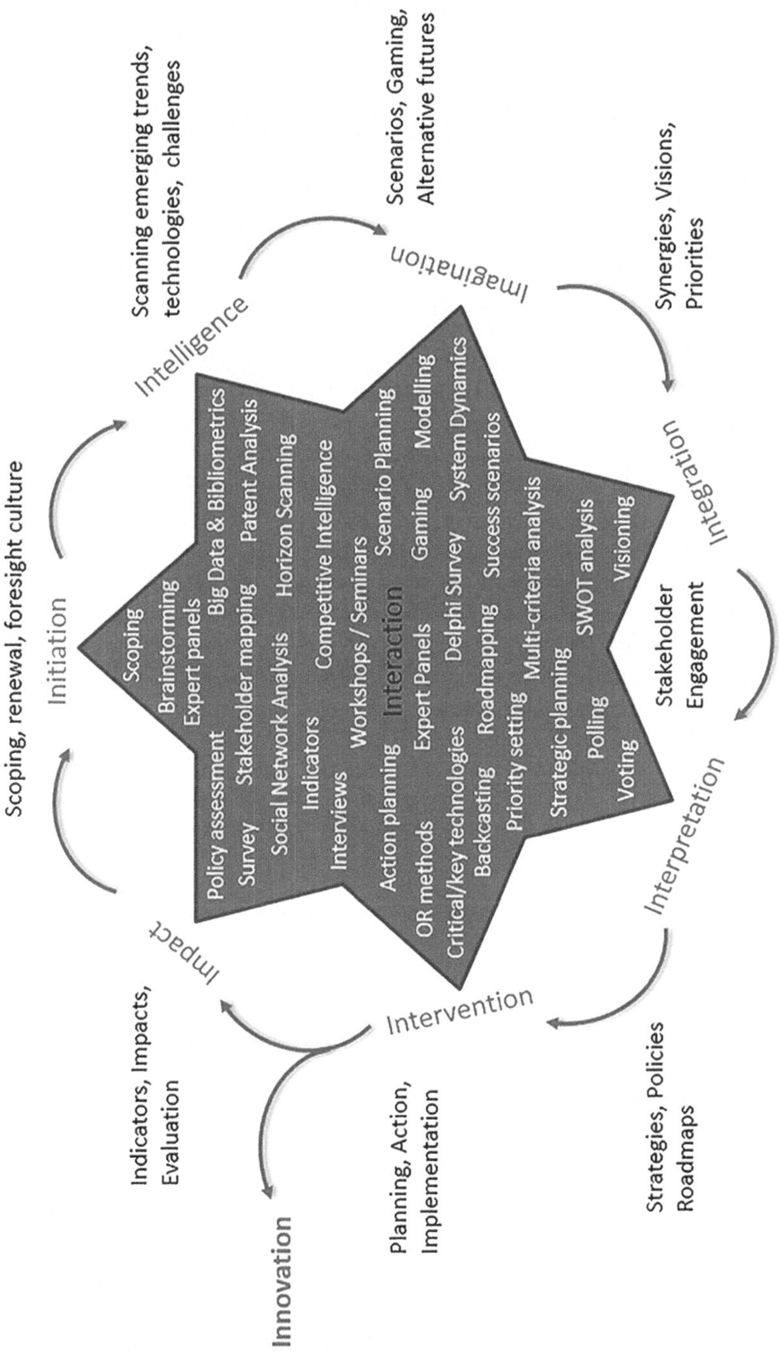

Fig. 1 Systemic Foresight Methodology, its phases, and frequently used methods in each phase

who consider the potential consequences of these technologies and their future implications.

Increasing "**futures and data mastery**" is another significant element in Foresight. We must ensure that people understand what the future is, what futures exist, how to find data about the future, and how to interpret it as part of their daily activities if we are to assist these systems. Data is now widely available and constantly being produced. Because it provides us with a wealth of information on the direction that advances are taking, we should ensure that we can use and monitor this data. Because of this, policymaking cannot proceed in the same manner as in the past, examining only a small number of concerns, primarily economic, developmental, and so forth.

These days, our systems are becoming a lot smarter. This brings about a new concept called "**Collective Human Artificial Intelligence" (CHAI)**. Human intellect is undoubtedly crucial when it comes to Foresight. We are creative beings who employ both logic and emotion simultaneously. However, machines also allow us to analyze vast amounts of data, comprehend dynamics, and monitor and track all those changes. Not only should Foresight involve anticipating the future and emerging technologies, but it should also involve understanding the effects of those developments. Three questions must be answered at the outset of any foresight exercise to consider all its facets as we build our science and determine whether it is viable. The first one is "What is possible?" Nature and science will tell us what is possible. The second question is, "What is feasible?" Technology and the economy will tell us what is feasible. And the third question is, "What is desirable?" Society, culture, and politics will tell us what is desirable.

Acknowledgments This chapter was prepared within the framework of the Basic Research Program at the HSE University and supported within the framework of the subsidy granted to the HSE by the Government of the Russian Federation for the implementation of the Global Competitiveness Program.

References

Burmaoglu, S., Saritas, O., & Sesen, H. (2020). IdeaChain: A conceptual proposal for blockchain-based STI policy development. *Foresight, 22*(2), 189–204.

Miles, I., Saritas, O., & Sokolov, A. (2016). *Foresight for science, technology and innovation*. Springer Verlag.

Saritas, O. (2006). Systems thinking for foresight, Ph.D. Thesis, PhD thesis for Manchester Business School, The University of Manchester.

Saritas, O. (2013). In D. Meissner, L. Gokhberg, & A. Sokolov (Eds.), *Systemic foresight methodology, in science, technology and innovation policy for the future: Potentials and limits of foresight studies* (pp. 83–117). Springer Verlag.

Saritas, O., & Cheah, S. (2020). *Systemic foresight methodology in action, in strategic foresight: Accelerating technological change*. De Gruyter.

Saritas, O., & Smith, J. (2011). The big picture–trends, drivers, wild cards, discontinuities and weak signals. *Futures, 43*(3), 292–312.

Foresight for Science, Technology and Innovation Policy: Relevance and Uses for Emerging Economies

Alexander Sokolov

Abstract The chapter outlines the crucial role of Science, Technology and Innovation (STI) in global economic and societal development, emphasising their contributions to creating industries, generating employment and addressing Grand Challenges. Foresight, defined as a systematic, participatory future-intelligence-gathering process, is presented as an essential tool for supporting policy-making, particularly in developing countries. The chapter explores the impact of STI on economic and social development, underscoring the significance of technological advancements in addressing global challenges. The growing importance of innovation in the competitiveness of nations is highlighted, prompting the need for STI policies. The chapter then delves into the concept of Foresight for STI, emphasising its role in identifying trends, opportunities and challenges. The discussion expands to include the relevance of Foresight for emerging economies, aiding them in overcoming technological gaps and effectively allocating resources. The chapter also addresses the global context, linking STI to the Sustainable Development Goals (SDGs) and Grand Challenges. The aftermath of the COVID-19 pandemic is discussed, indicating its impact on achieving SDGs and the need for foresight studies to guide STI policies. International experiences and best practices in STI foresight, both by developed countries and international organisations, are presented as valuable resources for developing nations. The text concludes with examples of successful STI foresight activities in developing countries and in Africa.

1 Introduction

Science, technology and innovation (STI) play a pivotal role for global economic and societal development. They contribute to creating new industries, generating employment opportunities and responding to Grand Challenges including poverty,

A. Sokolov (✉)
National Research University Higher School of Economics, Moscow, Russia
e-mail: sokolov@hse.ru

O. Saritas et al. (eds.), *21st Century Foresight*, Science, Technology and Innovation Studies, https://doi.org/10.1007/978-3-031-69557-5_3

hunger and diseases. The rapid pace of technological development holds the promise of revolutionising various sectors of the economy, bringing about transformative changes for humanity in the near future. International competitiveness of nations today strongly depends on their ability to deal with advanced technologies. Foresight as "the application of systematic, participatory, future-intelligence-gathering and medium-to-long-term vision building process to informing present-day decisions and mobilising joint actions" (Miles & Keenan, 2002) serves as an efficient tool to support policy-making. It helps to focus limited resources on most promising areas of science and technology, which is extremely important for developing countries.

This chapter presents a broad introduction of the role of Foresight for STI with a particular focus on the effects it can bring to developing countries. It also describes relevant activities of international organisations and best experience of Foresight projects implemented in a number of developing countries.

2 Science, Technology and Innovation as Drivers of Social and Economic Development

Science and Technology (S&T) play a significant role in social and economic development across the world. Technological advancements have dramatically changed various aspects of human life, including communication, transportation, healthcare and entertainment. Science has contributed to the development of newer, greener and more sustainable technologies that enhance the quality of life and improve living standards.

In terms of economic development, technology has created new industries and jobs, stimulated economic growth through innovation and provided greater access to markets and resources. S&T have led to significant breakthroughs in various fields such as agriculture, engineering, medicine and energy, providing solutions to some of the world's most pressing challenges, including poverty, hunger and disease. Fast development of information technologies and artificial intelligence, in particular, its penetration to all sectors of the economy, promises revolutionary changes for all human beings in the near future.

The impact of S&T on social development is also significant. Advancements in healthcare technology have improved life expectancy rates, as well as the quality of life for aged people and individuals with disabilities. Modern technologies provide new opportunities for learning and broaden access to education. Social media and information technology have revolutionised communication, making it easier and faster to connect with people in any region of the world. Overall, S&T play a critical role in social and economic development, providing solutions to many of the world's most pressing problems and improving the quality of life for people worldwide.

Technology plays a crucial role in the competitiveness of nations as it drives innovation, productivity and economic growth. In today's globalised economy, countries must compete on the basis of knowledge, innovation and technology development. Technology enables businesses to produce goods and services more efficiently and effectively, leading to higher productivity, better quality and lower costs. It can also create new industries and new jobs, spurring economic growth. Technological advancements can help countries to create resources, acquire technical expertise and attract foreign investment. They can also improve a nation's supply chain management, transportation and logistics capabilities. Furthermore, the adoption of the latest technologies helps to reduce the production cost and increase the quality of products, thus providing the basis for expanding the nation's niches in the global market.

During the last decades, science, technology and innovation (STI) have been a major source of new knowledge and have been increasingly considered as the key drivers of economic growth and social development. S&T achievements play an important role both for the future of humankind and for meeting the current, practical needs of the economy and society and responding to longer-term Grand Challenges. The evidence of this trend is confirmed by increasing the knowledge intensity of national economies, higher acceptance of science by society, international S&T cooperation and other phenomena (UNESCO, 2021).

Therefore, the better a country is at harnessing technology and the more advanced its technology is, the more competitive it can be in the global marketplace. This leads to constant growth of expenditure on research and development (R&D) both by governments and by businesses. The gross expenditure on R&D–GERD (in constant prices) in 2000–2019 has grown in the USA almost threefold, and in China, it has grown more than 15 times (OECD, 2022). In a number of developed/industrialised countries, GERD exceeds 3% of GDP (USA, Japan, Germany, Sweden, etc.) or even 4% (Israel and Republic of Korea). For most developing countries, this indicator is much lower.

All this calls for formulation and implementation of STI policies at the national level. In developed countries, formulation of such policies already has a rather long tradition, which is related to a large extent to application of the approach based on the concept of National Innovation Systems (NIS) introduced in the 1990s (Freeman, 1987; Nelson, 1993). At the same time, for individual countries, in particular developing ones, relevant institutional settings related to economic development, knowledge production and distribution can be fully different and have to be taken into account (Lundvall, 1992).

Research and innovation capabilities make economies and societies more resilient, but they require long-term investment in R&D, skills and infrastructure (OECD, 2023). Under growing uncertainties and geopolitical turbulence, emerging technologies, which bring both new opportunities and new risks, require well developed STI policies taking account of peculiarities of individual countries. The best practices that proved to be successful in one country could not always be applied in another one. Moreover, different policies can contradict each other; therefore, a coherent policy mix addressing the most essential, systemic failures and targeting economic growth and urgent societal problems is important (Gokhberg & Meissner, 2016). Thus, the "no-one-size-fits-all" approach has to be applied to STI policy development, in particular for developing countries. Moreover, governing emerging technologies require a more general and anticipatory framework addressing concerns through open and inclusive processes, and align innovation trajectories with societal goals, engaging stakeholders in the innovation process at its initial stage.

Under these conditions, an important task for any nation is to identify areas of development and practical application of science and technology that are promising in the long term, ensuring the use of the competitive advantages of domestic producers and solving the most pressing social problems. Taking into account the sharp increase in resource constraints, e.g. ones associated with the global economic crises, the presence of clear guidelines for the S&T development in the country in the medium and long term becomes especially relevant.

In many countries, both developed ones and emerging economies, governments select S&T priorities to ensure the efficient allocation of resources and the achievement of key development goals (Georghiou & Harper, 2011). By selecting S&T priorities, they can focus research and investments on areas where they have a comparative advantage and/or where they can address their most pressing development challenges. This can lead to the development of new technologies, products and services that can enhance economic competitiveness, create jobs and improve quality of life. Additionally, setting priorities can help countries build a critical mass of expertise in key areas and facilitate collaborations among researchers, businesses and other stakeholders. Overall, prioritising science and technology can enable countries to leverage their strengths, stimulate innovation and achieve sustainable economic and social development.

3 Foresight for Science, Technology and Innovation

Since the late 1990s, foresight has been increasingly acknowledged as a valuable instrument of S&T policies addressing key problems of national innovation systems. It helps to identify emerging S&T trends, potential market opportunities and societal challenges, which can be used to drive innovation policies and investment decisions. By undertaking a broad-based and inclusive process, foresight can bring together diverse stakeholders from academia, industry, government and civil society to foster a shared vision for innovation and build consensus on strategic priorities. This can help to align research and innovation agendas with broader societal goals, and promote the efficient use of resources to foster inclusive and sustainable growth. Furthermore, by engaging in foresight activities at the national level, countries can learn from international best practices and develop a shared understanding of global trends and issues that could affect their innovation systems in the future.

The ultimate goal of national foresight exercises is to co-ordinate STI agendas across public and private organisations, industrial and service sectors, and academic disciplines by developing new alliances between the producers and the consumers of knowledge (Meissner & Sokolov, 2013). These projects consider and make visible the processes by which research agendas and priorities are established, the degree and nature of autonomy in the practices of scientists and engineers, the relations of academic disciplines to each other and industrial knowledge, and the ends to which S&T are directed.

Foresight has proved to be an efficient tool to select S&T priorities because it allows countries to anticipate and prepare for the future by identifying potential emerging trends, opportunities and challenges (Miles et al., 2016). Foresight studies engage a broad range of stakeholders to provide a critical assessment of the current state and future prospects of S&T, and identify the areas of focus that offer the greatest potential for growth and development. This approach allows countries to make informed choices about where to invest their resources, and to prioritise R&D areas based on their potential social and economic impact. By using foresight to guide their S&T priorities, countries can maximise returns on investments, promote innovation and competitiveness and position themselves to address emerging challenges and future needs.

4 STI Foresight for Emerging Economies

Developing economies and emerging markets are expected to continue growing relatively fast, given their increasing labour force and expanding markets potential. Therefore, the future vision, better understanding of trends that will drive new markets and technologies that will be the basis of new global competitiveness is extremely important for these countries. Emerging economies need foresight to identify critical technologies that are essential for their development and competitiveness in the global market. Foresight helps less developed countries to anticipate future changes and emerging trends, to develop and implement appropriate policies and strategies to deal with them. Foresight helps emerging economies to overcome the technological and knowledge gaps they face by providing insights into technology trends and potential breakthroughs. It allows them to build on their strengths and identify new areas for investment and innovation.

Moreover, emerging economies often face challenges such as limited resources, weak institutional arrangements and political instability. Foresight can help address these challenges by enabling them to prioritise their needs and allocate resources effectively. By providing a long-term view of technological developments and potential scenarios, foresight can also inform decision-making on policies related to education, innovation and entrepreneurship that will impact the future growth and competitiveness of emerging economies.

In many developing countries, the emphasis of long-term socio-economic development plans is on restructuring the domestic economy towards an innovative development path, increasing competitiveness and contributing significantly to socio-economic growth through science and technology. The growing role of innovation and technological advancements in global socio-economic development poses new challenges for these countries. These challenges include modernising traditional economy sectors by deploying globally oriented, specialised industries, as well as establishing high-tech industries focused on international markets for high-tech products and intellectual services.

The 2030 Agenda for Sustainable Development,[1] adopted by all United Nations member states in 2015, identifies 17 Sustainable Development Goals (SDGs), which require a global partnership in order to provide peace and prosperity for people and the planet, now and into the future. These goals shared by all UN member states address a wide range of Grand Challenges including global hunger, poverty, exploitation of labour, illiteracy, infectious diseases, corruption, racism, migration of the productive workforce, inequality, gender biases, increased toxic emissions, biodiversity loss and climate change. Grand Challenges can be defined as "highly significant yet potentially solvable problems, such as urban poverty, insect borne disease, and global hunger that affect vast numbers of individuals, in often profound ways, and are typically complex with unknown solutions and intertwined technical and social elements" (Eisenhardt et al., 2016). Global problems, such as environment and climate change or ageing population, require a complex and multi-disciplinary approach in finding possible solutions. Being of global nature, they engage many countries, multiple actors from different sectors; therefore, addressing Grand Challenges is impossible without coordinated and coherent policies at different levels—from regional to national and international ones (Amanatidou, 2011). At the same time, developing countries are still lagging behind the developing countries by key development indicators—GDP per capita, infant mortality rate, life expectancy, etc. For example, GDP per capita in North America in 2020 was 59.2 thousand USD, whereas this indicator for East Asia and Pacific was 18.2, for Middle East and North Africa was 15.5, for South Asia was 5.8 and for Sub-Saharan Africa was only 3.6.[2]

The aftermath of COVID-19 has resulted in a significant slowdown in progress towards achieving the Sustainable Development Goals, as reported by the UN (Sachs et al., 2021). The UN also revealed that, due to the pandemic, the global poverty rate is anticipated to reach 7% by 2030, impeding the goal of eradicating poverty. The pandemic has led to the loss of equivalent to 255 million full-time jobs, and child labour has risen to 160 million—the first increase in two decades. Furthermore, almost half of the worldwide population (3.7 billion people) remain offline (Sachs et al., 2021).

Foresight studies play an important role in identifying promising areas of STI development as an integral component of the government STI policy worldwide. Therefore, it is advisable to use the experience gained by developed and, in particular, by industrialised countries and their best practices (both in methodologies and organisation of foresight studies) in developing countries. At the same time, these practices should be critically evaluated and focused on the most important challenges distinctive for the developing world. On this basis, relevant government agencies can form scientific and technical programmes, select priorities for research

---

[1] https://sdgs.un.org/goals.

[2] https://ourworldindata.org/human-development-index#hihd-vs-gdp-per-capita.

and development, determine strategic directions for the development of STI, thus increasing national competitiveness and improving life of citizens.

All this calls for formulation and implementation of STI policies at the national level. In developed countries, formulation of such policies already has a rather long tradition which is related to a large extent to application of the approach based on the concept of National Innovation Systems (NIS) introduced in the 1990s (Freeman, 1987; Lundvall, 1992; Nelson, 1993). A common feature of many developing countries, in particular the BRICS states, is the proliferation of new institutions and innovation systems resulting from a transitional economy (Chan & Daim, 2012).

Grand challenges highlight the need for multi-disciplinary research, engagement of a wide range of stakeholders to examine current issues and proposing relevant responses, coordination of different policies across different countries/regions and sectors. Therefore, opportunities for development require an integrated approach that should deal with both technological solutions and relevant social changes and economic transformations. In the context of STI, ensuring that benefits reach marginalised or impoverished groups is a critical challenge. Doing so would contribute to inclusive objectives that development policies aim to achieve, a goal shared among most developing countries. This objective is reflected in the political ambitions of these nations. STI foresight under these conditions has to address both inclusiveness and economic competitiveness that should be combined under a common vision (Aguirre-Bastos & Weber, 2018). This means that innovation policies have to combine both inclusive innovation and innovation for inclusiveness.

5 International Experience

Developing countries have already a solid record of successful implementation of foresight studies, e.g. in the field of science, technology and innovation. These activities are supported by a number of international organisations, which are among key players in promoting foresight worldwide. They provide a platform to bring together government, industry, academia and other stakeholders to identify and address key challenges and opportunities in different sectors. For example, the United Nations Development Programme (UNDP) and United Nations Industrial Development Organisation (UNIDO) help governments to develop foresight strategies that can enhance their socio-economic development. The Organisation for Economic Co-operation and Development (OECD) has also been actively involved in promoting foresight for inclusive and sustainable growth. The European Union's framework programmes for research and innovation (Horizon Europe and earlier framework programmes) have also emphasised the importance of foresight activities to inform research and innovation policies. These organisations provide technical expertise, funding support, training and capacity building to help countries develop their own foresight capabilities and prioritise their S&T activities. Among other international players in this area, there could be mentioned UN Food and

Agriculture Organisation, World Intellectual Property Organisation and International Energy Agency (FAO, 2022; WIPO, 2021; IEA, 2021, 2023).

5.1 *Best Practices of STI Foresight in International Organisations*

5.1.1 United Nations Development Programme (UNDP)

UNDP plays an important role in promoting S&T foresight in developing countries (GCPSE, 2014). It collaborates with national governments and other partner organisations to strengthen innovation systems and promote the use of foresight studies to inform science and technology policies. UNDP offers technical assistance to countries in conducting foresight studies, analysing results and developing strategies for implementation. The organisation also provides training and capacity building support to government officials, researchers and other stakeholders in S&T foresight methodologies.

Through its partnerships with other UN agencies, such as UNESCO and UNIDO, as well as regional organisations, UNDP helps countries to connect with global networks of experts and best practices in S&T foresight. The organisation also works to promote awareness of the importance of foresight and innovation for sustainable development, through publications, events and other communication activities.

It helps developing countries in, when it comes to allocating resources and capabilities, to prioritise longer-term issues (UNDP, 2021, 2022), taking account of limited capacities in resources available, skills, knowledge, etc., via cultivating the capacity to create sustainable long-term development policies and pushing the future agenda.

5.1.2 United Nations Industrial Development Organisation (UNIDO)

The UNIDO Technology Foresight Programme started in 1996. It focused on assisting developing countries to "upgrade their industrial sectors from resource-based to technology-based capacity in order to better integrate their products in the regional or global economy". The programme supported national technology foresight studies in emerging economies, in particular countries of Central and Eastern Europe under transition to market economy; former Soviet republics (CIS countries); Latin America (in 2000, UNIDO launched an ambitious programme of Technology Foresight for Latin America and the Caribbean), etc. UNIDO published methodological guides (UNIDO, 2005), organised numerous trainings and provided methodological support to individual countries and regions. Particular feature of this support was their connection with relevant local initiatives and the focus on cost effectiveness, awareness building, training activities and networking. Some projects supported by UNIDO targeted less developed or small countries in order to provide information on global technology, global challenges and potential promising areas for industrial development for their economies.

UNIDO promoted technology foresight at the regional level—mainly by building on national exercises already in progress. Its practical contribution was focused on the following issues: identifying favourable policy frameworks; strengthening interaction, communication and cooperation between stakeholders of the NIS; facilitating identification of priority areas in selected industrial sectors to increase national competitiveness and provide better access to the regional and the global markets; building capacities to identify risks and opportunities on how to improve value chain competitiveness through the application and use of technology foresight analysis and methodologies.

5.1.3 Organisation for Economic Cooperation and Development (OECD)

The OECD provides a platform for member-states and other countries to exchange best practices, methodologies and data for foresight exercises. Its basic principle states "in times of increasingly rapid change, growing complexity, and critical uncertainty, responsible governance requires preparing for the unexpected". The organisation has been involved in various foresight studies in different areas, including science and technology, space, environment, education and healthcare. Its main goal is to promote evidence-based policy-making and to support countries in addressing future challenges by developing strategic roadmaps and exploring potential scenarios. The OECD also provides useful guidance and recommendations on how to design and implement foresight exercises, such as involving stakeholders, defining the scope of the study and using a participatory approach.

The OECD Strategic Foresight Unit established in 2013 by the OECD Secretary General, is the successor to the International Futures Programme. Its activity targets at increasing the use and impact of strategic foresight in OECD policy expertise and policy-making by governments. The Unit works with governments and organisations facing different strategic challenges. Its work involves supporting OECD directorates and committees as well as governments of OECD member states and other countries in preparing for the future.

OECD implements projects and analysis by expanding thinking on emerging and disruptive developments of relevance for different policy topics (see for example OECD, 2021); facilitating workshops, delivering training and guidance on how to incorporate foresight into policy work, convening internal communities of interest to exchange futures insight. The organisation also helps governments to build foresight capacity by advice and guidance on national foresight systems as well as convening the Government Foresight Community—an international network of expertise, which includes experienced foresight experts from all over the world regularly gathering for discussing urgent issues and future trends.

5.1.4 United Nations Educational, Scientific and Cultural Organisation (UNESCO)

UNESCO's role in foresight studies is primarily focused on promoting the use of sciences and technology for sustainable development. UNESCO encourages foresight studies and anticipatory thinking in the areas of education, culture, sciences and communication to support national and regional development strategies. The organisation also supports foresight studies that take into account ethical, cultural and social implications of emerging technologies, as well as their potential impacts on sustainable development goals. Through its programmes and initiatives, UNESCO seeks to enhance the capacity of developing countries to undertake foresight studies, build institutional frameworks for technology assessment and promote the integration of foresight methodologies into national policy-making processes. Among the most important activities of UNESCO is the UNITWIN programme launched in 1992, which involves over 850 institutions in 120 countries, promotes international inter-university cooperation and networking to enhance institutional capacities through knowledge sharing and collaborative work. The programme supports, among others, the establishment of UNESCO Chairs in the field of futures literacy, thus promoting dissemination of foresight methodologies all over the world, with a special focus on developing countries. The Futures Literacy Laboratories designed by UNESCO help people "to make their anticipatory assumptions explicit and thereby reveal not only the determinants of the futures they imagine but also the attributes of the anticipatory systems and knowledge creation processes" (Miller, 2018).

5.1.5 World Economic Forum (WEF)

WEF organises discussions among global leaders from government, business and civil society on emerging trends, risks and opportunities that may affect the global economy and society (WEF, 2020). The forum uses strategic foresight methodologies to explore various possible scenarios and to identify potential solutions to address challenges and leverage opportunities. The World Economic Forum's foresight studies cover a wide range of topics, including geopolitical risks, technological disruptions, environmental sustainability, and social and economic inequalities. The goal of such studies is to shape a more sustainable and inclusive future by fostering global dialogue and collaboration.

5.1.6 Asia-Pacific Economic Cooperation

Asia-Pacific Economic Cooperation (APEC) and its Industrial Science and Technology Working Group (ISTWG) set up, through the support of the Thai government, the APEC Center for Technology Foresight (APEC CTF). The APEC CTF provides a number of foresight activities including international conferences, trainings as well as foresight studies (APEC, 2008). It assisted many countries of South-East Asia in organising S&T foresight studies (see for example DEPA, 2019). Among recently implemented projects, there are foresight studies in such areas like nanotechnology,[3] climate change[4] and human health.

5.1.7 Millennium Project

The Millennium Project is a global participatory think tank established in 1996, which has over 70 Nodes around the world. It connects futurists around the world and collaborates to improve global foresight on behalf of humanity and building a better future. The project team regularly publishes "State of the Future" reports, future visions like Work/Technology 2050: Scenarios, scanning global challenges, organises surveys, conferences and other activities.

5.2 *Best Practices of STI Foresight in Developing Countries*

Among developing countries, the most active in S&T foresight are new industrialised countries, first of all—the Republic of Korea and China as well as former countries of the Soviet Bloc—Russia and most of the countries of Central and Eastern Europe. Some examples of forward-looking activities in the developing world are presented below.

5.2.1 The Republic of Korea

One of the main drivers of technology foresight in South Korea has been the Ministry of Science and ICT (MSIT) and the Korea Institute of S&T Evaluation and Planning (KISTEP). The MSIT has been responsible for mapping out long-term technology development plans, while KISTEP has conducted foresight studies, evaluated R&D projects and provided policy support.

The Korean government has developed several programmes to support technology foresight. The most important of these is the Technology Foresight Programme launched in the 1990s (Choi & Choi, 2015). The programme includes a range of activities, such as technology roadmapping, expert panels and scenario development. The programme has been successful in identifying key technologies and trends, as well as informing policy decisions. KISTEP provides regular technology foresight studies (KISTEP, 2022a) that are used as a background for establishing

[3] https://apecctf.org/publications/volume-i-summary-report-of-a-foresight-project-nanotechnology-the-technology-for-the-21st-century-january-2002/.

[4] https://apecctf.org/publications_type/climate-change/.

national research priorities and identifying the most promising emerging technologies (KISTEP, 2022b). The results of foresight studies are closely integrated with the 5-year S&T Basic Plans and help provide a systemic basis for national level S&T planning.

The importance attributed to Korea's technology foresight has significantly grown. These studies have been increasingly focused on future social development and factors of demand to identify future technologies that could address acute societal problems. They consistently provide background information that feeds into policies and medium and long-term S&T strategies in different sectors.

5.2.2 China

The foresight activities in China began in 1992, when a study on "National key technologies selection" was launched. Technology foresight in China has been developing rapidly since then. In the mid-1990s, the Chinese government actively promoted the use of technology foresight to help identify key technological areas that should be developed to promote economic growth and overall development. In 1997, the State Scientific and Technological Commission initiated "Technology foresight in priority areas of the country". It identified three priority areas: agriculture, information and advanced manufacturing. In the early 2000s, the organisational structure and methodological system of national S&T foresight studies were gradually established. Several studies were implemented. In 2001, the Chinese Ministry of Science and Technology issued guidelines for conducting national technology foresight studies. Since then, it has launched several national-level foresight projects, covering areas such as energy, new materials, agriculture and information technology. The technology foresight survey based on new methodology started in 2002 in order to "clarify major technologies for China's socio-economic development for adoption in the National Guidelines for Medium- and Long-term Plans for Science and Technology Development and the 11th Five-Year Plan" (Tsujino & Yokoo, 2006). In addition, many provinces, cities and industries in China have conducted their own technology foresight studies. The latest national technology foresight launched in 2019 was aimed at the assessment of current state of national S&T and formulation of major goals of its development, identifying key technology areas and directions that will be relevant to China's long-term development in the next 15 years, and providing support for the decision-making and deployment of major S&T tasks and the preparation of medium- and long-term S&T development plans by the State Council.

As a new tool for strategic analysis and integration, China's technology foresight has accumulated rich practical experience, gradually forming a set of organisational systems and methodology with Chinese context (Like et al., 2023). These studies propose lists of technology projects, technology roadmaps and the priorities and serve as a timely arrangement of the future S&T development and a mechanism for formulation of medium- and long-term S&T plans. Technology foresight studies are considered at the macro, meso and micro levels, they highlight relationships between the planning objectives, S&T development objectives and thus help to clarify technical projects, which can promote the formulation of national S&T development plans and specific programmes to be more targeted and focused. The methodological background has been also rapidly developing and now includes dozens of modern foresight tools (Li et al., 2016). The Chinese government has also established a number of institutions and organisations to support technology foresight studies. For example, the National Technology Foresight Expert Advisory Committee provides strategic guidance and expert advice on foresight studies. The Chinese Academy of Science and Technology for Development conducts research on technology foresight methods and tools.

5.2.3 Russia

The first attempts of forward-looking activities related to S&T in Russia go back to the beginning of the Soviet era when long-term planning was introduced, the S&T plans were also developed for large-scale national projects (like atomic bomb or space). First fully-fledged S&T foresight projects started only in the late 1990s (Gokhberg et al., 2017). In 1996, Priority Areas for S&T Development and a list of 70 National Critical Technologies were formally approved by the Government. These lists were revised several times—in 2001–2015—and served a background for developing national S&T programmes (Sokolova et al., 2018; Grebenyuk & Shashnov, 2012).

In 2006, the first fully-fledged national foresight programme started with the goal to create a solid foundation for future-oriented S&T and innovation policies. The programme identified key areas of research that can bring the most substantial benefits in the longer term, to assess potential demand for innovation, and to understand which policy instruments could be most relevant to promote innovation activities. The time horizon for the programme was 2025. It covered a macroeconomic forecast, S&T Delphi and assessment of market demand for new technologies. The second cycle of the national S&T foresight (2009–2010) was largely focused on crisis-related issues. The results of the first cycle laid the foundation for the analysis of particular S&T-intensive clusters that could provide a momentum to innovation. The policy issues being addressed by foresight referred to a need in understanding the likely transformations of the global (and Russian) economy, new (collaborative and user-driven) ways of innovating, convergence among different technological platforms, etc. The national S&T foresight studies (Gokhberg et al., 2017) also considered grand challenges, their impact on Russia's socio-economic development.

5.2.4 India

India has initiated several foresight exercises and reports to shape its technology and innovation policy. The Technology Policy Statement issued by the Indian government in 1983 highlighted the need for undertaking technology forecasting and assessment studies on a systematic and continuing basis. In 1988, Under the Department of Science and Technology (DST), the Technology, Information, Forecasting and Assessment Council (TIFAC) was established with a mission to assess and develop roadmaps for future technological options in crucial sectors of socio-economic importance—formulation of strategy and nucleation of technology programmes. In 1996, TIFAC developed the Technology Vision (TV) 2020, which assessed prospects and potential pace of technological development for major sectors of the Indian economy. The TV2020 was a path breaking exercise because it spurred further forward-looking activities, in particular in different sectors of the economy. The next large-scale foresight study, also performed by TIFAC, was Technology Vision for 2035 (TIFAC, 2015). It identified basic needs for India by 2035 and key technologies that could help to meet these needs. It described key prerogatives like clean and potable water, food security, healthcare, energy supply, etc. For each of them, key relevant technologies were identified. Technology-related capabilities and constraints, grand challenges and actions to be taken were discussed.

The TV 2035 has been further developed in the form of several roadmaps covering different sectors: Education; Smart Advanced Manufacturing and Rapid Transformation Hub; Healthcare—Automated alerting system for medical emergencies; ICT-driven rural healthcare delivery; ICT—National Policy on Electronic 2019; Mission on Cyber Physical System and Quantum Computing; Water Roadmap; etc. Overall, technology foresight is seen as an essential tool for India to achieve its goals in becoming a knowledge-based economy and a global leader in innovation.

5.2.5 Latin America

In last 20 years, technology foresight has made a great progress in Latin America. Many countries in the region were implementing foresight exercises aimed at identifying opportunities and future S&T trends, as well as defining strategic priorities for innovation and economic development. In the late 1990s and 2000s, foresight studies were implemented in Argentina (Strategic Mid-Term Plan of STI for 2015), Chile ("E-ducation industry", sectoral studies for biotech, aquaculture and software industries), Cuba (technology watch and competitive intelligence), Venezuela (Technology Foresight Programme) as well as in Mexico, Bolivia, Uruguay and other countries (Georghiou et al., 2008). There were also several international foresight exercises, for example a study on fishery industry on the Pacific Coast of the South America supported by UNIDO in cooperation with the Government of Spain.

Brazil is another country in the region that has made significant progress in technology foresight. The government established a national foresight programme in the late 1990s, and since then a number of foresight projects have been implemented. The first of them was Brazil 2020 which was followed by the Prospectar Programme and Brazilian Technology Foresight Programme. These activities involved thousands of experts and complex methodologies: Delphi survey, scenarios, etc. (Cagnin, 2014). On top of that, several exercises focused on particular areas such as biotechnology, agriculture, nanotechnology, information and communication technologies, etc.

Another notable example is Colombia. The Colombian Technology Foresight Programme was launched in 2003 and implemented in two stages—until 2008. It was the first national foresight Programme in Latin America that passed a full-scale international evaluation (Popper et al., 2010).

Overall, technology foresight has been rapidly developing in Latin America—both in organisational capacities and methodology (Pinto & Medina, 2020). It has become one of key tools for promoting innovation and economic development in Latin America, and many countries in the region are investing in this area to ensure they remain competitive in the global economy.

5.3 *Foresight in Africa*

There are numerous foresight studies devoted to Africa, which cover social and economic issues like human development, economic growth and transformation (Cilliers et al., 2011), global governance, climate change and other UN sustainable development goals (Coulibaly, 2019). Most of them touch, to some extent, STI problems but cannot be considered as fully-fledged STI foresight studies.

The most notable STI foresight activities in the African continent were implemented in South Africa. The first National Research and Technology Foresight in the late 1990s organised by the Department of Arts, Culture, Science and Technology covered a number of areas sectors of S&T policy: Agriculture and Agro-processing; Biodiversity; Crime Prevention, Criminal Justice and Defence; Energy; Environment; Financial Services; Health; ICT; Manufacturing and Materials; Mining and Metallurgy; Tourism; Youth. It included a large-scale Delphi survey followed by the selection of 20 critical technologies based on an open-ended questionnaire filled in by several key experts (Pouris & Raphasha, 2015). Michael Kahn concluded that STI foresight had limited influence on innovation policy. Nevertheless, it contributed to building trust in the scientific community, raising awareness and increasing public participation. Many black researchers who have led or served on expert panels occupied important positions in the system of science, which is recognised as a success in terms of increasing inclusion. However, in identifying new technological targets, success has been much more modest (Kahn, 2008). There were several other foresight initiatives focused on problems of different sectors (Day et al., 2009).

The presented book is devoted to the STI foresight exercise for South Africa. It was implemented in 2018–2019, and its first results were discussed at the Science Forum South Africa (NACI, 2019). The project can serve as an example how a fully-fledged foresight study could be organised and implemented at the national level in a developing country, which methodologies can be used, which challenges the study could address and how the results obtained can help policymakers to improve the national innovation system.

6 Conclusions

Science and technology foresight in developing countries has some specific features due to their unique socio-economic and political context. First, there is often R&D capacity in developing countries, and consequently, there is limited scientific knowledge to draw upon in the foresight process. This may require building research capacity and engaging in technology transfer activities with developed countries.

Secondly, there is a need to consider the challenges faced by developing countries, such as poverty, disease and environmental degradation, as well as the opportunities for development and progress. Foresight activities in developing countries should aim to identify and prioritise areas of research and innovation that have the potential to address these challenges and leverage opportunities.

Finally, engagement with stakeholders, including policymakers, industry, academia and civil society, is critical in developing countries. This is important for ensuring that the foresight process is inclusive and that the recommendations are relevant and actionable.

Overall, science and technology foresight in developing countries should not only focus on technological advancements but should also address the broader socio-economic context and aim to achieve sustainable development goals.

Acknowledgements The chapter was prepared within the framework of the Basic Research Program at the HSE University.

References

Aguirre-Bastos, C., & Weber, M. (2018). Foresight for shaping National Innovation Systems in developing economies. *Technological Forecasting and Social Change, 128*, 186–196.

Amanatidou, E. (2011). Grand challenges—A new framework for foresight evaluation. EU-SPRI conference. https://euspri-forum.eu/wp-content/uploads/2019/07/amanatidou_paper_sept2011.pdf

APEC. (2008). *A decade of foresight. Technology foresight in Asia-Pacific*. APEC TFC.

Cagnin, C. (2014). STI Foresight in Brazil. *Foresight-Russia, 8*(2), 46–55.

Chan, L., & Daim, T. (2012). Exploring the impact of technology foresight studies on innovation: Case of BRIC countries. *Futures, 44*(6), 618–630.

Choi, M., & Choi, H. (2015). Foresight for science and technology priority setting in Korea. *Foresight and STI Governance, 9*(3), 54–67.

Cilliers, J., Hughes, B., & Moyer, J. (2011). African futures 2050: the next forty years. Institute for Security Studies, Monograph 175.

Coulibaly, B. (Ed.) (2019). Foresight africa. top priorities for the continent in 2019. Africa Growth Initiative, 2019.

Day, B., Greenwood, P., & Karuri-Sebina, G. (2009). *Thinking about tomorrow today. An introduction to using foresight in South Africa*. COFISA, Ministry of Foreign Affairs of Finland, Department of Science and Technology, Republic of South Africa.

DEPA. (2019). Thailand digital technology Foresight 2035.

Eisenhardt, K. M., Graebner, M. E., & Sonenshein, S. (2016). Grand challenges and inductive methods: Rigor without rigor mortis. *Academy of Management Journal, 59*(4), 1113–1123.

FAO. (2022). *The future of food and agriculture. Drivers and triggers for transformation*. Food and Agriculture Organization of the United Nations.

Freeman, C. (1987). *Technology policy and economics performances: Lessons from Japan*. Pinter.

GCPSE. (2014). *Foresight as a strategic long-term planning tool for developing countries*. UNDP Global Centre for Public Service Excellence.

Georghiou, L., & Harper, J. C. (2011). From priority-setting to articulation of demand: Foresight for research and innovation policy and strategy. *Futures, 43*, 243–251.

Georghiou, L., Harper, J. C., Keenan, M., Miles, I., & Popper, R. (Eds.). (2008). *The handbook of technology Foresight*. Edward Elgar.

Gokhberg, L., & Meissner, D. (2016). Seizing opportunities for national STI development, in: deploying foresight for policy and strategy makers: creating opportunities through public policies and corporate strategies in science, technology and innovation. / Ed. by L. Gokhberg, D. Meissner, A. Sokolov. Netherlands: Springer, 2016, 267–273.

Gokhberg, L., Sokolov, A., & Chulok, A. (2017). Russian S&T Foresight 2030: Identifying new drivers of growth. *Foresight, 19*(5), 441–456.

Grebenyuk, A., & Shashnov, S. (2012). Critical technologies in Russia: Identifying and implementation. *Maintenance Problems, 4*, 105–116.

IEA. (2021). *An energy sector roadmap to carbon neutrality in China*. International Energy Agency.

IEA. (2023). *Energy technology perspective 3 2023*. International Energy Agency.

Kahn, M. (2008). Foresight in South Africa: Estimating results through the prism of time. *Foresight and STI Governance, 2*(2), 56–65.

KISTEP. (2022a). *The 6th science and technology foresight (2021–2045)*. KISTEP.

KISTEP. (2022b). *10 emerging technologies*. KISTEP.

Li, N., Chen, K., & Kou, M. (2016). Technology foresight in China: Academic studies, governmental practices and policy applications. *Technological Forecasting and Social Change, 119*, 246–255.

Like, Y., Fei, X., & Sokolov, A. (Eds.). (2023). *Foresight and STI planning in China*. Springer (in print).

Lundvall, B. A. (1992). *National innovation system, toward a theory of innovation and interactive learning*. Pinter Publishers.

Meissner, D., & Sokolov, A. (2013). Foresight and science, technology and innovation indicators. In *Handbook of innovation indicators and measurement (Chap. 16)* (pp. 381–402). Edward Elgar publishing.

Miles, I., & Keenan, M. (Eds.). (2002). *Practical guide to regional foresight in the United Kingdom*. European Commission.

Miles, I., Saritas, O., & Sokolov, A. (2016). *Foresight for science, technology and innovation*. Springer.

Miller, R. (2018). Futures literacy laboratories (FLL) in practice. In R. Miller (Ed.), *Transforming the future. Anticipation in the 21st century*. Routledge.

NACI. (2019). *South Africa foresight exercise for science, technology and innovation*. Synthesis Report. Nacional Advisory Council on Innovation.

Nelson, R. R. (1993). *National innovation systems: A comparative analysis*. Oxford University Press.

OECD. (2021). *Global scenarios 2035: Exploring implications for the future of global collaboration and the OECD*. OECD.

OECD. (2022). *Main science and technology indicators*. OECD.

OECD. (2023). *Science, technology and innovation outlook 2023*. OECD.

Pinto, J. P., & Medina, J. (2020). Hybrid processes for a new era of strategic foresight. *Foresight, 22*(3), 287–307.

Popper, R., Georghiou, L., Keenan, M., & Miles, I. (2010). *Evaluating foresight. Fully-fledged evaluation of Colombia technology foresight Programme*. Universidad del Valle.

Pouris, A., & Raphasha, P. (2015). Priorities setting with foresight in South Africa. *Foresight and STI Governance, 9* (3), 66–79.

Sachs, J. D., Kroll, C., Lafortune, G., Fuller, G., & Woelm, F. (2021). *Sustainable development report 2021. The decade of action for the sustainable development goals*. Cambridge University Press.
Sokolova, A., Grebenyuk, A., & Sokolov, A. (2018). Twenty years of S&T priority setting in Russia: Lessons learned. *Foresight, 20*(5), 449–466.
TIFAC. (2015). *Technology vision 2035*. TIFAC.
Tsujino, T., & Yokoo Y. (2006). Technology foresight surveys in China. *Science and Technology Trends, Quarterly Review*, No 20, July 2006.
UNDP. (2021). *Horizon scanning: The future of 21st century governance*. Oslo Governance Centre.
UNDP. (2022). *UNDP RBAP: Foresight playbook*. UNDP.
UNESCO. (2021). UNESCO Science Report. The race against time for smarter development. https://unesdoC.unesco.org/ark:/48223/pf0000377433.locale=en
UNIDO. (2005). *UNIDO technology Foresight manual. Organization and methods* (Vol. 1). UNIDO.
WEF. (2020). *Critical frontier: Leveraging technology to combat COVID-19*. White Paper. World Economic Forum.
WIPO. (2021). *WIPO technology trends 2021: Assistive technology*. World Intellectual Property Organization.

Foresight Methodology and Commonly Used Methods in the New Generation Foresight

Ozcan Saritas, Natalia Veselitskaya, and Alexander Sokolov

Abstract This chapter provides an overview of the Foresight methods, which were used in the South African Foresight for STI (SAForSTI) process. The SAForSTI 2030 exercise brought together a variety of qualitative and quantitative Foresight methods with innovations in their use. Over years, Foresight has embraced a spectrum of quantitative and qualitative methods from various disciplines to establish consensus among key stakeholders, identify global challenges and opportunities, explore technologies and markets, and develop alternative scenarios, visions, priorities, policies, strategies, and action. The chapter delves into the 15 most commonly used methods in Foresight, exploring their advantages, limitations, and practical applications.

1 Introduction

The overall Foresight process and phases draw upon a variety of qualitative and quantitative methods, and their integrated use. Methods are the tools to be used as part of the means to explore ideas and acquire information and data to clarify situations and to negotiate a solution. There are many methods used in Foresight as described in presentations. The selection of methods from this pool is based on the objectives of the exercise and resources available including skills to use methods, time, money, the availability of participants, and other human resources as facilitators. It should be remembered methods are quite flexible tools. They can be attributed different functions and can be integrated in various ways in line with the

O. Saritas (✉)
Rochester Institute of Technology Dubai, Dubai, UAE

National Research University Higher School of Economics, Moscow, Russia
e-mail: oxscad1@rit.edu; osaritas@hse.ru

N. Veselitskaya · A. Sokolov
National Research University Higher School of Economics, Moscow, Russia
e-mail: nveselitskaya@hse.ru; sokolov@hse.ru

O. Saritas et al. (eds.), *21st Century Foresight*, Science, Technology and Innovation Studies, https://doi.org/10.1007/978-3-031-69557-5_4

objectives and methodology of the Foresight exercise. New methods can also be created during the process. Describe here are the types of methods that you plan to use for your Foresight exercise.

Whether quantitative or qualitative, methods are not departure points of Foresight. Suitable methods are combined in line with the context and content of the Foresight exercise as well as available resources to constitute a full Foresight methodology. The overall methodology aims at developing and supporting understanding of situations in their temporal nature, enabling divergent thinking within a defined time horizon, prioritising issues, and setting up visions, as well as converging those visions into policies and strategies. Certain methods during the process help to enable networking, mutual learning, collective visioning, and action-oriented thinking.

This chapter describes the most used 15 methods in Foresight with their advantages and limitations as well as their practical applications. Figure 1 highlights those quantitative and qualitative methods, which were used in the SAForSTI 2030 exercise in a combined way.

2 Commonly Used Methods in Foresight

Foresight methodology has been developing during the last decades, in particular with respect to STI Foresight. The key reasons for that were ever-increasing uncertainty and turbulence, high economic, political, social and environmental risks, growing interdisciplinarity of S&T, and dispersion of expert knowledge. Identification of strategic development alternatives in such an environment requires achieving consensus among key stakeholders. Over time, Foresight has absorbed a wide range of quantitative and qualitative methods from various disciplines, which help to provide communications between experts, integrate opinions of experts, obtain an integrated vision of the future development of the area under study, and formulate actions to be performed in order to reach a better future. Foresight methods are used to identify global challenges, threats, and windows of opportunity in the area under study; explore promising technologies, markets, innovative products; identify critical technologies and priorities; develop alternative scenarios; and for many other purposes.

Among the latest methods involved in Foresight studies, the key role is played by evidence-based methods—modelling, statistical, bibliometric and patent analysis, and the most recently—methods based on artificial intelligence including semantic analysis of big data. Different methods are often used together at different stages of Foresight (see Miles et al., 2016). Such linkage between the frequently used methods is illustrated in Fig. 2. The combination of quantitative evidence-based methods with expert knowledge brings much more reliable and well-grounded results. Below, one can find a structured description of most advanced methods, many of which were used in the study described in the book.

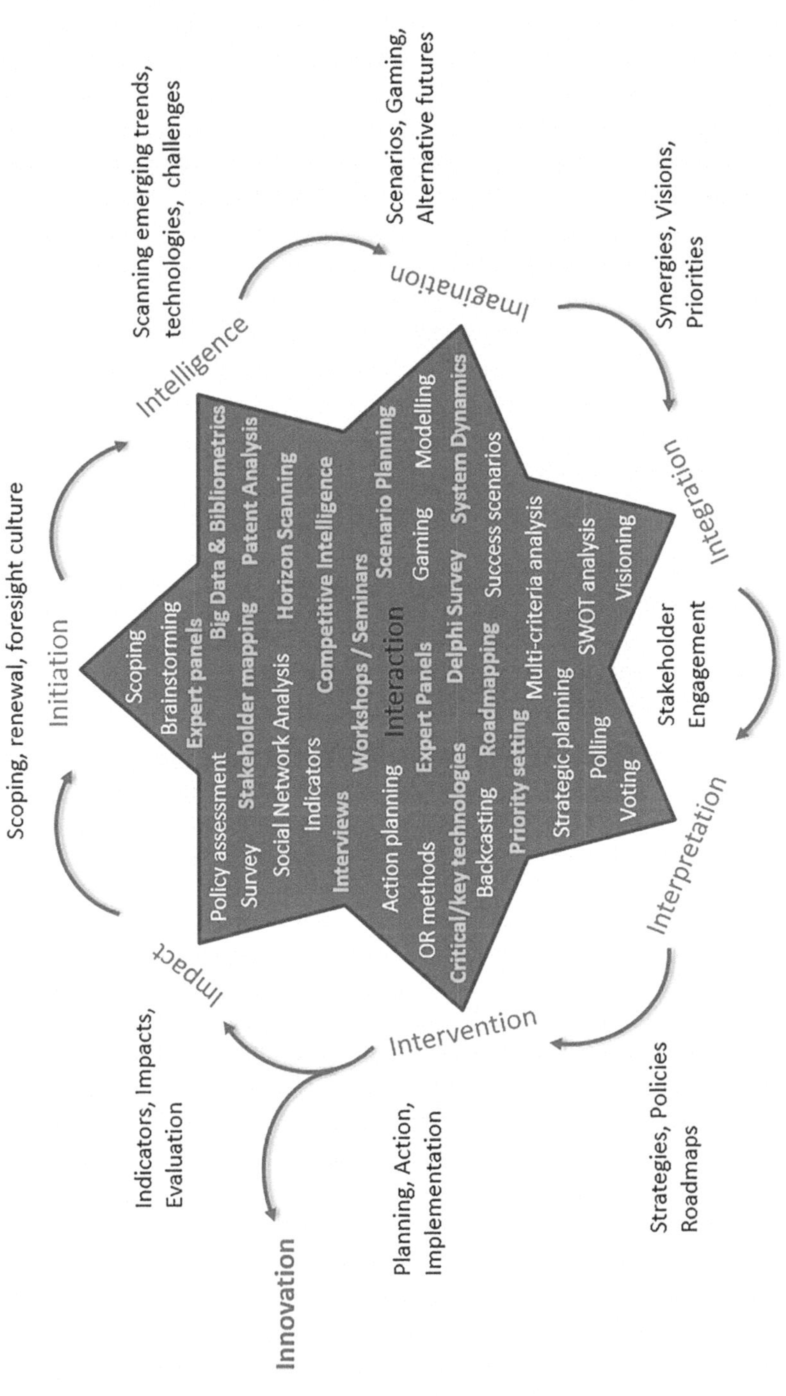

Fig. 1 Methods used in the STIForSA 2030 exercise

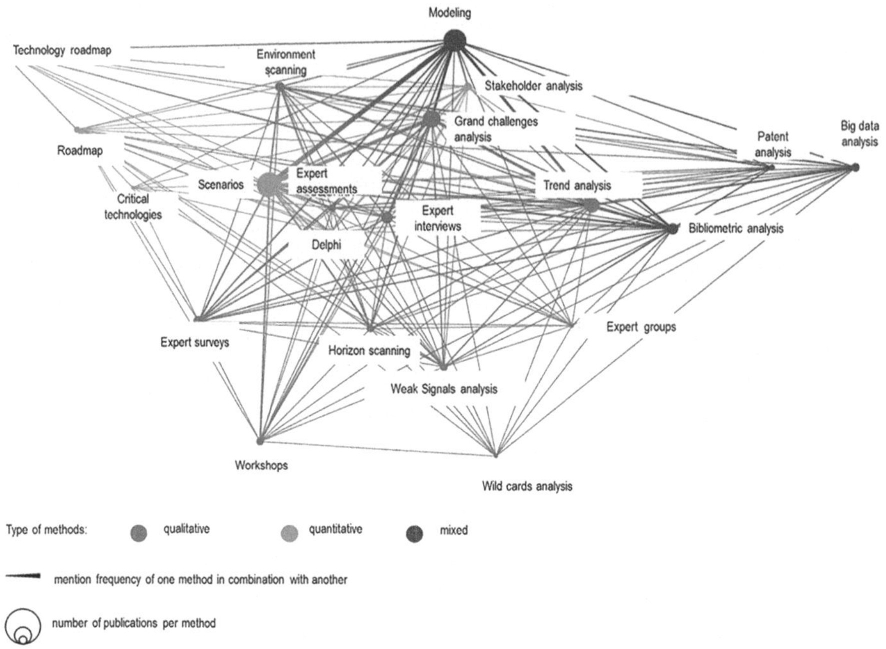

Fig. 2 Correlation between frequently used methods of Foresight. Source: Authors' calculations based on 4960 publications from Web of Science (2016–2021)

Search query: Foresight OR "Technolog* Forecast*" OR "Future* planning" OR "Futur*-orient* Technolog* Analy*" OR "Futur* orient* Technolog* Analy*" OR "Scenario* planning" OR "Long-range planning" OR "long range planning" OR "forward-looking analysis".

2.1 The Delphi Survey

Definition: A Delphi survey is an anonymous questionnaire-based survey of experts conducted in two or more rounds, providing the respondents access to the results of the previous round. Delphi surveys are conducted to summarise the opinions of a large number of experts (from several hundred to tens of thousands).

Advantages

- Delphi survey results can serve as a basis for in-depth analysis.
- It allows one to obtain opinions of a wide range of relevant experts.
- The questionnaire can be adjusted to precisely match the study objectives (assess the market, trends, technologies, policies).

Limitations

- It is a labour-intensive process (developing a questionnaire, organising the actual survey, involving external experts), with high time and financial costs.
- There is a risk that creative solutions suggested by a small number of experts will be ignored.

Delphi surveys have been actively applied in Japanese foresight studies conducted every 5 years since 1971. In the scope of the most recent 11th Japanese S&T Foresight (NISTEP, 2019[1]) study, a Delphi survey on S&T development prospects until 2050 was conducted with the participation of 5300 experts, covering seven science and technology areas and 702 topics. Delphi surveys were also carried out in the framework of the Russian S&T Foresight 2025 project (HSE, 2007–2008[2]), involving more than 2000 experts.

2.2 Scenarios

Definition: Scenarios describe the future states of an object based on studying, structuring, and ranking various external and internal factors which determine alternative development paths.

Scenarios are built to accomplish the following:

- Facilitate long-term planning, strategy development, and decision-making based on analysing a range of likely future development alternatives.
- Consider alternative future states of the system in a highly uncertain and turbulent situation (taking into account wildcard events), for management purposes.

All scenarios must be realistic, mutually exclusive, and collectively exhaustive.

Advantages:

- Scenarios allow one to take into account a variety of structured external and internal factors affecting the development of the environment when preparing development strategies and other documents.
- Using scenarios allows one to manage an organisation more flexibly in case of sudden changes in the environment, by considering various possible developments in advance.
- Scenarios allow one to move from traditional planning based on extrapolating current trends to building multiple visions of the future.

Limitations:

- Building reliable scenarios is a complex task.

---

[1] NISTEP (2019) An Overview of the 11th Foresight Survey in Japan // Innovation and Development Policy. No. 3. P. 79—90.

[2] Sokolov A., Chulok A. Russian Science and Technology Foresight – 2030: Key Features and First Results // Foresight and STI Governance, v. 6, No. 1.

- There is a risk of limiting the range of possible future dynamics (between two and five scenarios are typically considered, but there can be many more development options).
- Cognitive inertia, reliance on limited rationality, and biased assertions can hinder the objectivity of scenario planning, and lead experts to dismiss unlikely scenarios without due consideration.

The best-known scenario building cases include the practices of Shell:[3] Since the early 1990s, the company regularly prepares development scenarios for various areas, ranging from the energy sector's transformation to post-COVID-19 social development and the sustainable development of specific countries. In the scope of Russian S&T Foresight 2030, the Technological Adaptation and Technological Leap scenarios were built. They were presented as integrated paths; in addition to the base scenario, an accelerated development scenario was prepared. This allowed for avoiding the need to build a third one, which could be seen as transitional from the base to the accelerated development scenarios.

2.3 *Technology Roadmaps*

Definition: A technology roadmap is a summary document describing a multi-level strategic development system in the area under consideration, based on a single time scale.

Technology roadmaps (TRMs) are applied to:

- Match demand for new products and services with the relevant production potential, by presenting emerging and developing markets, and prospective technologies that make it possible to produce competitive products and services, on a single timeline,
- Provide an analytical base to support strategic decision-making,
- Set priorities and suggest measures to implement them,
- Define a long-term sequence of actions to meet the demand and provide a sufficient supply of resources at any given time,
- Identify potential research and development areas and obtain an understanding of possible development trends in the area under consideration.

TRMs can be developed for various target objects such as technology areas, product groups, companies, and sectors of the economy.

Advantages:

- TRMs describe alternative development paths for technologies, products, and markets using a single timeline.
- They allow one to assess specific features of a technology area or a market.

[3] Shell (2008) Scenarios: An Explorer's Guide. Hague: Shell International BV.

- TRMs help manage large projects.
- Allow one to identify stakeholders' needs.
- Support strategic and investment decision-making.

Limitations:

- High resource intensity.
- The need to regularly monitor the external environment and adjust the TRM accordingly.
- The effects from the use of TRMs are delayed.

Several thousand industry and corporate TRMs are currently in use around the world in various areas, such as, e.g. the Motorola TRM (Willyard & McClees, 1987[4]), the 2006 EU Nanomaterials TRM,[5] etc. An interesting Russian case is the TRM System for S&T Development of the Civil Aircraft Industry until 2030 and the Integrated RM (2014–2015[6]). TRMs are also developed at the international, national, and regional levels.

2.4 Priorities and Critical Technologies

Definition: The critical technologies (CT) method is applied to set S&T development priorities and identifies the most promising technologies that require priority support. Various sets of criteria are used to assess the "criticality" (importance) of a particular technology.

The need to set priorities is due to the fact that even the richest countries cannot finance R&D in all areas. Given the limited availability of resources (financial, human, infrastructural, etc.), one has to focus on the most promising research areas.

Critical technologies are selected based on the following principles:

- Potential to contribute to accomplishing major socio-economic objectives,
- Medium-term planning horizon (typically 10 years),
- Allow one to make use of existing competitive advantages (the country's territory, available resources, etc.)
- Assessment of feasibility: availability of necessary resources (financial, human, materials and equipment, S&T groundwork, etc.)
- Links to science, technology, and innovation policy tools,

---

[4] Willyard C., McClees C. (1987) Motorola's Technology Roadmap Process // Research Management. Vol. 30. No. 5. P. 13—19.

[5] European Commission (2006) Roadmaps at 2015 on nanotechnology application in the sectors of: materials, health & medical systems, energy. https://www.nanowerk.com/nanotechnology/reports/reportpdf/report74.pdf (accessed on: 15.03.2023).

[6] HSE (2015) Roadmaps for Science and Technology Development of the Aircraft Industry until 2030. M: HSE.

Advantages:

- Lists of priorities and critical technologies help one develop clear policy recommendations to support and promote the development of the S&T sphere and focus on accomplishing major socio-economic objectives.
- They also help identify key areas for the practical implementation of R&D results, taking into account the existing competitive advantages.

Limitations:

- A relatively small number of experts are typically involved in CT selection, which may affect the quality of the policy recommendations they develop and implies the need to be particularly careful in setting up the expert working groups.
- Implementation tools need to be carefully designed, including the provision of financial support, so that CTs do not remain priorities only "on paper".
- There are risks of certain groups lobbying for their vested interests due to insufficiently formalised CT selection procedures.

Many countries prepare lists of CTs. For example, the USA has regularly published them since 1989,[7] France since 1995,[8] and Russia since 2002.[9] In all these countries, such lists are regularly updated.

2.5 Stakeholder Analysis

Definition: Stakeholder analysis involves identifying, assessing, and prioritising stakeholders who influence decision-making in the area under consideration.

Stakeholder analysis is conducted to accomplish the following:

- Assign and assess stakeholders' roles in the project.
- Motivate stakeholders to accomplish the set objectives and prepare final recommendations.
- Assess the impact of stakeholders' possible actions on the project results.

Advantages:

- Stakeholder analysis allows one to take into account the opinions of representatives of the government, academia, business, and the public to improve the project's effectiveness.

---

[7] National Science and Technology Council (2022) Critical and Emerging Technologies. List Update, Published in the United States of America.

[8] Ministre de l'Économie (2020) Préparer l'industrie du future. https://www.entreprises.gouv.fr/files/files/directions_services/politique-et-enjeux/innovation/technologies-cles-2020/technologies-cles-2020.pdf (accessed on: 25.02.2023).

[9] Grebenyuk A., Shashnov S. (2012) Critical technologies in Russia: identifying and implementation // Maintenance Problems. Vol. 4. P. 105—116.

- This technique helps to identify and prevent conflicts as well as to minimise their consequences.

Limitations:

- Involving specific stakeholder groups may not be easy.
- Experts' opinions may influence stakeholders and vice versa.

Stakeholder analysis is a key element of a foresight project, both at the preliminary and final stages. Examples of the practical application of this method abroad include Future State 2030 (European Commission, 2019[10]), Technology and Society (European Commission, 2022[11]), Roadmap for Healthcare Development in Iran (Iran, 2019[12]), and in Russia—Nanotechnology Market Forecast (HSE, 2007–2011[13]), Russian S&T Foresight 2030 (RF Government, 2014[14]), and Aviation Industry S&T Foresight Until 2030 and Beyond (TsAGI, HSE, 2014[15]).

2.6 Horizon Scanning, Analysis of Challenges and Trends

Definition: This is a method of a systemic search for and the identification of signs of potentially important events, development areas, emerging challenges, trends, etc. This technique is applied to identify global-, regional-, national-, industry-, and corporate-level socio-economic and S&T challenges and trends, and produce evidence-based short-, medium-, and long-term forecasts.

Advantages:

- Horizon scanning allows one to build visions of the future in various fields.
- Covers a wide range of time horizons.
- Improves decision-making.

Limitations:

- The need for constant monitoring of the external environment.
- Hard-to-predict wildcard events may occur, capable of drastically changing the trend.

[10] Vesnic L., Scapolo F. (2019) The Future of Government 2030+: Policy implications and recommendations, EUR 29853 EN, Luxembourg: Publications Office of the European Union.

[11] European Commission (2022) Societal engagement with key enabling technologies (SocKETs).

[12] Sajadi S., Majdzadeh R., Yazdisadeh B., Mohtasham F., Mohseni M., Doshmangir L., Lavis J. (2019) A roadmap for strengthening evidence informed health policy-making in Iran: protocol for a research programme // Health Research Policy and Systems. May.

[13] HSE (2014b) Russian S&T Foresight: 2030. New Materials and Nanotechnology. M.: HSE.

[14] RF Government (2014) Russian S&T Foresight: 2030. http://government.ru/news/9800/ (accessed on: 18.02.2023).

[15] TsAGI, HSE (2014) Aviation Industry S&T Foresight Until 2030 and Beyond. M.: TsAGI.

This method is actively applied by international organisations (e.g. UNDP[16]), R&D centres (such as, e.g. RAND[17] or Fraunhofer Institute for Technological Trend Analysis[18]), consulting agencies (e.g. Gartner[19]), and companies (e.g. DHL[20]) in their predictive and monitoring studies.

2.7 *Wildcard Events and Weak Signals*

Definition: A wildcard event is a poorly predictable event with significant potential effects if it occurs. A weak signal is unimportant in itself but is an emerging event that indicates radical possible transformations in the future.

An analysis of wildcard events and weak signals helps to accomplish the following:

- Identify new opportunities and strengthen one's competitiveness.
- Analyse the consequences of current strategies and already made decisions.
- Obtain a clearer understanding of the company's future activity areas.
- Analyse risks.
- Prepare for possible negative developments.
- Predict changes, reduce uncertainty.
- Identify barriers hindering market entry and exit.
- Identify trends affecting the business environment.

Advantages:

- An analysis of wildcard events and weak signals helps make the business more flexible and resilient, by developing strategies to adapt to radical changes in the system.
- It also helps reduce potential losses from wildcard events.
- It allows one to predict the behaviour of complex systems in a dynamic environment.

Limitations:

---

[16] UNDP (2021) Horizon Scanning: The Future of 21st Century Governance. Norway: Oslo Governance Centre.

[17] RAND (2022) Skills needs in selected occupations over the next 5—10 years. Research report https://assets.publishing.service.gov.uk/government/uploads/system/uploads/attachment_data/file/1095989/Skills_needs_in_selected_occupations_over_the_next_5–10_years_.pdf (accessed on: 15.03.2023).

[18] Fraunhofer Institute for Technological Trend Analysis (2015) Models of Horizon Scanning https://www.isi.fraunhofer.de/content/dam/isi/dokumente/ccv/2015/Models-of-Horizon-Scanning.pdf (accessed on: 15.03.2023).

[19] Gartner (2018) Scan, Assess, React: A 3-Phased Approach to Help Tech CEOs Respond to Disruptive Forces https://www.gartner.com/en/documents/3894665 (accessed on: 15.03.2023).

[20] DHL (2022) The logistics trend radar https://www.dhl.com/content/dam/dhl/global/csi/documents/pdf/csi-logisticstrend-radar-6-dhl.pdf (accessed on: 15.03.2023).

- Identifying wildcard events and assessing the probability of their occurrence are not easy tasks.
- The assessment of wildcard events' significance is subjective.
- Wildcard events are affected by subjective factors that cannot be quantitatively measured.

International initiatives involving the use of this technique include the iKNOW Project[21] (development of early warning systems in the EU by identifying and analysing weak signals and wildcard events, 2009–2013) and the analysis of weak signals in the social sphere in the scope of the Canadian Policy Horizons initiative (2020[22]). Among the Russian projects, the setting up of trend and wildcard event radars in various industries in 2022–2024 should be noted.

2.8 Foresight Seminars

Definition: A foresight seminar is a group session held to build a vision of the future and draw up alternative development scenarios. The difference between a foresight session and other seminars (strategic, expert, etc.) is the focus on working with the future, the wide coverage of considered trends, and the objective to support managerial decision-making.

Foresight seminars are held to accomplish the following:

- Gather information from key experts on the topic under study.
- Develop a shared expert community position.
- Collectively generate new ideas.
- Test and validate the results of the pre-foresight study.

Advantages:

- Foresight seminars allow one to take a broader look at the area under consideration.
- They help obtain insights from experts that are otherwise difficult to get using other methods.
- An opportunity to summarise expert community opinions.
- Results are obtained quickly due to the joint work of experts and beneficiaries in the same group.

Limitations:

---

[21] European Commission (2011a) Final Report Summary. iKNOW: interconnecting Knowledge for the early identification of issues, events and developments. Luxembourg: Publications Office of the European Union.

[22] Government of Canada (2020) Exploring Social Futures, Policy Horizon Canada. https://horizons.gC.ca/en/2020/03/20/exploring-social-futures/ (accessed on: 08.12.2022).

- High requirements for moderator's qualifications to build a constructive discussion.
- The need to involve all stakeholders in the area under consideration.
- There are risks of participants' low activity, a formal approach to conducting the foresight session, and obtaining unverifiable results.

Foresight sessions are held in the scope of most foresight projects, e.g. the Japanese S&T Foresight (NISTEP, 2019[23]) and Russian S&T Foresight 2030 (RF Government, 2014[24]).

2.9 Expert Panels

Definition: An expert panel involves periodically surveying a group of experts participating in the foresight project regarding a certain topic. The term originates from the Latin word *expertus* (experienced) and the English word *panel* (list).

Expert panels are particularly effective in accomplishing the following objectives:

- Monitor the changes in the area under study.
- Obtain consistent peer reviews from qualified professionals throughout the project.

Advantages:

- Continuous involvement of experts in the foresight process.
- Participation of experts from different fields.
- More accurate forecasts based on tacit expert knowledge.

Limitations:

- Risks of insufficiently substantive expert evaluation or of the expert group collapsing during the foresight process.
- Risks of psychological incompatibility of expert panel participants: polarisation, group thinking, cognitive distortions, etc.
- Risks of biased expert assessments due to inadequate organisation of the group work.

Expert panels are used in most large-scale foresight projects, e.g. in the Sixth S&T Foresight of the Republic of Korea 2021–2045 (2022[25]), in the development of

[23] NISTEP (2019) An Overview of the 11th Foresight Survey in Japan // Innovation and Development Policy. No. 3. P. 79—90.

[24] RF Government (2014) Russian S&T Foresight 2030. http://government.ru/news/9800/ (accessed on: 18.02.2023).

[25] KISTEP (2022) The sixth Science and Technology Foresight (Republic of Korea). https://www.kistep.re.kr/board.es?mid=a20401000000&bid=0046&act=view&list_no=42923 (accessed on: 09.02.2023).

the EU New Horizons S&T Policy Programme (2018[26]), the List of Critical Technologies of the Russian Federation (2011[27]), and so on.

2.10 Expert Interviews

Definition: Expert interviews are conducted to obtain implicit expert knowledge via a structured conversation with highly qualified professionals or opinion leaders.

Expert interviews are aimed to accomplish the following:

- Obtain an expert opinion on the current and target state of the research subject.
- Formalise experts' empirical knowledge about the area under consideration.

Advantages:

- Rapid immersion in the relevant professional context.
- A chance to obtain a "genius forecast" about the future development paths for the subject area.

Limitations:

- Risks of the respondents' lobbying for their own vested interests.
- The need for additional validation of expert assessments.
- Risks of involving insufficiently competent experts.

Expert interviews are one of the most common foresight techniques. The following international cases are worthy of note: the EU New Horizons S&T Policy Programme;[28] the South African Science, Technology, and Innovation Foresight (NACI, 2021[29]); and the Thailand Digital Technology Foresight 2035 (2019[30]). In Russia, this method was also seen as fundamental in the implementation of most foresight projects; e.g. expert interviews were conducted in the scope of such studies as Russian S&T Foresight 2030 (RF Government, 2014[31]), Roadmap for S&T Development of the Civil Aircraft Industry Until 2025, and others.

---

[26] Gheorghiu R., Dragomir B., Andreescu L., Cuhls K. M. (2017) New horizons: Data from a Delphi Survey in Support of Future European Union Policies in Research and Innovation. Luxembourg: Publications Office of the European Union.

[27] Grebenyuk A., Shashnov S. (2012) Critical technologies in Russia: identifying and implementation // Maintenance Problems. Vol. 4. P. 105—116.

[28] Gheorghiu R., Dragomir B., Andreescu L., Cuhls K. M. (2017) New horizons: Data from a Delphi Survey in Support of Future European Union Policies in Research and Innovation. Luxembourg: Publications Office of the European Union.

[29] NACI (2019) South Africa Foresight Exercise for Science, Technology and Innovation. Synthesis Report. Pretoria: Nacional Advisory Council on Innovation.

[30] DEPA (2019) Thailand Digital Technology Foresight 2035. Frost & Sullivan.

[31] RF Government (2014) Russian S&T Foresight 2030. http://government.ru/news/9800/ (accessed on: 18.02.2023).

2.11 Competitive Intelligence

Definition: Competitive intelligence is "the process of collecting and disseminating practical information for the development and implementation of strategic programmes" (Calof, 2020[32]).

Three levels of competitive intelligence are commonly identified, which correspond to the organisation management levels:

- Strategic: the company's external and internal environments are analysed to identify opportunities and risks for long-term goal setting, prepare forecasts, and develop recommendations.
- Tactical: events and information related to the company's daily operations (competitors, partners, suppliers) are analysed to clarify its short-term goals and objectives (for up to 3 years).
- Operational: unforeseen changes, sudden risks and threats which can affect the company's activities in the short term are analysed to quickly adapt to the current conditions (Grosheva, 2020).

Advantages:

- Competitive intelligence allows one to monitor industry trends, expectations, and technologies, and forecast future developments based on a large amount of data.
- It helps increase the return on investments in decision-making by reducing the number of initiatives (prioritising strengths).
- It allows one to obtain operational information about political, regulatory, or other changes potentially capable of affecting the company/industry and adapt to them.
- It allows one to predict competitors' behaviour, use their experience and positions to increase the competitiveness of one's own products, services, and research.
- It allows one to more quickly bring products and services to the market through in-depth trend analysis and the adoption of clear performance indicators (KPIs).

Limitations:

- Collecting the correct information for analysis may turn out to be difficult: one must be aware of all important events, gather information in the "field", and use reliable sources.
- Data quickly becomes obsolete and must be constantly updated. Implementing automated data collection tools partially removes this limitation.

Many large European and US companies conduct competitive intelligence on an ongoing basis and have integrated these activities into their organisational structure.

[32] Calof J. (2020) Competitive Intelligence through the Eyes of the Professionals. Форсайт, № 14(3), pp. 6–13.

In particular, Motorola, Xerox, Avnet, IBM, Procter & Gamble, LexisNexis, and others have relevant divisions.

2.12 Big Data Mining

Definition: Big data mining is a set of methods, tools, and approaches for working with large arrays of heterogeneous (unstructured) information. Typically, such approaches include collecting, structuring, and processing data, which allows the organisation to "extract" additional value from the information it gathered.[33]

Big data mining one allows to:

- Obtain new knowledge from a large "layer" of data.
- Automate analysis, making it more standardised and objective.
- "Substantiate" previously made decisions with a large amount of information, its quick collection, and a wide variety of big data processing tools
- Improve the efficiency of specific analytical operations, and the project as a whole.

Advantages:

- The possibility of using expert big data analysis to develop corporate strategies and "predict" the results of their implementation.
- More valid decisions due to reduced subjectivity.
- Increased "flexibility" of organisations (including public sector ones) that mine big data online (streaming analytics).

Limitations:

- Implementing changes related to the use of big data in organisations' business processes and IT infrastructure (including public sector ones) is a complex task.
- Big data mining requires a lot of financial and human resources.

Examples of the use of big data for predictive research purposes include "Analysis of limitations and opportunities for the development of the European transport sector using big data" (2020[34]); "Semantic data analysis to set S&T development priorities for South Africa for 2020–2030" (2020–2021[35]). In Russia, big data mining methodology was applied in dozens of projects to analyse technology trends and identify new technologies, products, and markets.

---

[33] https://azure.microsoft.com/ru-ru/resources/cloud-computing-dictionary/what-is-big-data-analytics.

[34] Western Norway Research Institute (2021) Leveraging big data to manage transport operations: LeMo report. https://www.bigdata.vestforsk.no/completed (accessed on: 08.02.2023).

[35] NACI (2019) South Africa Foresight Exercise for Science, Technology and Innovation. Synthesis Report. Pretoria: Nacional Advisory Council on Innovation.

2.13 Bibliometric Analysis

Definition: Bibliometric analysis involves the application of mathematical and statistical methods to analyse publication activity.

This is conducted to accomplish the following:

- Obtain objective information about the R&D sector's performance.
- Benchmark specific countries, universities, R&D organisations, and individual scientists by their publication activity.
- Analyse the trends in specific knowledge areas.
- Identify promising research areas.
- Identify leading competence centres.

Advantages:

- Objective and measurable indicators.
- Reliable data (verifiable information).

Limitations:

- High cost of database access.
- Incomplete coverage of non-periodical publications.

Bibliometric analysis is an essential element of most foresight studies. Examples of international projects include "Digitalisation of Science, Technology and Innovation" (OECD, 2020[36]), "The Development Strategy of China's Engineering Science and Technology for 2035" (2019[37]), the Japanese 11th S&T Foresight (2019[38]), and Russian ones—"Frontiers of Science" (HSE, 2022[39]) and others.

2.14 Patent Analysis

Definition: Patent analysis involves analysing data on new technological solutions.

This is carried out to accomplish the following:

- Assess the effectiveness of S&T activities.

[36] OECD (2020) The Digitalisation of Science, Technology and Innovation. Key developments and policies. https://www.oecd-ilibrary.org/sites/b9e4a2c0-en/1/2/2/index.html?itemId=/content/publication/b9e4a2c0-en&_csp_=e6f980ccaddecf7f87720ff84a3888b2&itemIGO=oecd&itemContentType=book (accessed on: 25.02.2023).

[37] Chinese Academy of Engineering Science (2019) The Development Strategy of China's Engineering Science and Technology for 2035. Beijing: China Science and Technology Publishing and Media Co., Ltd. (на китайском языке).

[38] NISTEP (2019) An Overview of the 11th Foresight Survey in Japan // Innovation and Development Policy. No. 3. P. 79—90.

[39] HSE (2022b) Frontiers of Science M.: HSE.

- Evaluate the quality of S&T results and the level of technology readiness.
- Analyse the development of specific technology areas and identify emerging technologies.
- Assess the market potential for technology developers.
- Identify key developers of technological solutions and patent activity leaders.

Patent analysis uses the following indicators: number of patent applications, inventive activity rate, number of patent applications by technology area; number of issued (received) patents, number of valid patents, number of intellectual property items applied at one's own production facilities, technological dependence rate, and technological specialisation index.

Advantages:

- Patent analysis objectively reflects the creation of new patentable solutions.
- It allows one to use patent information to accomplish various research objectives.

Limitations:

- High cost of database access.
- A wide range of alternative intellectual property protection methods are not covered by patent analysis.
- The descriptive nature of this technique, the need to combine it with other approaches to enhance the analytical potential.

Patent analysis is carried out over the course of desk research at the preliminary stage of a foresight study. This method was applied in such international projects as "The Development Strategy of China's Engineering Science and Technology for 2035" (2019[40]) and "Patent Analysis of the Diffusion of Artificial Intelligence Technologies" (US Patent Office, 2020[41]). Among the Russian projects, one could note the "Analysis of patent activity in high-technology areas" (HSE, 2021–2023[42]).

[40] Chinese Academy of Engineering Science (2019) The Development Strategy of China's Engineering Science and Technology for 2035. Beijing: China Science and Technology Publishing and Media Co., Ltd. (на китайском языке).

[41] USPTO (2020) Public views on artificial intelligence and intellectual property policy. https://www.dwt.com/-/media/files/blogs/artificial-intelligence-law-advisor/2021/09/uspto%2D%2Dai-report%2D%2D10-2020.pdf (accessed on: 25.02.2023).

[42] HSE (2022a) Development of certain high-technology areas. White paper. M.: HSE.

2.15 System Dynamics

Definition: System dynamics (SD) can be described as a method for modelling and simulating complex physical and social systems[43] (Forrester, 1958[44]).

The SD approach is applicable to dynamic problems arising in complex social, managerial, economic, or ecological systems—literally any dynamic system characterised by interdependence, information feedback, and circular causality.

The SD approach is based on continuous top-down analysis, which makes it particularly useful for managing long-term strategic projects.

Advantages:

- It helps one understand patterns in dynamic systems and their underlying structures.
- This is a comprehensive methodology capable of crystallising systemic thinking in realistic models.

Given these strengths, the method has found application in the exact and natural sciences and humanities. The use of SD has especially grown in corporate, social, and environmental projects.

Limitations:

- Lack of a methodological basis for analysing and modelling relationships between organisational subjects, which can be provided by organisational cybernetics or the living systems theory. System dynamics cannot explain how organisations grow from a strategic point of view. Similarly, system dynamics cannot predict future development if any kind of random events or qualitative changes in the environment occur, such as, e.g. technological revolutions or economic crises.
- This method cannot take into account organisational diversity[45] (complexity). Though it does allow for comprehensively assessing the issue under consideration, it has no potential for designing a model structure that would take into account the complexity surrounding the studied system.
- System dynamics modelling is based on an inductive approach which is inconvenient for models describing the individual behaviour of a large number of agents. When it comes to issues SD cannot handle on its own, it must be supplemented by other methods.

[43] Complex systems are a subset of the systems theory that focuses on group behaviour of interacting objects, but covers a much broader class of systems including non-complex ones, or which simplified conventional models can still be viable.

[44] Forrester J. W. (1971) World dynamics, Cambridge Mass.

[45] Diversity is a technical term for complexity which refers to a large number of potential system states or behaviours (Ashby, 1956; Beer, 1966).

There are two particularly popular application areas for the SD approach in foresight studies: system dynamic market modelling and quantitative assessment of growth areas (or rather, product and service innovations).

As to the cases of practical usage of this method, a project commissioned by the Russian Ministry of Education and Science to design a system forecasting Russian enterprises' demand for highly skilled personnel should be noted,[46] in the scope of which dynamic predictive models were developed for the Russian Federation as a whole and for the country's 83 regions in particular.

Acknowledgements The chapter was prepared within the framework of the Basic Research Program at the HSE University.

[46] https://www.anylogic.ru/resources/case-studies/forecasting-human-resources-demand-in-russian-economy/.

Part II
Using Foresight for Science, Technology and Innovation: The Case of South Africa

South African Foresight for STI (SAForSTI): Background, Rationales, and Objectives

Mlungisi Cele, Ozcan Saritas, and Alexander Sokolov

Abstract This chapter provides a concise overview of the South Africa Foresight Exercise for Science, Technology, and Innovation (SAForSTI). The primary objective of this exercise was to explore the future landscape of science, technology, and innovation (STI), with a focus on leveraging STI to enhance the well-being of all South Africans by addressing socio-economic challenges inclusively and sustainably. The outcomes of SAForSTI have played a pivotal role in shaping the priorities outlined in the new 10-year plan for STI leading up to 2030. When determining the STI areas to be scrutinised and analysed through the foresight process, several factors were considered, including: (i) Established areas of STI functioning effectively with minimal expected change, (ii) Areas demonstrating high potential for growth, and (iii) Emerging STI areas within the South African national innovation system. SAForSTI employed diverse methodologies informed by both global best practices and local context, with the analysis phase benefiting from big data analytics. A key aspect of SAForSTI was extensive stakeholder engagement. Ultimately, SAForSTI identified seven STI domains, each with its associated priorities or thrusts, including the Circular Economy, High-tech Industrialisation, Nutrition Security, ICTs and Smart Systems, Health Innovation, Sustainable Energy, and Education for the Future.

These domains are intricately linked to South Africa's prevailing challenges encompassing inequality, poverty, unemployment, food security, health, education, water, and climate, among others. Addressing these systemic societal challenges necessitates long-term planning and policy interventions spanning technological, economic, and social dimensions, transcending national boundaries. Concurrently,

M. Cele
National Advisory Council on Innovation, Pretoria, South Africa
e-mail: mlungisi.cele@dst.gov.za

O. Saritas (✉)
Rochester Institute of Technology Dubai, Dubai, UAE
e-mail: oxscad1@rit.edu

A. Sokolov
National Research University Higher School of Economics, Moscow, Russia
e-mail: sokolov@hse.ru

O. Saritas et al. (eds.), *21st Century Foresight*, Science, Technology and Innovation Studies, https://doi.org/10.1007/978-3-031-69557-5_5

South Africa finds itself amidst an era of unparalleled technological advancements permeating various facets of human and planetary existence. This includes remarkable progress in information and communications technology (ICT), biotechnology, nanotechnology, the Internet of Things (IoT), robotics, artificial intelligence (AI), machine learning, blockchain technology, and additive manufacturing, among others. Like other developing nations, South Africa is navigating ways to harness the opportunities presented by these technologies while also preparing to mitigate unintended negative consequences. A youthful population can serve as a catalyst and significant asset amidst the rapid acceleration of technological innovation and socio-technological transformations associated with emerging technologies.

This chapter draws insights from the final synthesis report of the South Africa Foresight Exercise for Science, Technology, and Innovation 2030 published by the National Advisory Council on Innovation, Department of Science and Innovation, Republic of South Africa (NACI, 2019, South Africa foresight exercise for science, technology and innovation 2030, synthesis report, National Advisory Council on Innovation, Department of Science and Innovation, Republic of South Africa. Retrieved June 3, 2024, from https://www.naci.org.za/wp-content/uploads/2020/07/South-African-Foresight-Exercise-For-Science-Technology-and-Innovation-2019.pdf).

1 Introduction and Background

Foresight is "a systematic, participatory, future-intelligence-gathering and medium to long-term vision-building process aimed at enabling present-day decisions and mobilising joint action" (Miles & Keenan, 2002). Foresight, and ForSTI, when it is applied to issues of STI: (a) considers multiple futures, which may include possible, plausible and desirable futures; (b) is a participative process and (c) is action-oriented (Miles et al., 2016). Foresight is a process that seeks to create shared "visions" of the future, appraisals that stakeholders are willing to endorse by the actions they choose to take today. Foresight empowers decision-makers and policy planners to use new ways of thinking and talking about the unfolding future and implementing strategic plans that are compatible with such a future. The premise of foresight is that the future is still in the making and can be actively influenced or shaped, rather than already decided by past thinking and actions, and therefore to be merely awaited and accepted (NACI, 2019).

Foresight empowers the governments of developing countries to create their own visions of the future, rather than depending on the perspectives of highly developed nations, which may not be applicable to their specific contexts. This approach allows public service organisations to better anticipate future policy environments and provide decision-makers with a broader range of options for achieving inclusive growth and social justice. By incorporating participatory Foresight, countries can reduce their dependence on foreign or local technical experts and actively involve citizens in shaping discussions and decisions about their future.

Foresight has become a well-established component of strategy, decision-making, and advocacy in both public and private sectors in many countries. This widespread adoption and institutionalisation of foresight capabilities can be seen in countries like the USA, Finland, the United Kingdom, the European Union, South Korea, Japan, and Russia. It is also prevalent in organisations such as the United Nations and the Organisation for Economic Co-operation and Development (OECD), as well as in most multinational corporations and civil society groups focused on issues like environmental or technological activism (NACI, 2019).

In South Africa, the period from 2014 to 2019 was marked by critical reflection on the implementation of Science, Technology, and Innovation (STI) policy and its influence on the country's future. The Department of Science and Innovation (formerly the Department of Science and Technology) undertook several initiatives, including a review of the 1996 White Paper on Science and Technology, a situational and performance analysis of the national system of innovation, and the initial phase of the STI Institutional Landscape Review. Consequently, a new White Paper on STI was developed and approved by the Cabinet in March 2019. The objectives of this White Paper were as follows:

- "Improved coherence and coordination
- Increased partnering between business, academia, government and civil society in the national system of innovation (NSI).
- Strengthened and transformed NSI institutions.
- Increased human capabilities.
- Expanded research enterprise.
- Enhanced enabling environment for innovation.
- Improved funding across the NSI.

In pursuing these objectives, the aim was formulated as follows:

- To ensure full alignment with the NDP objectives.
- To expand on successes and propose new approaches where necessary.
- To take advantage of opportunities presented by megatrends.
- To promote transformation" (NACI, 2019, p.22).

Building on past successes and introducing new strategies where necessary, the White Paper on STI outlines a long-term policy framework for the South African government to enhance the role of science, technology, and innovation in fostering a more prosperous and inclusive society. It highlights inclusivity, transformation, and partnerships as central themes, and suggests various actions to improve policy coherence, develop human capacity, expand knowledge, boost innovation performance, and increase investment.

According to the White Paper on STI, the role and functions of NACI will be expanded to include monitoring and evaluation, coordination (through interministerial and Presidency-led mechanisms), and the enhancement of planning and agenda-setting via the institutionalisation of foresight exercises. Future Foresight exercises should build on and learn from previous initiatives since the adoption of the 1996 White Paper on Science and Technology. These efforts include the National

Research and Technology Foresight exercises of the 1990s and SAForSTI, which provide insights into potential futures for South Africa and priority options for the new STI decadal plan that will serve as an implementation tool for the White Paper on STI.

SAForSTI aimed to explore the future application of Science, Technology, and Innovation (STI) to tackle societal, sustainability, and economic challenges, foster inclusive and sustainable growth, and enhance quality of life and well-being in South Africa. The specific tasks of the foresight exercise included the following:

- "To identify local and global drivers or trends likely to have a major influence on STI and society in the next decades
- To analyse the performance of the NSI in order to identify its weaknesses and strengths, and to characterise the current policy mix.
- To identify opportunities and competitive advantages in the National System of Innovation.
- To provide an understanding of what is being achieved and assess whether changes should be made.
- To make recommendations on the following:
  - Possible visions or futures, using a time horizon of 2030.
  - Key underlying principles.
  - Policy options for future priorities/grand challenges for the new decadal plan" (NACI, 2019, p.23).

2 Methodology

The STI Foresight process developed for South Africa utilised a variety of data and information collection methods, including desktop studies, big data analyses, online voting, high-level stakeholder interviews, workshops, and expert contributions. Prior to the workshops and online voting, selected stakeholders participated in a foresight training course to establish a shared understanding of the foresight process. Figure 1 illustrates the overall process followed.

In summary, SAForSTI consisted of three phases:

1. During the preparation and initiation phase, several desk studies were undertaken, which served as inputs for the next phase. In addition, planning and preparation took place to facilitate the implementation of the remainder of the project.
2. During the STI domain identification and selection phase, the top-priority STI domains were identified, using workshops and online processes.
3. The STI thrust identification and prioritisation phase was centred on two stakeholder workshops, during which possible futures in each domain were envisaged, and related proto-scenarios were developed (NACI, 2019, p.32).

The output was a collection of proposals for STI thrusts. Both Phase 2 and Phase 3 involved interviewing a variety of STI stakeholders to gather input for identifying

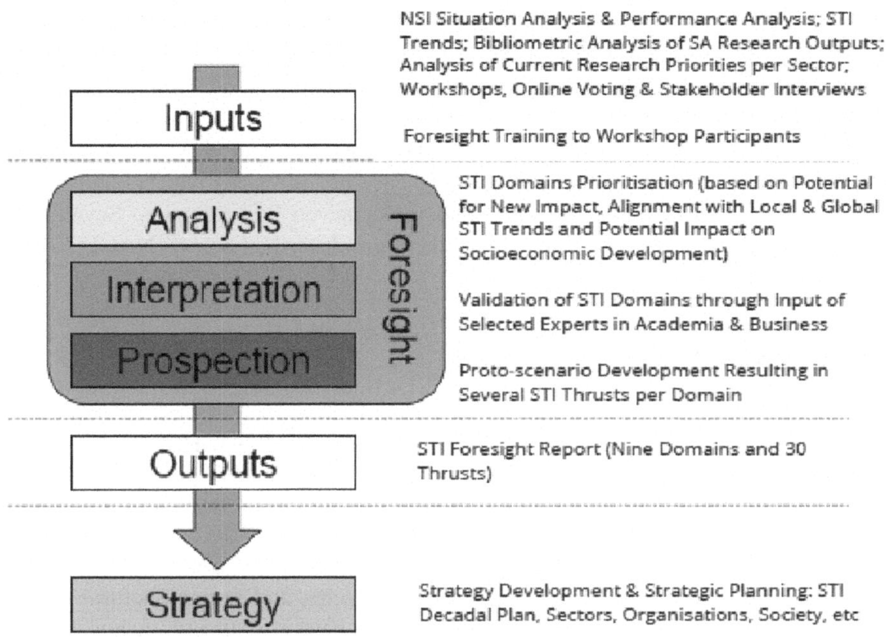

Fig. 1 Summary of the STI Foresight process for South Africa. Source: NACI (2019), p.32

key domains and thrusts. Big data analysis spanned these phases, encompassing bibliometric analyses of South Africa's scientific outputs and semantic analyses of extensive document sets to pinpoint emerging and evolving STI-related trends. The activities associated with these phases took approximately 17 months to complete.

3 Foresight Process

The National STI Foresight 2030 study identified seven STI domains and their associated thrusts. STI domains are broadly defined but focused areas related to science, technology, and innovation, as well as societal needs or issues. The study identified multiple thrusts within each STI domain. Thrusts are STI-related priority areas that outline South Africa's objectives for 2030. Throughout the Foresight process, various STI areas were identified and prioritised based on criteria such as their potential for new impact, alignment with global and STI trends, and contribution to socio-economic development. Initially, eight STI domains were selected through online and offline consultations with stakeholders in South Africa's STI system. This number was then reduced to seven by merging two of the domains.

Thrusts were determined through the following stages of the Foresight study. First, a long list of thrusts was proposed through a Foresight workshop. Then, these were shortlisted using a new set of prioritisation criteria by considering their importance and feasibility. The importance criteria included the (i) socio-economic impact; (ii) new impact; and (iii) strategic value of each thrust. The feasibility criteria included (i) availability of required knowledge and expertise; (ii) availability of institutional capacity; (iii) availability of infrastructure; (iv) required policy and regulatory environment in place; (v) social and ethical acceptability; (vi) amount of relevant funding currently allocated; and (vii) ease of addressing barriers and obstacles. A total of 25 thrusts were identified under seven STI domains. Seven STI domains and thrusts identified for each of them are described below (NACI, 2019).

3.1 Circular Economy

South Africa's economy currently follows a traditional, linear production model (make, use, dispose). In contrast, a circular economy keeps resources in use for as long as possible, maximising their value during use, and then recovering and regenerating products and materials at the end of their service life. This approach involves converting biological and non-biological waste into new resources and materials. Unlike the linear model, a circular economy recycles and remanufactures waste products, significantly reducing the amount of waste requiring disposal. The Circular Economy domain focuses on creating products that are restorative and regenerative by design, allowing them to circulate through the economy repeatedly, thus minimising waste. This includes converting waste into new resources and materials, as well as restoring and protecting biodiversity (NACI, 2019). Four thrusts were identified in the Circular Economy domain, including:

(a) CE1: Reduce, reuse and recycle waste.
(b) CE2: Ensuring sustainable water, energy and food (agriculture) security.
(c) CE3: Low-carbon and climate-resilient economy.
(d) CE4: Smart connectivity and mobility in communities.

While numerous nations are actively pursuing the development of a circular economy, South Africa finds itself on a fresh trajectory in this regard. A comprehensive waste strategy, founded on innovation, transdisciplinarity, and science, technology, and innovation (STI), is imperative to curtail expenses, foster job creation and micro-industries, and bolster the economy, public health, environmental integrity, and biodiversity preservation.

Analysing big data has revealed that South Africa possesses considerable and pertinent research capabilities to tackle circular economy challenges, particularly in agricultural and biological sciences, environmental science, and earth sciences. Sustained funding for research in these domains is crucial. Moreover, there are several areas warranting further investment to fortify research capacities, such as engineering, materials science, chemistry, chemical engineering, biochemistry, genetics and molecular biology, energy, and the social sciences.

The execution of these initiatives could entail the establishment, through strategic collaborations, of pilot sites in rural, peri-urban, and urban settings to showcase possibilities and facilitate broader implementation. Such endeavours necessitate political endorsement to commence, necessitating the formation of a coordinating and implementing body within a designated government department.

3.2 High-Tech Industry

South Africa must swiftly adopt cutting-edge manufacturing methods to enhance its competitiveness and counteract the looming threats of job losses and industrial decline associated with the onset of the Fourth Industrial Revolution (4IR). Within the realm of High-tech Industrialisation lie advanced manufacturing techniques and technologies such as robotics, artificial intelligence (AI), the Internet of Things (IoT), and additive manufacturing. Although South Africa boasts pockets of excellence in advanced manufacturing, particularly in additive manufacturing, there's a pressing need to elevate these capabilities. Identifying niche sectors where global competitiveness is feasible and directing efforts towards leveraging advanced manufacturing systems for the beneficiation of natural resources is essential. The integration of advanced manufacturing holds the promise of fostering the emergence of new Small, Medium, and Micro Enterprises (SMMEs) in South Africa (NACI, 2019). Three STI thrusts for the High-tech Industrialisation domain were identified and elaborated on during SAForSTI:

(a) HT1: Enabling small business to adopt high tech.
(b) HT3: New thinking for new industries.
(c) HT4: New thinking for old industries.

The High-tech Industrialisation domain aims to support small businesses in penetrating the high-tech market and leveraging technology for their growth, aligning with the National Development Plan's (NDP) emphasis on job creation and employment expansion. Simultaneously, fostering new high-tech sectors and revitalising existing industries are crucial for enhancing the competitiveness and efficiency of the country's industrial landscape through innovative combinations of traditional, contemporary, and advanced technologies. Despite noticeable growth in scientific research focus within this domain, particularly in recent years, South Africa's research output remains notably limited, with substantial deficiencies in research capacity, particularly in engineering, materials science, computer science, mathematics, and decision sciences. Rectifying this requires significant investment and support to catalyse change (NACI, 2019).

3.3 Nutrition Security

The Nutrition Security domain explores how agriculture, particularly precision agriculture, can contribute to enhancing nutrition security in the country while also mitigating the effects of climate change. Adequate nutrition is fundamental for alleviating poverty and fostering healthy populations, especially for disadvantaged South Africans. Unfortunately, malnutrition and stunting afflict 25% of children under the age of five, detrimentally impacting their cognitive and physical development and exacerbating existing socio-economic disparities. Concurrently, there is a concerning rise in the prevalence of overweight and obese adults and children, affecting over 60% of South African women. Integrated and transdisciplinary solutions are imperative to tackle both nutritional challenges and agricultural reforms in response to climate change (NACI, 2019). Four STI thrusts for the Nutrition Security domain were identified and elaborated on during SAForSTI:

(a) NU1: Zero impact agriculture.
(b) NU2: Use and acceptance of modern biotechnology.
(c) NU3: Personalised information for healthy nutrition for all.
(d) NU4: Precision and big data in agri-businesses.

South Africa demonstrates notable strengths in research output, particularly within the Nutrition Security domain, where there is a growing scientific focus. It ranks second highest among BRICS countries in terms of publications per GDP and per population size. Nevertheless, further investment is essential to enhance specialisation in key disciplines relevant to this domain, such as biochemistry and energy, and to elevate the quality of research outputs, especially in fields like the social sciences (NACI, 2019).

3.4 ICTs and Smart Systems

This thrust seeks to ensure that broadband Internet access is both affordable and readily available to everyone, while also fostering the creation of innovative applications and services rooted in the Fourth Industrial Revolution (4IR). It tackles the pressing challenge of lowering the considerable expenses associated with ICT infrastructure and rectifying the existing disparities in its distribution, which perpetuate the rural/urban digital gap. Furthermore, it aims to address the deficiency in digital skills, particularly in rural and marginalised communities (NACI, 2019). Four thrusts have been identified under this domain:

(a) IT1: Checks and balances for a digital future.
(b) IT2: ICT infrastructure and Internet access.
(c) IT3: Big data, data analytics and decision support.
(d) IT4: Smart and sustainable municipal service delivery.

One of the foremost challenges lies in effectively coordinating research endeavours and mobilising pertinent stakeholders, spanning from national to local government, industry, NGOs, and communities. Establishing partnerships with industry players, adhering to global standards, and collaborating with regulatory bodies are imperative to ensure the interoperability and compatibility of ICT systems. Adequate allocation of radio spectrum, along with mechanisms for spectrum sharing, is essential. The envisioned outcome is a seamless ICT infrastructure that extends affordable broadband Internet access to all, bridging the digital divide and bolstering South Africa's GDP. This will foster improved communication and information exchange among stakeholders within the ICT ecosystem—government, industry, and society—while nurturing a burgeoning Small, Medium, and Micro Enterprises (SMMEs) sector comprising network and ICT service providers. Despite the increasing scientific focus on the ICT domain in South Africa, the publication output remains notably low (0.06% of global output), as does the quality of research outputs in most relevant areas. Hence, strategic investment in research capacity pertinent to the ICT domain warrants consideration (NACI, 2019).

3.5 *Health Innovation*

Health stands as a pivotal domain in South Africa's Science, Technology, and Innovation (STI) landscape. Rather than simply adopting and modifying health solutions tailored for affluent First World populations, the focus should be on addressing the needs of local marginalised communities. This domain is intricately linked to the country's challenges regarding poverty, where substandard living conditions invariably lead to compromised health outcomes. Furthermore, there is a pressing need to deepen our understanding of the impact of climate change on human health, particularly among marginalised groups. By fostering innovative local solutions, South Africa has the potential to not only address its own health concerns but also to transfer these solutions to other African nations, thereby creating employment opportunities and generating revenue. Overall, there's a necessity to optimise the South African health system to enhance healthcare delivery, drug development, and diagnostic and treatment services (NACI, 2019). Three STI thrusts were identified and elaborated on for the Health domain during SAForSTI, including:

(a) HE1: Optimisation of health systems.
(b) HE2: Improving the quality of healthcare.
(c) HE3: Digitisation of health systems.

The three focal points within the Health domain aim to enhance healthcare efficiency by ensuring quality support systems and fair allocation of resources, facilitated by improved supply and procurement mechanisms and swift diagnostic systems. This encompasses the establishment of an integrated national patient record system accessible to all healthcare providers, including those in remote areas. Advancements in telemedicine and remote diagnostics systems will enable healthcare workers to expand their roles, fostering the potential for increased self-directed healthcare at home in the long run. The integration of Information and Communication Technology (ICT) into healthcare endeavours will contribute to cultivating a healthier populace, benefiting society.

While there has been a notable expansion in scientific research output within the Health domain in South Africa, there is a call for greater collaboration between the natural and social sciences to comprehensively grasp the drivers and challenges associated with the adoption of proposed healthcare systems. An investment strategy is imperative to enhance the quality of pertinent social science research outputs, and further research is warranted to elucidate the change management processes and educational needs for optimal utilisation and adoption of technologies aimed at bolstering healthcare services (NACI, 2019).

3.6 Sustainable Energy

As society progresses, energy has emerged as a fundamental requirement for sustaining a productive life. While major urban centres in South Africa typically enjoy a stable energy supply, rural and peri-urban marginalised communities often lack access to reliable energy sources. Leveraging clean and affordable energy options such as solar, wind, and bioenergy presents an opportunity for these communities to embrace new technologies without the necessity of replacing existing ones (NACI, 2019). Four STI thrusts for the Energy domain were identified and elaborated on during the second phase of SAForSTI:

(a) EN1: Clean, affordable and sustainable energy for all.
(b) EN2: Renewable energy sources and technologies.
(c) EN3: Energy efficiency solutions for industry plus household use.
(d) EN4: Distributed energy generation and storage.

Considering the persistent energy crisis in South Africa and its widespread ramifications on the economy and society, exploring sustainable energy production alternatives has become imperative. There is a noticeable surge in scientific focus on the Energy domain within South Africa, with publication rates outpacing those in other Science, Technology, and Innovation (STI) domains. However, the specialised expertise in crucial fields like energy, chemistry, chemical engineering, and biochemistry remains inadequate and will necessitate substantial additional investment to meet the escalating energy needs of a growing population and the imperative for sustainable economic advancement (NACI, 2019).

3.7 Education for the Future

The Education for the Future domain focuses on expanding learning access and enhancing learning outcomes by leveraging technology and integrating relevant findings from neuroscience research. It underscores the pressing necessity to cultivate educators, students, and graduates capable of navigating the demands of the Fourth Industrial Revolution (4IR) and tackling inclusivity and sustainability issues in education. South Africa is tasked with ensuring that quality education is accessible to all citizens as a fundamental human right, serving as the bedrock for equitable society and economic prosperity (NACI, 2019). These objective gains particular significance amid the evolving skill sets demanded by the onset of the 4IR. Three STI thrusts were identified and expanded on for the Education domain during the second phase of SAForSTI:

(a) ED1: Skills for the fourth Industrial Revolution.
(b) ED2: Inclusive innovation and development.
(c) ED3: Curriculum development 2030.

The Education for the Future domain presents a comprehensive array of initiatives aimed at preparing individuals for upcoming challenges. These include fostering technical skills essential for emerging fields like Artificial Intelligence (AI), as well as nurturing soft skills such as critical and creative thinking. Lifelong learning is emphasised through programmes like Early Childhood Development (ECD), Adult Basic Education and Training (ABET), and artisan development, all tailored to industry demands. Curriculum reform is proposed to integrate entrepreneurship education across disciplines, supplemented by research and development efforts to pioneer new teaching methodologies. However, bibliometric analysis reveals a notable shortage of research in STEM education fields within South Africa. While existing endeavours predominantly focus on arts, social sciences, and psychology, some attention is directed towards neuroscience. Increased funding for STEM educational research is imperative to identify and implement resilient strategies that equip individuals for the challenges of the 4IR (NACI, 2019).

4 Results and Discussion

The advancement of globally competitive science and technology is a fundamental driver of national wealth creation and societal progress, representing a key prerequisite for success for major nations like South Africa. The strategic identification of promising research fields and relevant applications of scientific discoveries is pivotal in shaping national science and technology policy. The "South Africa Science, Technology, and Innovation (STI) Foresight for 2030" delineates seven STI domains and associated priorities (thrusts) to guide national innovation strategies in the coming decade.

The evaluation of current national capacities within these domains and thrusts, as presented in the foresight study, relies on a comprehensive semantic analysis of big data and an examination of publication trends. SAFEST offers detailed insights into the country's research landscape, strengths and weaknesses, comparisons with global leaders and premier research institutions, scientific specialisations, and international collaborations. A plethora of visual aids, including presentations, graphs, and tables, provide a comprehensive overview of national science and technology capacities, while also spotlighting key economic and societal challenges that can be addressed through these capacities.

New technologies must align with societal needs, and in turn, shape societal evolution. The analysis of STI domains underscores the crucial co-evolutionary relationship between science, technology, and society. While South Africa shows strong positioning in certain STI areas, such as ICTs, there are notable gaps in others, highlighting the need for strategic interventions to ensure balanced progress across research, capacities, skills, and infrastructure development. The SAForSTI underscores the feasibility of STI goals and targets while emphasising the importance of integrated strategies to address cross-cutting topics that span multiple domains, often articulated in various ways. Some of these significant topics are elaborated below.

There are many topics that were mentioned or highlighted in multiple domains, sometimes articulated in different ways. Some of the more important of these are elaborated below (NACI, 2019).

4.1 Climate Change

Climate change denotes a substantial, protracted alteration in the anticipated patterns of typical weather conditions within a region. Its repercussions are far-reaching and have the potential to impede the attainment of established priorities across various domains. Consequently, addressing the challenge posed by climate change is imperative across numerous sectors. For instance, within the Circular Economy domain, a critical impetus is to mitigate the effects of climate change by curbing the current pace of environmental degradation through transitioning to a low-carbon economy, thereby striving for a more sustainable climate equilibrium. Moreover, the circular economy endeavours to adapt to the inevitable impacts of climate change by fostering a more resilient economic framework. In the Nutrition Security domain, research efforts will be directed towards developing drought-resistant crops, enhancing production systems, combating diseases, and managing water and energy resources efficiently, all geared towards alleviating the adverse effects of climate change (NACI, 2019).

4.2 Inclusive Access to ICTs

Improved and inclusive access to connectivity, along with the foundational ICT infrastructure, is indispensable for effectively addressing numerous priorities across the seven domains. For instance, endeavours such as ensuring water security, implementing precision farming techniques, facilitating technology-mediated learning, and establishing a comprehensive national digital health system all hinge on widespread ICT access among citizens.

The advent of the 4IR necessitates a well-educated workforce equipped to navigate the demands of technological advancements such as Artificial Intelligence (AI), the Internet of Things (IoT), and automation. This entails ensuring that recent graduates enter the workforce with the requisite skills and providing substantial retraining and skill enhancement opportunities for existing employees. Given the scale of skills training required, technology-mediated approaches will play a pivotal role, contingent upon widespread connectivity and accessibility to such technology (NACI, 2019).

4.3 Big Data Analytics

Most of the aforementioned domains necessitate the capacity for both intensive and extensive utilisation of data and data analysis to fulfil their defined objectives. For instance, the establishment of a comprehensive, digitised health system serves as a foundational requirement for achieving various goals, including remote diagnosis and more accurate epidemiological studies. Moreover, the dependence on data generated for monitoring diverse activities, such as advanced manufacturing processes, patient health record systems, and the data-driven implementation of precision farming to enhance agricultural productivity and crop nutritional quality, underscores the importance of big data and data analytics capabilities.

Enhanced proficiency in data analytics, characterised by a greater number of individuals possessing deeper expertise, will be imperative across numerous domains, including Water Security, the Circular Economy, Health, High-tech Industrialisation, ICTs, and Nutrition Security (NACI, 2019).

4.4 Entrepreneurship

Entrepreneurship has historically been intricately intertwined with past industrial revolutions, and the 4IR is no different—it remains a pivotal catalyst for these transformative periods. However, on a broader scale, entrepreneurial endeavours serve as the sustainable pathway for navigating the uncharted territories that individuals and society must traverse, spanning diverse realms such as high-tech industrialisation,

ICTs, and nutrition security. It is through the successful entrepreneurial ventures, particularly in small-scale yet highly efficient agricultural production setups, that nutrition security can be achieved while concurrently alleviating prevalent levels of poverty and unemployment.

In both established and emerging industries, fostering innovative thinking necessitates robust entrepreneurial support to harness the potential advantages of high-tech industrialisation. This underscores the importance of embedding entrepreneurship across all educational curricula, as emphasised within the Education domain (NACI, 2019).

4.5 Transport

Efficient transportation of people and goods is a fundamental aspect of a thriving economy. As such, transportation is a pervasive element that must be considered across various domains. The distribution of goods, both to urban and rural regions, holds significance for health products within the Health domain and agricultural products within the Nutrition Security domain. A robust circular economy hinges on effectively coordinated inter and intra-sectoral transportation and logistics connections. The true potential of a smart city can only be unlocked if smart transportation systems are integrated into the comprehensive smart city solution (NACI, 2019).

4.6 Knowledge, Research, and Skills

Many of the proposed thrusts underscored the imperative for augmenting knowledge, skills, and human capacities. The nature of the required knowledge, skills, and human capacities varies somewhat across domains, albeit with identifiable synergies. For instance, the Circular Economy domain necessitates the development of expertise in water, energy, and agricultural technologies, pertinent to the Water Security, Sustainable Energy, and Nutrition Security domains respectively. Moreover, there is a demand for expertise concerning the interplay between these domains.

Another instance pertains to the necessity for understanding social processes and fostering cohesion, particularly emphasised in the Circular Economy and the Future of Society domains. However, this requirement extends to other domains as well, facilitating the human-centric application of STI to effectively address social challenges in South Africa, rather than exacerbating them.

A common theme across domains is the imperative to bolster capacity to elevate expertise levels across various knowledge domains, as envisaged in the Education domain. Most thrust proposals underscore the necessity for additional research pertinent to their respective domains, albeit without delving into specific research areas extensively. Institutions equipped with research capabilities must be adequately resourced to undertake requisite research endeavours. This also entails ensuring a sufficient pool of researchers with the requisite expertise, necessitating training for additional researchers where needed (NACI, 2019).

4.7 Collaboration and Partnerships

A prominent recurring theme among the thrust proposals emphasises the necessity for collaboration among institutions and individuals in diverse capacities. This entails fostering heightened collaboration between government departments and science councils, fostering public-private partnerships, nurturing international collaborations, and enhancing engagement with both national and global experts, as well as civil society.

It is imperative to cultivate a culture of collaboration and partnership actively to expedite progress and achieve more substantial outcomes. Any obstacles to collaboration or disincentives hindering cooperation should be identified and mitigated to the greatest extent feasible. In their stead, mechanisms incentivising extensive collaboration should be instituted (NACI, 2019).

4.8 Funding Instruments

Another ubiquitous resource category emphasised across the 30 thrust proposals is the availability of funding sources. Funding is imperative for the execution of all initiatives. While the necessity for funding is universal, certain domains may identify specific options for acquiring funds, along with specialised funding mechanisms pertinent to a particular domain. For instance, redirecting some or all of the existing fossil fuel subsidies could serve as a potential funding source for the Energy domain. Conversely, financial resources are essential in the High-tech Industrialisation domain for experimental development and the commercialisation of technologies (NACI, 2019).

5 Conclusions

Foresight endeavours are typically undertaken with specific objectives, such as influencing policy, shaping strategic planning, or informing decision-making processes. The SAForSTI initiative aligns with this paradigm. Initially, its purpose was to contribute insights for formulating a 10-year plan to enact the 2019 White Paper on Science, Technology, and Innovation (STI). However, the outcomes of SAForSTI extend beyond this scope, potentially informing strategic planning endeavours conducted by both public and private sector entities. Furthermore, the knowledge gained from executing the initiative, along with the involvement of STI stakeholders in its activities, can catalyse momentum towards realising the goals outlined in the White Paper on STI. Participating in foresight exercises can yield additional benefits, such as fostering networking relationships that transcend conventional boundaries, spanning sectors, disciplines, and institutions. Consequently, leveraging multisectoral and interdisciplinary resources and expertise may foster more innovative approaches to implementing the White Paper (NACI, 2019).

Following the SAForSTI exercise, there are additional insights to be gleaned or observations to be noted. While some of these are already apparent from the description of the SAForSTI process and its outcomes, several more noteworthy points merit discussion:

- "On the whole, participation in SAForSTI by STI stakeholders was enthusiastic and substantial. While there is certainly scope to reach even greater levels of participation, the wholehearted involvement of STI stakeholders augers well for future foresight processes.
- The results of reviews and analyses of data relevant to SAForSTI formed key inputs to the exercise, both at its initiation, and at various points during the process. However, more extensive reviews and analyses that could have focused on the early outputs of the process, such as the selected STI domains, would likely have further improved the quality of the final outputs of the process.
- The analysis of the STI domains and thrusts highlighted the importance of having a reliable evaluation of the capacity in South Africa for research and development in different disciplines and in interdisciplinary areas. Such information facilitates the selection of realistic priorities that are attainable, given the resources available. It also aids in targeting resources to build needed capacity effectively and efficiently.
- Collaboration with international and local partners in the design and implementation of SAForSTI resulted in significant additions to the process and its outputs. An international partner provides an external perspective, apart from potential additional capabilities, which lends credibility to the exercise.
- The 2019 White Paper on STI calls for additional institutional capacity to enable regular foresight exercises to be undertaken, for example, to aid in directing the expansion of the NSI to meet national objectives, and in the formulation of successive decadal plans. The experience of undertaking SAForSTI serves to underscore the benefit of holding regular foresight exercises, so that foresight expertise may be extended and enhanced amongst the STI stakeholder community, resulting in higher-quality results from such exercises" (NACI, 2019, pp.137–138).

It is anticipated that SAForSTI's contribution to formulating the initial decadal plan for implementing the 2019 White Paper on Science, Technology, and Innovation will shape a brighter future for South Africa by leveraging the potential of science, technology, and innovation (NACI, 2019).

Acknowledgements This chapter draws on the final synthesis report of the South Africa Foresight Exercise for Science, Technology and Innovation 2030 published by National Advisory Council on

Innovation, Department of Science and Innovation, Republic of South Africa (NACI, 2019). The chapter was prepared within the framework of the Basic Research Program at the HSE University.

References

Miles, I., & Keenan, M. (2002). *Practical guide to regional foresight in the UK*. European Communities.

Miles, I., Saritas, O., & Sokolov, A. (2016). *Foresight for science, technology and innovation*. Springer.

NACI. (2019). South Africa foresight exercise for science, technology and innovation 2030, synthesis report, National Advisory Council on innovation, Department of Science and Innovation, Republic of South Africa. Retrieved June 3, 2024, from https://www.naci.org.za/wp-content/uploads/2020/07/South-African-Foresight-Exercise-For-Science-Technology-and-Innovation-2019.pdf

Research Landscape of South Africa: Current Trends in Research Output, Thematic Focus and Scientific Partnerships

Ozcan Saritas and Maxim Kotsemir

Abstract This chapter conducts a comprehensive analysis of the research landscape in South Africa, focusing on various aspects such as the country's publication activity, contribution to global knowledge generation, thematic structure of publications, scientific specializations, article quality measured by citation indicators and international research collaboration profiles. The analysis utilizes the Scopus database, employing a range of bibliometric indicators to examine key trends in South Africa's research development. The findings are presented in terms of publication activity, thematic structure, citation impact and international collaboration, providing insights into South Africa's position in the global research landscape. Additionally, the chapter compares these findings with the priorities identified in the "South Africa Science and Technology Foresight for 2030", highlighting matches and mismatches between scientific capacity and strategic priorities. The information presented is crucial for developing strategies to support critical science and technology areas, aligning them with the country's vision for 2030. The methodology involves deriving key indicators from the Scopus database and UNESCO Institute of Statistic and OECD Main Science and Technology Indicators databases, offering a detailed examination of South Africa's research landscape and its relative position in global research and development.

O. Saritas (✉)
Rochester Institute of Technology Dubai, Dubai, UAE

National Research University Higher School of Economics, Moscow, Russia
e-mail: oxscad1@rit.edu; osaritas@hse.ru

M. Kotsemir
National Research University Higher School of Economics, Moscow, Russia
e-mail: mkotsemir@hse.ru

O. Saritas et al. (eds.), *21st Century Foresight*, Science, Technology and Innovation Studies, https://doi.org/10.1007/978-3-031-69557-5_6

1 Introduction

This chapter provides a comprehensive analysis of research landscape in South Africa in terms of: level of the country's publication activity and its contribution to the global process of knowledge generation; thematic structure of publications, their scientific specializations; quality of articles measured by citation indicators; similarity of thematic structures of publications; international research collaboration profiles; and finally closeness and relative influence of each country in collaboration with other countries.

The findings presented here should be considered in conjunction with the outputs of the "South Africa Science and Technology Foresight for 2030", which identified seven science and technology domains and thrusts (priorities) associated to them. Matches and mismatches between the priorities and scientific capacity should be assessed to develop strategies for supporting some of the critical science and technology areas in order to achieve 2030 visions of the country.

Scopus database was used for analysis with the timespan of 2000–2019 that allows demonstrating some of the key trends in the development of research landscape in South Africa. A wide range of bibliometric indicators was used for analysis including citation indicators, indices of structural difference and indicators of scientific collaboration. Among the key outputs produced are:

- Publication activity of South Africa, the country's contribution to the global process of knowledge generation;
- Thematic structure of publications of South Africa, its scientific specialization and the quality of articles in the country's specialization areas;
- The similarity of the thematic structure of publications and,
- Research collaboration profiles, collaboration closeness and the dominance level of South Africa in comparison with other countries including BRICS, the USA and the European Union (EU) 28.

Different approaches are used to visualize data in form of illustrative tables and charts. In the analysis, the following types of documents are considered (unless otherwise indicated): articles, conference proceedings and reviews. A publication is considered to belong to a certain country if at least one of its authors is affiliated with this country.

2 Methodology

The comprehensive scientometric analysis of research landscape in South Africa is presented in terms of: level of the country's publication activity and its contribution to the global knowledge generation; thematic structure of publications, their scientific specializations; quality of articles measured by citation indicators; similarity of thematic structures of publications; international research collaboration profiles;

and finally closeness and relative influence of each country in collaboration with other countries.

Key indicators of financial and human resources of R&D sector in South Africa were derived from UNESCO Institute of Statistic database (section "Science, technology and innovation"). Last update of UNESCO dataset is in June 2020. For OECD and EU-28 countries in total, the data was derived from OECD Main Science and Technology Indicators (OECD MSTI) database (updated in August 2020). The following indicators were used for the analysis:

- Number of researchers in full-time equivalent (measured in thousand person-years);
- Volume of gross expenditures on research and developments (GERD) (measured in billions US dollars (USD) in purchasing power parity (PPP));
- Ratio of GERD to GDP (measured in per cents);
- Volume of GERD per one researcher FTE (measured in thousand USD PPP).

For the four above-mentioned indicators, South Africa was compared with leading countries in statics for the year 2018 (or the last available year with data) and in dynamics with Sub-Saharan Africa countries, global level and upper middle-income countries. Key indicators of financial and human resources of R&D sector in South Africa are analysed in Sect. 3.

Building on the analysis for our previous report "South Africa Science and Technology Foresight for 2030", we take Scopus as a database for our bibliometric analysis. Scopus is one of the largest science citation databases worldwide. As of the start of February 2021, Scopus indexed more than 81.5 million documents. Timespan of the analysis covers the period of 2000–2019 since the year 2020 is not yet complete in Scopus database. As publications in all our calculations, we take (unless otherwise indicated) documents like articles, reviews and conference papers indexed in Scopus. A publication is considered to belong to a certain country if at least one of its authors is affiliated with this country. Different approaches are used to visualize data in form of illustrative tables and charts. A wide range of bibliometric indicators was used for analysis that allows demonstrating some of the key trends in the development of research landscape in South Africa.

For the comparison of publication activity of South Africa with other countries, we use the following list of indicators:

- Number of publications of a country in Scopus, thousand units;
- Share of a country in the global number of publications, %;
- Position of a country in the global rankings of countries by number of publications, place.

Analysis of basic indicators of publication activity of South Africa is done in Sect. 4.

Citation indictors were derived from Scopus SciVal—electronic analytical tool based on Scopus data.[1] In our analysis, we used citation indicators like:

- Field-Weighted Citation Impact (FWCI), points;
- Share of publications in Q1 Journal Quartile by CiteScore in all publications of a country, %;
- Share of publications in Top 1% citation percentiles in all publications of a country, %.

Section 5 is devoted to the analysis of citation indicators of South Africa in Scopus database.

According to methodology, derived by Scopus, Field-Weighted Citation Impact (FWCI) for country indicates how the number of citations received by a country's publications compares with the average number of citations received by all other similar publications in the data universe: how do the citations received by this country's publications compare with the world average? Field-Weighted Citation Impact of "World", or the entire Scopus database, is 1.00.[2] FWCI of more than 1.00 indicates that the country's publications are cited more than would be expected based on the global average. For example, FWCI 1.65 for country A means that country's A publications are cited 65% more than the world average. FWCI of less than 1.00 indicates that the country's publications are cited less than would be expected based on the global average. For example, FWCI 0.63 for country A means that country's A publications are cited 27% less than the world average.

Publications in Q1 Journal Quartile by CiteScore are publications in journals that are included in top-25th percentile (top 25%), first quartile or Quartile 1 (Q1) of journals by the value of CiteScore in at least one of Scopus subject category of a journal.[3] CiteScore according to Scopus methodology "is a simple way of measuring the citation impact of serial titles such as journals. Serial titles are defined as titles, which publish on a regular basis (i.e. one or more volumes per year). Calculating the CiteScore is based on the number of citations to documents (articles, reviews, conference papers, book chapters and data papers) by a journal over four years, divided by the number of the same document types indexed in Scopus and published in those same four years".[4] CiteScore 2019, the most recent CiteScore metrics for Scopus-indexed journals calculated in May 2020, counts the citations received in 2016–2019 to articles, reviews, conference papers, book chapters and data papers published in 2016–2019, and divides this by the number of these documents published in 2016–2019.

Publications in top 1% citation percentiles in SciVal indicate the extent to which a country's publications are presented in the most-cited percentiles of all

[1] See more on SciVal on: https://www.elsevier.com/solutions/scival.

[2] More description of methodology of FWCI calculation as well as formulas of FWCI can be found at: https://service.elsevier.com/app/answers/detail/a_id/28190/supporthub/scival/p/10961/.

[3] See more on breaking of publications by Journal Quartiles in Scopus at: https://service.elsevier.com/app/answers/detail/a_id/29508/supporthub/scival/.

[4] Detailed description of CiteScore methodology.

publications in Scopus: how many publications of a country are in the top 1% of the most-cited publications.[5]

Indicators of thematic structure for South Africa include:

- Number of South Africa publications in specific Scopus subject area in 2019, units;
- Share of a specific Scopus subject area in all South Africa publications in 2019 and for 2015–2019, %.
- Position of South Africa in the global ranking of countries by number of publications in a specific Scopus subject area in 2019;
- Ratio of publications of South Africa to the number of publications of global leader by number of publications in specific Scopus subject areas in 2019, %;
- Share of a specific Scopus subject area in the total number of publications of the world (all Scopus-indexed publications) for 2015–2019, %;
- Relative comparative advantage (RCA) index value of South Africa for 2015–2019, points.

Relative comparative advantage (RCA) index was calculated for South Africa that reflects the degree of its S&T specialization in a given Scopus subject area.

For a given country (j) and a given subject area (i), revealed comparative advantagesindex(RCA)indexiscalculatedasfollows: $RCA_{subjarea\,i\,country\,j} = \frac{Share_{subjarea\,i\,country\,j}}{Share_{subjarea\,i\,world.}}$ where $Share_{subjarea\ i\ country\ j}$ is the share of publications on subject area i ($i = 1, \ldots, 27$) in the total number of publications of a specific country j in the Scopus database; $Share_{subjarea\ i\ world}$ is the share of publications on subject area i ($i = 1, \ldots, 27$) in the global number of publications in the Scopus database.

Subject areas where RCA index value is higher than 1.0 are considered as areas of research specialization of South Africa. Subject areas with the RCA index above 1.5 are considered as the key areas of scientific specialization of the South Africa. As shown in Shashnov and Kotsemir (2018), in bibliometric studies, relative (revealed) comparative advantages index is also called as "scientific specialization index" or "index of scientific specialization" or "relative (scientific) specialization index". Thematic structure of Scopus-indexed publications of South Africa is studied in Sect. 6.

Finally, indicators of international research collaboration include:

- Number of internationally collaborated publications (ICPs) in South Africa, units;
- Shares of ICPs in the total number of publications of South Africa, % (for all publications of South Africa and for all Scopus subject area);
- Number of joint publications of South Africa with a specific country, units;
- Share of a country in all (ICPs) of South Africa, %;
- Growth of number of joint papers of South Africa with a specific country in 2019 to 2000, times;
- Number of internationally collaborated publications of South Africa in a specific Scopus subject area, units;

[5] https://service.elsevier.com/app/answers/detail/a_id/28193/supporthub/scival/p/10961/.

- Share of a specific Scopus subject area in all internationally collaborated publications of South Africa, %.

Section 7 focuses on the analysis of international research collaboration of South Africa in Scopus.

Bibliometric analysis on four domains of S&T development of South Africa (Health Innovation, Circular Economy, Education, High-tech Industry) was derived from Microsoft Academic Graph (MAG) database. As of start of February 2021, Microsoft Academic Graph indexes more than 250 million of publications from 49 thousand journals and 4500+ conferences.[6] Microsoft Academic Graph (MAG) is a heterogeneous information storage platform containing both the texts of scientific publications and the citation relationships between these publications, as well as metadata: names of authors, institutions, journals, conferences and topics of research (the total number of topic in MAG is currently over 739 thousand). Data on citations in MAG is derived from Crossref metadata search mechanism. Further, the data on publications and citations from MAG was processed by iFORA to receive time series on number of publications and citations on each of four South Africa domains on country level. Corpuses of publications on each of four domains in MAG database were derived by search of keywords related with each specific domain in titles and abstracts of MAG-indexed publications. Processing of MAG-indexed publications by iFORA was performed in December 2019 so the data for 2018 and 2019 in tables is incomplete. For each of four South Africa S&T development domains, we calculate the following indicators from Microsoft Academic Graph Database:

- Number of publication of a country in a specific domain, thousand units;
- Share of country in the global number of publications in a specific domain, %;
- Position of a country in the global ranking by number of publications in a specific domain (among top-50 countries by number of publications for 2000–2019);
- Share of a country in the global number of citation on publications in a specific domain, %;
- Position of a country in the global ranking by number of citations received on publications in a specific domain (among top-50 countries by number of publications for 2000–2019);
- Citation impact of country in a specific domain relative to world, points.

For each specific domain, we show publication activity indicators and citation indicators for leading countries by number of publications/citation received in 2019 and South Africa.

Citation impact of country relative to world is a ratio of citation impact of a specific country to world average citation impact in the corpus of MAG-indexed publications on a specific domain. Citation impact of country is an average number of citations received by publications of this country in a specific year. Tables with

[6] See more on Microsoft Academic Graph on https://academic.microsoft.com/home

bibliometric analysis of MAG-indexed publications of South Africa in four domains of S&T development of a country (Health Innovation, Circular Economy, Education, High-Technology Industry) are provided in Sect. 8.

3 Financial and Human Resources of R&D Sector in South Africa

When South Africa's overall scientific and technological potentials are assessed, it is seen that the country still has considerable progress to make (Fig. 1). There is an urgent need to increase the R&D intensity of the country. By number of researchers (29.5 thousand person-years in full-time equivalent (FTE) in 2017), South Africa is far behind leading countries (compared to 1866.1 in 2018 (thousand person-years) for China, 1434.4 in 2017 for the USA, 678.1 for Japan in 2018). Meanwhile, South Africa is among the leaders in Africa by number of researchers: 67.6 thousand person-years in Egypt in 2017 and 33.9 thousand person-years in Algeria in 2017.

By volume of gross expenditures on research and development (GERD), 6.4 billion USD PPP in South Africa also stays far behind from the leading countries (581.6 in the USA in 2018, 554.3 in China in 2018). By GERD volume, South Africa is among the leaders in African continent. The volume of GERD in Egypt was 8.8 billion USD PPP in 2017, and in Algeria, it was 3.4 billion USD PPP.

By one of key indicators for R&D, gross expenditure on research and development (GERD) as percentage of GDP South Africa is also far behind leading countries—0.83%, whereas Israel has 4.95% and Korea Rep has 4.81%. GERD-to-GDP ratio in South Africa is much lower than in OECD (2.38%), EU28 (2.03%) and global level (1.73%) and upper middle income (1.48%) but much higher than the total level for Sub-Saharan Arica (0.38%).

Volume of GERD per one researcher FTE in South Africa (216.4 thousand USD PPP in 2017) was comparable with leading countries, EU-28 (221.6), upper middle-income countries (226.2) slightly lower that global level (242.6) and OECD level (269.8) and higher than in Sub-Saharan Africa countries (Table 1).

South Africa in 1997–2017 took 25%–33% of all researchers from Sub-Saharan Africa countries. Table 1 presents the "contribution" of South Africa to the total volume in comparison with Sub-Saharan Africa, upper-middle income countries and the World. The data indicates:

- High contribution of South Africa to the total volume of GERD of Sub-Saharan Africa
- By GERD as a percentage of GDP South Africa is much higher than Sub-Saharan Africa but much lower than Upper middle-income countries and the global level
- By GERD per researcher, FTE South Africa is higher than Sub-Saharan Africa and comparable with Upper middle-income countries and the global level

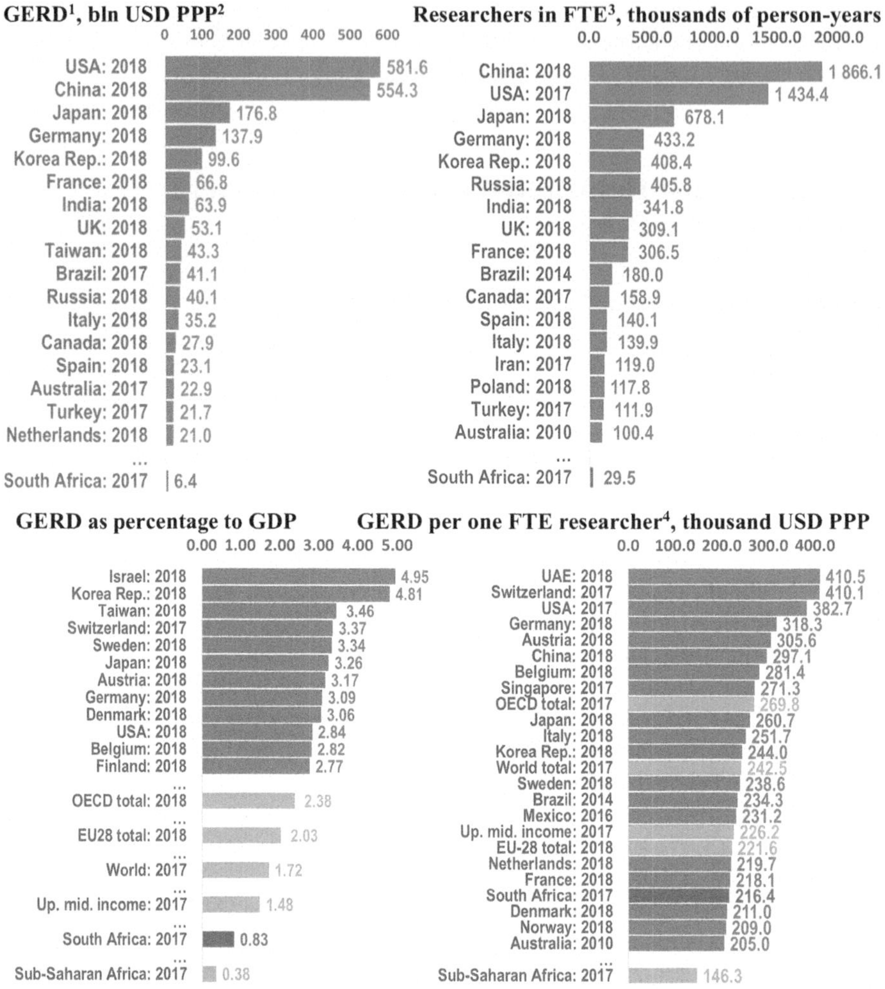

Fig. 1 Key R&D expenditures and R&D personnel indicators for South Africa in comparison with best benchmark countries in 2018 or last year with available data. Notes. 1. GERD is gross expenditures on research and development; 2. PPP is purchasing power parity. 3. FTE is full-time equivalent. 4. GERD per one FTE researcher is calculated for countries with at least 10,000 person-years of researchers in full-time equivalent. 5. To maintain continuity with our previous analysis, we take EU-28 (taking UK as a member of EU). 6. For China, we take data for Mainland China. Source: Derived from (Shashnov & Kotsemir, 2018) and updated with the latest available data with focus on South Africa. Calculated by authors using UNESCO Institute of Statistic database (section "Science, technology and innovation"). Last update: June 2020

- By "intensity" of Researchers per million inhabitants South Africa is much higher that Sub-Saharan Africa countries but more than two time lower than Upper middle-income countries and the global level.

Table 1 Key R&D expenditures and R&D personnel in South Africa in comparison to Sub-Saharan Africa countries, upper middle-income countries and the world in 1997–2019

Indicators	1997	2001	2005	2010	2015	2016	2017
Number of researchers (FTE) in South Africa	8533.0	14,182.0	17,303.0	18,719.6	26,159.4	27,656.2	29,515.2
Share of South Africa in the total number of researchers in all countries from Sub-Saharan Africa, %	25.2	33.0	33.5	27.6	29.1	29.2	29.5
GERD in billion current PPP$ in South Africa	1.80	2.61	4.05	4.43	5.83	6.10	6.39
Share of South Africa in the total GERD volume in all countries from Sub-Saharan Africa, %	53.2	60.3	66.3	48.5	46.0	45.6	45.8
GERD as a percentage of GDP							
South Africa	0.58	0.72	0.86	0.74	0.80	0.82	0.83
Africa (Sub-Saharan)	0.31	0.35	0.37	0.36	0.37	0.38	0.38
Upper middle-income countries	0.57	0.70	0.84	1.13	1.41	1.45	1.48
World	1.42	1.54	1.53	1.62	1.69	1.69	1.72
GERD per researcher, FTE (in '000 current PPP$)							
South Africa	211.1	183.9	234.0	236.5	223.0	220.4	216.4
Africa (Sub-Saharan)	104.4	110.2	132.1	138.9	148.6	147.6	146.3
Upper middle-income countries	48.8	69.6	97.7	152.8	205.7	212.9	226.2
World	129.9	154.6	170.5	202.7	228.7	233.3	242.5
Researchers per million inhabitants (FTE)							
South Africa	198.5	311.2	361.4	365.5	472.3	492.0	517.7
Africa (Sub-Saharan)	57.3	65.4	70.8	81.2	93.9	96.0	99.0
Upper middle-income countries	536.9	559.9	670.9	873.5	1071.3	1112.3	1135.5
World	775.2	818.9	903.3	1018.4	1154.3	1173.6	1198.0

Source: Calculated by authors using UNESCO Institute of Statistic database (section "Science, technology and innovation"). Last update: June 2020

4 Publication Activity

Figure 2 shows the dynamics of publications in South Africa in Scopus for 2000–2019. The figure indicates that in average the highest year-to-year growth rate of number of publications was registered in 2008–2014. The highest growth of the share of South Africa in the global number of publications was registered in 2009–2016.

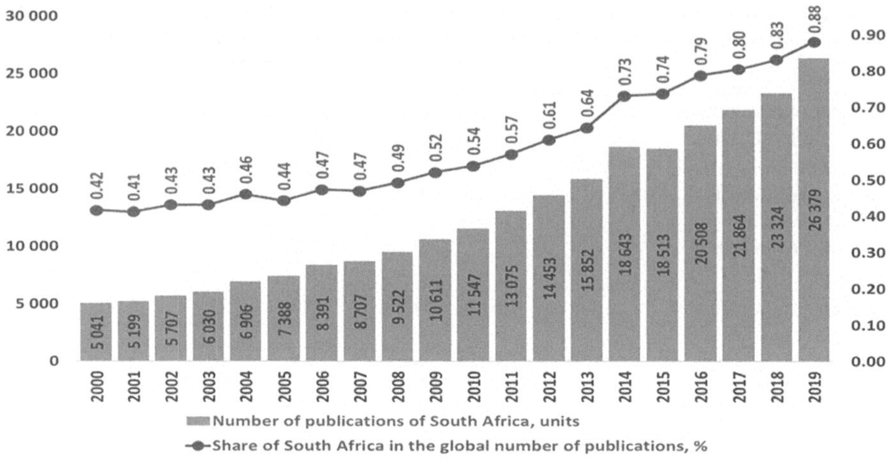

Fig. 2 Dynamics of publications in Scopus (2000–2017) (vertical axis indicates 1000 publications). Source: Authors' calculations from data taken from Scopus database. Types of publications include articles, reviews and conference papers. Data is current as of May 2020

Table 2 shows the number of publications in different countries their shares in the global number of publications and positions in global science. It is seen that China[7] has made a significant progress in terms of the number of publications (13.04 times growth for the studies period) and jumped from sixth place in 2000 to the first place in 2014. The number of publications from South Africa remained rather modest.

Further details are shown on Table 2, which, in addition, illustrates the share of South Africa in global scientific publication output. In the global ranking on the total number of publications, South Africa is among the top-30 countries. According to the table, South Africa has moved from 36th place in 2000 to 28th in 2019. However, the growth of publications of South Africa is quite slow in comparison to fast-growing countries from top 50 by number of publications in 2019. Countries like Portugal, Iran, Saudi Arabia, Malaysia and Indonesia had lower number of publications than South Africa in the year 2000, but in 2019, they outperformed South Africa by number of publications. Pakistan and Egypt that started with lower number of publications in 2000 came closer to South Africa by 2019.

On the other hand, Finland and Israel, that had in 2000 much higher number of publications than South Africa has lower number of publications than South Africa in 2019. Furthermore, Austria and Denmark that have much higher number of publications than South Africa in 2000 can only slightly outperform South Africa by number of publications in 2019.

South Africa is the leader by number of Scopus-indexed publications among all African countries. In 2019, South Africa holds 28th position in global rankings with

[7] In all our calculations, we take China as Mainland China since Hong Kong and Macao are counted as separate countries in the Scopus database.

Table 2 Key statistics of publication activity of leading countries in Scopus in 2000–2019

	Number of publications in Scopus, '000						Share in global number of publications in Scopus						Position in ranking number of countries by number of publications						Growth of the number of publications 2019 to 2000, times
Country	2000	2005	2010	2015	2018	2019	2000	2005	2010	2015	2018	2019	2000	2005	2010	2015	2018	2019	
China	51.7	170.2	339.7	450.1	591.6	673.5	4.26	10.17	15.85	17.91	21.09	22.49	6	2	2	2	2	1	13.04
United States	353.6	463.9	517.1	567.6	592.8	606.0	29.14	27.72	24.13	22.58	21.14	20.23	1	1	1	1	1	2	1.71
United Kingdom	98.3	118.5	141.2	166.1	179.0	188.1	8.10	7.08	6.59	6.61	6.38	6.28	3	4	3	3	3	3	1.91
India	24.5	38.9	75.8	131.8	163.7	174.9	2.02	2.33	3.54	5.24	5.84	5.84	13	12	9	5	5	4	7.13
Germany	86.3	114.6	137.4	156.5	164.8	167.2	7.12	6.85	6.41	6.23	5.88	5.58	4	5	4	4	4	5	1.94
Japan	103.2	123.4	123.6	118.7	124.8	124.0	8.50	7.37	5.77	4.72	4.45	4.14	2	3	5	6	6	6	1.20
Italy	43.1	60.6	78.3	99.7	107.9	113.4	3.55	3.62	3.66	3.97	3.85	3.79	8	8	8	8	8	7	2.63
Russian Federation	34.5	39.6	39.5	66.2	100.8	110.0	2.85	2.36	1.84	2.63	3.60	3.67	9	11	15	13	9	8	3.19
France	61.3	78.4	97.4	108.8	110.5	108.4	5.05	4.68	4.55	4.33	3.94	3.62	5	6	6	7	7	9	1.77
Canada	43.4	64.9	81.0	92.2	99.6	104.9	3.58	3.88	3.78	3.67	3.55	3.50	7	7	7	9	10	10	2.42
Australia	29.2	42.4	60.9	83.4	92.9	100.3	2.41	2.53	2.84	3.32	3.31	3.35	11	10	11	10	11	11	3.43
Spain	30.0	45.9	66.8	81.1	88.2	93.1	2.47	2.74	3.12	3.23	3.15	3.11	10	9	10	11	12	12	3.10
South Korea	18.3	37.4	59.4	78.7	82.9	86.6	1.51	2.24	2.77	3.13	2.96	2.89	15	13	12	12	13	13	4.74
Brazil	15.2	26.2	48.6	66.0	77.9	81.1	1.26	1.57	2.27	2.63	2.78	2.71	17	15	13	14	14	14	5.32
Iran	1.9	8.2	29.4	44.1	57.0	63.0	0.15	0.49	1.37	1.75	2.03	2.10	48	33	20	16	15	15	33.94
Netherlands	24.7	34.1	44.5	52.2	56.1	58.7	2.04	2.04	2.08	2.08	2.00	1.96	12	14	14	15	16	16	2.37

(continued)

Table 2 (continued)

Country	Number of publications in Scopus, '000						Share in global number of publications in Scopus						Position in ranking number of countries by number of publications						Growth of the number of publications 2019 to 2000, times
	2000	2005	2010	2015	2018	2019	2000	2005	2010	2015	2018	2019	2000	2005	2010	2015	2018	2019	
Poland	15.0	24.4	30.2	41.5	47.9	47.6	1.23	1.46	1.41	1.65	1.71	1.59	18	18	19	17	17	17	3.18
Turkey	7.7	20.0	31.6	40.3	41.7	47.1	0.64	1.20	1.47	1.60	1.49	1.57	25	20	18	19	19	18	6.11
Indonesia	0.8	1.2	2.6	8.1	33.9	44.9	0.06	0.07	0.12	0.32	1.21	1.50	60	62	57	48	22	19	59.45
Switzerland	17.7	24.5	32.1	40.7	43.8	44.6	1.46	1.46	1.50	1.62	1.56	1.49	16	17	17	18	18	20	2.51
Sweden	19.1	23.3	28.1	36.4	39.4	41.4	1.58	1.39	1.31	1.45	1.40	1.38	14	19	21	21	20	21	2.16
Taiwan	13.5	25.4	39.4	36.8	34.7	36.8	1.12	1.52	1.84	1.46	1.24	1.23	19	16	16	20	21	22	2.72
Malaysia	1.7	3.4	15.4	26.9	32.5	35.6	0.14	0.20	0.72	1.07	1.16	1.19	49	46	28	23	23	23	20.72
Belgium	13.1	19.1	24.6	30.2	31.8	32.3	1.08	1.14	1.15	1.20	1.13	1.08	20	21	22	22	24	24	2.47
Denmark	9.9	12.3	16.7	24.3	26.6	27.9	0.82	0.73	0.78	0.97	0.95	0.93	22	25	24	24	25	25	2.82
Saudi Arabia	2.0	2.4	6.2	19.0	22.4	26.8	0.16	0.14	0.29	0.75	0.80	0.89	47	50	46	29	31	26	13.52
Portugal	4.4	8.2	15.0	21.9	24.1	26.6	0.36	0.49	0.70	0.87	0.86	0.89	37	32	31	27	27	27	6.02
South Africa	**5.0**	**7.4**	**11.5**	**18.5**	**23.3**	**26.4**	**0.42**	**0.44**	**0.54**	**0.74**	**0.83**	**0.88**	**36**	**37**	**36**	**32**	**30**	**28**	**5.23**
Mexico	6.6	11.0	15.2	19.9	23.8	25.9	0.54	0.66	0.71	0.79	0.85	0.87	29	28	29	28	29	29	3.96
Austria	9.0	13.5	18.0	22.8	24.1	25.4	0.74	0.81	0.84	0.91	0.86	0.85	24	23	23	26	26	30	2.82
Egypt	3.3	4.7	9.3	16.2	20.9	24.6	0.27	0.28	0.43	0.65	0.74	0.82	39	42	41	37	34	31	7.41
Czech Republic	6.5	10.4	16.5	23.1	23.8	24.5	0.54	0.62	0.77	0.92	0.85	0.82	30	29	25	25	28	32	3.75
Pakistan	1.2	2.8	7.3	12.0	19.5	23.6	0.10	0.17	0.34	0.48	0.70	0.79	55	49	44	42	37	33	19.12
Norway	6.7	9.7	14.3	18.3	21.9	23.4	0.55	0.58	0.67	0.73	0.78	0.78	28	31	34	34	32	34	3.51
Hong Kong	7.6	11.9	14.3	17.5	21.3	23.1	0.62	0.71	0.67	0.70	0.76	0.77	26	27	33	35	33	35	3.06

Singapore	6.0	10.4	14.5	18.8	20.7	21.6	0.49	0.62	0.68	0.75	0.74	0.72	32	30	32	30	35	36	3.63
Israel	12.0	14.3	16.0	18.6	20.1	21.3	0.99	0.86	0.75	0.74	0.72	0.71	21	22	27	31	36	37	1.77
Finland	9.6	12.6	15.1	18.5	19.5	20.9	0.79	0.75	0.71	0.74	0.70	0.70	23	24	30	33	38	38	2.18
Thailand	2.2	4.9	9.5	12.4	17.7	19.0	0.19	0.30	0.44	0.49	0.63	0.63	43	40	40	40	39	39	8.46
Greece	6.8	11.9	16.3	16.9	17.6	18.8	0.56	0.71	0.76	0.67	0.63	0.63	27	26	26	36	40	40	2.77

Notes. 1. For China, we take data for Mainland China since Macao and Hong Kong are presented in Scopus as separate countries. 2. Forty countries with the highest number of publications in 2019 in Scopus are shown in this table

Source: Derived from (Shashnov & Kotsemir, 2018) and updated for 2019. Authors' calculations from data are taken from Scopus database. Types of publications include articles, reviews and conference papers. Data is current as of May 2020

26.4 thousand publications. Egypt takes the second place with number of publications comparable to South Africa and 31st place by number publications in 2019 (24.6 thousand publications). Other African countries stay far behind South Africa and Egypt. In 2019, Nigeria holds 51st position in global rankings by number of Scopus-indexed publications (10.9 thousand publications); Algeria holds 54th position (8.1 thousand publications); Tunisia holds 57th position (7.8 thousand publications); Ethiopia holds 66th position (4.2 thousand publications), etc.

5 Citation Analysis

Table 3 shows key citation indicators for South Africa, top-10 countries by number of publications in 2019, benchmark countries with very high citation indicators (Singapore, Hong Kong, Denmark, Switzerland, Netherlands) and in BRICS, EU-28, OECD and SADC countries and countries of Africa for 2000–2019.

Field-Weighted Citation Impact (FWCI) of South Africa shows that publications of South Africa since 2005 are cited more actively than the world average. With this, the FWCI level of South Africa is higher than in BRICS countries and African countries, which is comparable with the OECD, EU-28 and SADC countries level, and only lower that in some top-10 countries (UK, USA, Germany, Italy, France, Canada) by number of publications in 2019 and much lower than in benchmark countries with very high citation indicators.

In the period of 2000–2019, 39%–43% of all publications of South Africa are publications in Q1 journals. Share of publications in Q1 journals by CiteScore in all publications of South Africa is higher than in BRICS countries and African countries. This figure is comparable with SADC countries level, slightly lower than in OECD, EU-28 and much lower than in some top-10 countries by number of publications in 2019 (like USA, UK, Germany, France, Canada) and in benchmark countries with very high citation indicators. Share of publications in top 1% citation percentiles in all publications of South Africa jumped from 0.54% in 2000 to 1.36% in 2010, and since 2010, it was on the level of 1.36%–1.50%.

By the value of this indicator, South Africa is much better than African counties in total and is slightly better than BRICS, OECD, EU-28 and SADC countries. However, the share of publications in top 1% citation percentiles is far behind from benchmark countries with the highest values of citation indicators.

6 Thematic Structure of Publications

Table 4 shows the positions of South Africa in global science by Scopus subject areas in 2019. According to the table, South Africa lags other countries in global ranking on number of publications in Scopus. The country is placed usually below 20th in 27 Scopus subject areas. Among the "strongest" subject areas for South

Table 3 Key citation indicators of South Africa in comparison to other countries and country groups in Scopus

Country or Country group	Field-Weighted Citation Impact, points (world average level is 1.00)						Share of publications in Q1 Journal Quartile by CiteScore in all publications of a country, %						Share of publication in Top 1% Citation Percentiles in all publications of a country, % (world average level is 1.00)					
	2000	2005	2010	2015	2018	2019	2000	2005	2010	2015	2018	2019	2000	2005	2010	2015	2018	2019
South Africa	**0.87**	**1.07**	**1.21**	**1.28**	**1.24**	**1.16**	**39.2**	**39.4**	**39.0**	**40.8**	**43.1**	**42.2**	**0.54**	**1.01**	**1.36**	**1.37**	**1.48**	**1.42**
Top-10 countries by number of publications in Scopus in 2019																		
China	0.48	0.55	0.67	0.93	1.05	1.06	24.3	19.9	21.8	37.3	42.8	45.1	0.21	0.35	0.54	1.16	1.67	1.86
United States	1.50	1.50	1.52	1.49	1.41	1.40	55.3	51.1	52.2	54.8	56.0	55.6	2.06	2.07	2.20	2.16	1.94	1.87
United Kingdom	1.39	1.49	1.55	1.61	1.56	1.56	56.5	53.1	52.5	56.7	60.1	60.7	1.59	1.96	2.21	2.38	2.23	2.23
India	0.58	0.69	0.75	0.76	0.85	0.81	32.8	35.1	27.7	25.4	24.3	24.5	0.31	0.52	0.56	0.61	0.71	0.83
Germany	1.23	1.33	1.42	1.43	1.34	1.35	48.0	45.3	46.9	50.4	51.5	51.9	1.16	1.47	1.87	1.97	1.78	1.87
Japan	0.90	0.90	0.92	0.93	0.93	0.94	41.1	38.4	37.7	38.7	40.6	40.4	0.58	0.70	0.80	0.98	0.98	1.15
Italy	1.14	1.28	1.37	1.46	1.43	1.41	50.2	47.4	46.2	46.8	48.7	47.6	0.91	1.31	1.57	1.66	1.70	1.72
Russian Federation	0.50	0.53	0.55	0.74	0.76	0.81	20.3	19.4	18.7	18.1	15.6	15.9	0.22	0.29	0.37	0.41	0.42	0.49
France	1.19	1.27	1.34	1.35	1.30	127	49.3	47.4	47.3	49.6	51.7	52.7	1.09	1.35	1.63	1.79	1.74	1.78
Canada	1.40	1.46	1.50	1.56	1.45	1.47	54.5	50.8	51.8	55.9	57.3	57.9	1.58	1.74	2.00	2.15	1.98	2.18
Some benchmark countries with high citation indicators (from top-50 countries by number of publications in Scopus in 2019)																		
Singapore	1.03	1.43	1.65	1.81	1.85	1.82	43.6	45.5	49.6	57.1	59.0	59.2	0.66	1.40	2.40	3.55	4.10	4.26
Hong Kong	1.17	1.43	1.51	1.80	1.82	1.79	50.7	49.1	49.8	58.4	62.7	63.7	0.91	1.56	1.92	3.06	3.64	3.53
Denmark	1.55	1.72	1.78	1.93	1.77	1.76	58.9	58.2	58.1	60.9	63.6	64.6	1.75	2.43	2.69	2.98	2.73	2.74
Switzerland	1.61	1.82	1.86	1.94	1.78	1.73	54.6	54.4	55.5	58.8	61.9	61.8	2.06	2.55	3.05	3.43	3.25	2.99
Netherlands	1.55	1.67	1.81	1.84	1.73	1.72	59.8	56.7	58.3	62.9	63.8	64.4	1.74	2.34	2.83	2.97	2.68	2.69

(continued)

Table 3 (continued)

Country or Country group	Field-Weighted Citation Impact, points (world average level is 1.00)						Share of publications in Q1 Journal Quartile by CiteScore in all publications of a country, %						Share of publication in Top 1% Citation Percentiles in all publications of a country, % (world average level is 1.00)					
	2000	2005	2010	2015	2018	2019	2000	2005	2010	2015	2018	2019	2000	2005	2010	2015	2018	2019
Groups of countries																		
BRICS	0.55	0.60	0.69	0.86	0.95	0.96	27.0	23.7	23.3	32.7	35.9	37.6	0.27	0.39	0.52	0.91	1.23	1.39
Africa	0.72	0.81	0.87	0.94	1.00	1.00	38.0	35.7	31.3	32.5	34.6	34.3	0.32	0.54	0.71	0.83	0.87	0.97
OECD	1.20	1.22	1.23	1.20	1.15	1.16	49.2	45.3	44.7	46.9	48.8	49.0	1.26	1.31	1.36	1.31	1.24	1.25
EU28	1.15	1.20	1.23	1.22	1.18	1.19	48.7	45.2	43.9	46.3	48.3	48.6	1.05	1.21	1.31	1.32	1.24	1.27
SADC[a]	0.86	1.06	1.21	1.24	1.19	1.14	40.4	40.3	40.3	42.0	43.9	43.4	0.47	0.95	1.27	1.32	1.32	1.28

Notes. For China, we take data for Mainland China

[a]SADC is Southern African Development Community

Source: Authors' calculations from Scopus SciVal "Benchmarking" Toolbox. Types of publications include articles, reviews and conference papers. Scopus data was last updated in SciVal as of 13 January 2021

Table 4 Positions of South Africa in global science by Scopus subject areas in 2019

Subject Ara	Number of South Africa publications in subject area in 2019	Share of subject area in all South Africa publications in 2019	Position of South Africa in the global ranking of countries by number of publications in subject area in 2019	% of South Africa to global leader by number of publications in 2019	Global leader by number of publications
Agricultural and Biological Sciences	3668	13.9%	20	7.4%	China (since 2019)
Arts and Humanities	1654	6.3%	15	6.3%	USA
Biochemistry, Genetics and Molecular Biology	2217	8.4%	34	2.6%	China (since 2019)
Business, Management and Accounting	1705	6.5%	17	10.2%	USA
Chemical Engineering	972	3.7%	34	1.7%	China (since 2009)
Chemistry	1601	6.1%	35	1.7%	China (since 2007)
Computer Science	2608	9.9%	44	2.1%	China (since 2006)
Decision Sciences	644	2.4%	27	4.0%	China (since 2017)
Dentistry	51	0.2%	50	1.8%	USA
Earth and Planetary Sciences	1925	7.3%	22	4.2%	China (since 2015)
Economics, Econometrics and Finance	774	2.9%	19	6.5%	USA
Energy	1130	4.3%	36	2.1%	China (since 2005)
Engineering	3562	13.5%	42	1.6%	China (since 2007)
Environmental Science	2519	9.5%	25	4.1%	China (since 2016)
Health Professions	424	1.6%	26	3.8%	USA
Immunology and Microbiology	997	3.8%	23	4.8%	USA
Materials Science	1731	6.6%	39	1.3%	China (since 2007)

(continued)

Table 4 (continued)

Subject Ara	Number of South Africa publications in subject area in 2019	Share of subject area in all South Africa publications in 2019	Position of South Africa in the global ranking of countries by number of publications in subject area in 2019	% of South Africa to global leader by number of publications in 2019	Global leader by number of publications
Mathematics	1478	5.6%	43	2.1%	China (since 2012)
Medicine	5641	21.4%	27	3.0%	USA
Multidisciplinary	658	2.5%	27	4.3%	USA
Neuroscience	285	1.1%	41	1.1%	USA
Nursing	455	1.7%	26	3.0%	USA
Pharmacology, Toxicology and Pharmaceutics	691	2.6%	33	2.9%	China (since 2018)
Physics and Astronomy	2516	9.5%	38	2.2%	China (since 2013)
Psychology	826	3.1%	23	2.9%	USA
Social Sciences	4942	18.7%	15	6.9%	USA
Veterinary	264	1.0%	26	5.8%	USA
Total	**26,379**		**28**	**3.9%**	**China (since 2019)**

Source: Authors' calculations from data taken from the Scopus database. The types of publications include articles, reviews and conference papers. Data is current as of May 2020

Africa in Scopus in 2019 are: "Social sciences" (15th and 6.9% to the global leader); "Arts and humanities" (15th and 5.3%); "Business, Management and Accounting" (17th and 10.2%); "Economics, Econometrics and Finance" (19th, 6.5%), "Agricultural and Biological Sciences" (20th position and 7.4% of the global leader). In "Agricultural and Biological" and "Social Science" also take quite high share in the total number of publications of South Africa in Scopus in 2019. On the other hand, South Africa stands behind top-40 countries in 2019 in Scopus subject areas like "Neuroscience" (41st place), "Engineering" (42nd place), Mathematics (43rd place), "Computer Science" (44th place) and finally "Dentistry" (50th place).

Besides the number of publications, it is also important to understand the key areas of competence in scientific work. These are demonstrated by the thematic structure of publications of South Africa in Scopus (Table 5). In 2005–2009 and 2015–2019, 2.25% of Scopus-indexed publications of South Africa were concentrated in "Medicine". Among the other important areas in the thematic structure of publications of South Africa are "Social Sciences", "Agricultural and Biological Sciences" and "Engineering".

The key areas of scientific specialization of South Africa in Scopus in 2005–2009 and 2015–2019 are: "Social Sciences", "Agricultural and Biological Sciences" and

Table 5 Key indicators of thematic structure of publications of South Africa in Scopus in 2005–2009 and 2015–2019

	Relative comparative advantage (RCA) index value in South Africa		Share of subject area in the total number of publications of country			
			South Africa		World	
Subject area	2005–2009	2015–2019	2005–2009	2015–2019	2005–2009	2015–2019
Agricultural and Biological Sciences	2.72	1.83	19.1%	14.5%	7.0%	8.0%
Arts and Humanities	1.94	1.74	6.6%	6.8%	3.4%	3.9%
Biochemistry, Genetics and Molecular Biology	0.85	0.78	9.7%	8.8%	11.4%	11.2%
Business, Management and Accounting	1.09	2.08	2.7%	5.6%	2.5%	2.7%
Chemical Engineering	0.68	0.64	3.0%	3.3%	4.4%	5.2%
Chemistry	0.73	0.70	6.7%	6.4%	9.2%	9.1%
Computer Science	0.55	0.62	6.6%	8.7%	12.0%	14.1%
Decision Sciences	1.12	1.41	1.1%	2.4%	1.0%	1.7%
Dentistry	0.60	0.32	0.3%	0.2%	0.5%	0.5%
Earth and Planetary Sciences	1.93	1.47	9.1%	7.4%	4.7%	5.0%
Economics, Econometrics and Finance	1.44	1.60	2.0%	2.8%	1.4%	1.7%
Energy	0.80	0.78	2.2%	3.7%	2.8%	4.8%
Engineering	0.56	0.56	12.5%	12.4%	22.3%	22.2%
Environmental Science	1.81	1.34	8.3%	8.5%	4.6%	6.3%
Health Professions	0.88	1.32	0.9%	1.6%	1.0%	1.2%
Immunology and Microbiology	2.02	1.45	5.7%	3.9%	2.8%	2.7%
Materials Science	0.54	0.58	6.0%	6.7%	11.2%	11.5%
Mathematics	0.75	0.68	5.4%	5.4%	7.2%	7.9%
Medicine	0.86	0.95	21.5%	21.6%	24.9%	22.8%
Multidisciplinary	0.74	1.19	0.7%	2.6%	0.9%	2.2%
Neuroscience	0.44	0.50	1.1%	1.2%	2.5%	2.4%

(continued)

Table 5 (continued)

Subject area	Relative comparative advantage (RCA) index value in South Africa		Share of subject area in the total number of publications of country			
			South Africa		World	
	2005–2009	2015–2019	2005–2009	2015–2019	2005–2009	2015–2019
Nursing	0.87	1.02	1.5%	1.6%	1.7%	1.6%
Pharmacology, Toxicology and Pharmaceutics	0.74	0.85	2.3%	2.7%	3.1%	3.2%
Physics and Astronomy	0.63	0.79	8.8%	10.1%	14.1%	12.7%
Psychology	1.34	1.23	2.6%	2.8%	2.0%	2.3%
Social Sciences	1.95	2.01	13.8%	17.6%	7.1%	8.8%
Veterinary	1.94	1.45	1.8%	1.1%	0.9%	0.8%

Source: Authors' calculations from data taken from the Scopus database. The types of publications include articles, reviews and conference papers. Data is current as of May 2020

"Arts and Humanities". Values of revealed comparative advantages (RCA) index for these subject areas were higher than 1.70 in 2005–2009 and 2015–2019.

Also for some areas of Scientific specialization, we can see the fast increase of RCA index value in 2015–2019 in comparison to 2015–2019: "Business, Management and Accounting" from 1.09 to 2.08, for "Decision Sciences" from 1.12 to 1.41, for "Economics, Econometrics and Finance" from 1.44 to 1.60. "Health Professions", "Multidisciplinary" and "Nursing" became the areas of scientific specialization of South Africa in 2015–2019. Other areas of specialization of South Africa in 2005–2009 and 2015–2019 like "Earth and Planetary sciences", "Environmental science", "Immunology and Microbiology", "Psychology" and "Veterinary" show decrease of RCA value in 2015–2019 to 2005–2009.

"Dentistry" and "Neuroscience" show the lowest values of RCA index in the structure of publications of South Africa in Scopus in 2005–2009 and 2015–2019. In 2015–2019, subject areas like "Chemical Engineering", "Computer Science", "Engineering", "Material Science" and Mathematics" have quite low level of RCA index—0.58–0.70.

Figure 3 presents the specialization areas of South Africa and its thematic structure in 2015–2019 by Scopus subject areas in a more illustrative way. Figure illustrates that South Africa has very high level of RCA index in "BUSI – Business, Management and Accounting" (2.08), "SOCI—Social Sciences" (2.01), "AGRI—Agricultural and Biological Sciences" (1.83), "ARTS—Arts and Humanities" (RCA: 1.74) and "ECON—Economics, Econometrics and Finance" (1.60). Other areas of scientific specialization are also observed in "EART—Earth and Planetary Sciences", "IMMU–Immunology and Microbiology", "VETE—Veterinary", "DECI—Decision Sciences", "ENVI—Environmental Science", "HEAL—Health Professions", "PSYC—Psychology", "MULT—Multidisciplinary", "NURS—Nursing" ($1.0 < RCA < 1.47$). Low specialization is observed in some of the key areas like "MATH—Mathematics", "CENG—Chemical Engineering",

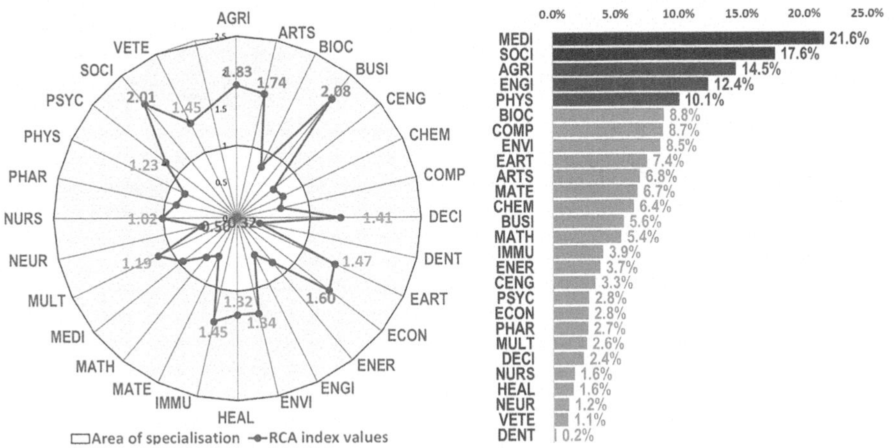

Fig. 3 Specialization areas of South Africa Subject structure of publications of South Africa by Scopus subject areas for 2015–2019. Twenty-seven Scopus subject areas are abbreviated as follows: AGRI—Agricultural and Biological Sciences; ARTS—Arts and Humanities; BIOC—Biochemistry, Genetics and Molecular Biology; BUSI—Business, Management and Accounting; CENG—Chemical Engineering; CHEM—Chemistry; COMP—Computer Science; DECI—Decision Sciences; DENT—Dentistry; EART—Earth and Planetary Sciences; ECON—Economics, Econometrics and Finance; ENER—Energy; ENGI—Engineering; ENVI—Environmental Science; HEAL—Health Professions; IMMU—Immunology and Microbiology; MATE—Materials Science; MATH—Mathematics; MEDI—Medicine; MULT—Multidisciplinary; NEUR—Neuroscience; NURS—Nursing; PHAR—Pharmacology, Toxicology and Pharmaceutics; PHYS—Physics and Astronomy; PSYC—Psychology; SOCI—Social Sciences; VETE—Veterinary. Source: Authors' calculations from data taken from the Scopus database. The types of publications include articles, reviews and conference papers. Data is current as of May 2020

"COMP—Computer Science", "MATE—Materials Science", "ENGI—Engineering", "NEUR—Neuroscience", "DENT—Dentistry" (RCA < 0.75). Considering the science and technology priority domains and priorities for South Africa, particularly Engineering and Computer Sciences areas need immediate attention for the development of the scientific capacity. Key areas of research in South Africa in 2015–2019 are "Medicine" and "Social sciences" and to a lesser extent "Agricultural and biological sciences", "Engineering" and "Physics and Astronomy".

7 International Collaboration

Another important aspect of scientific work is "international collaboration", which represents joint research and knowledge exchange between countries. Figure 4 shows dynamics of South Africa's research collaboration measured in join international publications. In 2004, the share of internationally collaborated publications in

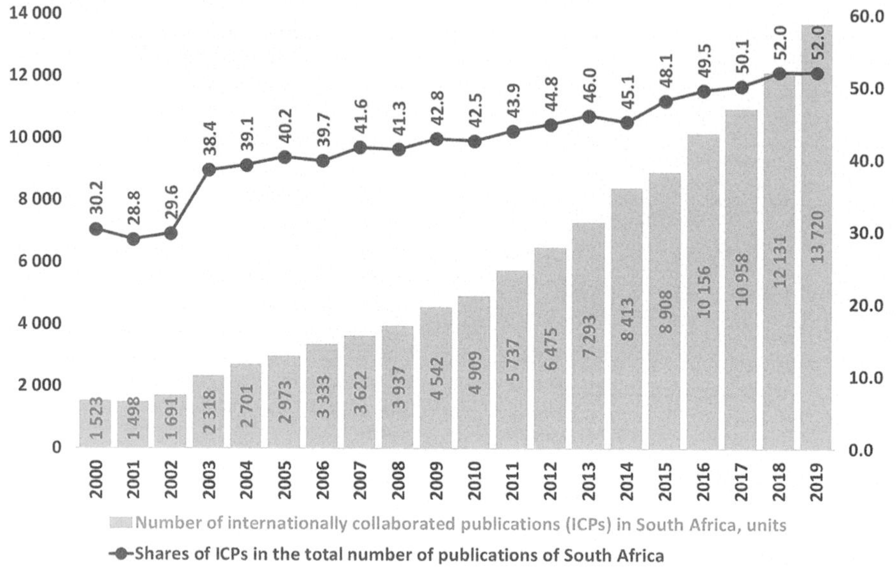

Fig. 4 Key indicators of international research collaboration of South Africa in Scopus in 2000–2019. Source: Authors' calculations from data taken from the Scopus database. The types of publications include articles, reviews and conference papers. Data is current as of May 2020

the total number of publications jumped from 29.6% to 38.4% and further by 2019 this indicator gradually increased to 52.0%.

Table 6 shows that the key scientific partners of South Africa in research collaboration in Scopus are the USA and the UK during 2000–2019. Australia and European countries like Germany, France and the Netherlands are among the top scientific partners of South Africa. Among BRICS countries India and China are among top 10 partner countries in Scopus in 2019. We should note the burst of joint publications of South Africa and Kenya—from 12 in 2000 to 1410 in 2019 so Kenya in 2019 became fifth most important partner of South Africa in Scopus. In this aspect (i.e. very fast growth of joint publications with South Africa for 2000–2019), we can see countries like Portugal, Ghana, Iran, Uganda, Romania, Thailand, Pakistan and Tanzania among the others.

Table 7 presents the thematic structure of internationally collaborated publications of South Africa in years 2005–2009 and 2015–2019, 2010 and 2017 by Scopus subject area. It is seen that subject areas like Business, Management and Accounting; Health Professions; Decision Sciences; Nursing; and Arts and Humanities are gaining momentum in terms of the growth of their share in total number of publications.

Table 6 Key scientific partners of South Africa in international research collaboration in Scopus in 2000, 2005, 2010, 2015 and 2019

Country	Number of joint publications with a country					Growth of joint papers 2019 to 2000, times	Share of a country in all internationally collaborated publications (ICP s) of South Africa				
	2000	2005	2010	2015	2019		2000	2005	2010	2015	2019
All ICP s of South Africa	**1523**	**2973**	**4909**	**8908**	**13,720**	**9.01**	**100.0%**	**100.0%**	**100.0%**	**100.0%**	**100.0%**
1. United States	492	994	1554	2937	3943	8.01	32.30%	33.43%	31.66%	32.97%	28.74%
2. United Kingdom	369	636	1121	2000	3054	8.28	24.23%	21.39%	22.84%	22.45%	22.26%
3. Australia	152	286	511	1104	1661	10.93	9.98%	9.62%	10.41%	12.39%	12.11%
4. Germany	187	340	558	1179	1591	8.51	12.28%	11.44%	11.37%	13.24%	11.60%
5. Nigeria	12	48	175	390	1410	117.50	0.79%	1.61%	3.56%	4.38%	10.28%
6. France	113	235	463	987	1235	10.93	7.42%	7.90%	9.43%	11.08%	9.00%
7. China	30	73	183	543	1177	39.23	1.97%	2.46%	3.73%	6.10%	8.58%
8. Netherlands	83	174	341	847	1138	13.71	5.45%	5.85%	6.95%	9.51%	8.29%
9. India	33	74	210	723	1065	32.27	2.17%	2.49%	4.28%	8.12%	7.76%
10. Canada	102	203	363	839	1038	10.18	6.70%	6.83%	7.39%	9.42%	7.57%
11. Italy	49	101	258	594	932	19.02	3.22%	3.40%	5.26%	6.67%	6.79%
12. Sweden	38	80	241	588	865	22.76	2.50%	2.69%	4.91%	6.60%	6.30%
13. Switzerland	55	105	241	614	844	15.35	3.61%	3.53%	4.91%	6.89%	6.15%
14. Spain	31	82	216	597	835	26.94	2.04%	2.76%	4.40%	6.70%	6.09%
15. Brazil	22	53	135	441	667	30.32	1.44%	1.78%	2.75%	4.95%	4.86%
16. Belgium	56	157	214	428	654	11.68	3.68%	5.28%	4.36%	4.80%	4.77%
17. Japan	48	65	166	391	558	11.63	3.15%	2.19%	3.38%	4.39%	4.07%
18. Russia	33	37	96	361	496	15.03	2.17%	1.24%	1.96%	4.05%	3.62%
19. Norway	19	57	161	405	472	24.84	1.25%	1.92%	3.28%	4.55%	3.44%
20. Poland	26	56	108	326	468	18.00	1.71%	1.88%	2.20%	3.66%	3.41%

(continued)

Table 6 (continued)

Country	Number of joint publications with a country					Growth of joint papers 2019 to 2000, times	Share of a country in all internationally collaborated publications (ICP s) of South Africa				
	2000	2005	2010	2015	2019		2000	2005	2010	2015	2019
Country	Number of joint publications with a country					Growth of joint papers 2019 to 2000, times	Share of a country in all internationally collaborated publications (ICP s) of South Africa				
	2000	2005	2010	2015	2019		2000	2005	2010	2015	2019
21. Denmark	26	61	158	391	423	16.27	1.71%	2.05%	3.22%	4.39%	3.08%
22. Kenya	20	48	110	256	421	21.05	1.31%	1.61%	2.24%	2.87%	3.07%
23. Portugal	4	24	68	286	406	101.50	0.26%	0.81%	1.39%	3.21%	2.96%
24. Turkey	10	19	60	253	377	37.70	0.66%	0.64%	1.22%	2.84%	2.75%
25. Austria	28	56	116	302	376	13.43	1.84%	1.88%	2.36%	3.39%	2.74%
26. Zimbabwe	28	36	72	190	356	12.71	1.84%	1.21%	1.47%	2.13%	2.59%
27. Finland	16	24	67	223	354	22.13	1.05%	0.81%	1.36%	2.50%	2.58%
28. Saudi Arabia	8	11	44	185	352	44.00	0.53%	0.37%	0.90%	2.08%	2.57%
29. Czech Republic	8	40	75	263	348	43.50	0.53%	1.35%	1.53%	2.95%	2.54%
30 Chile	13	15	69	267	347	26.69	0.85%	0.50%	1.41%	3.00%	2.53%
31. Ghana	1	7	41	138	329	329.00	0.07%	0.24%	0.84%	1.55%	2.40%
32. Iran	2	13	34	150	321	160.50	0.13%	0.44%	0.69%	1.68%	2.34%
33. New Zealand	48	62	103	225	306	6.38	3.15%	2.09%	2.10%	2.53%	2.23%
34. Greece	14	16	44	247	304	21.71	0.92%	0.54%	0.90%	2.77%	2.22%
35. Argentina	9	39	83	225	297	33.00	0.59%	1.31%	1.69%	2.53%	2.16%
36. Israel	33	42	77	232	290	8.79	2.17%	1.41%	1.57%	2.60%	2.11%
37. South Korea	6	14	46	143	269	44.83	0.39%	0.47%	0.94%	1.61%	1.96%
38. Mexico	16	32	86	159	260	16.25	1.05%	1.08%	1.75%	1.78%	1.90%
38 Romania	3	7	43	220	260	86.67	0.20%	0.24%	0.88%	2.47%	1.90%
40. Thailand	3	24	48	167	252	84.00	0.20%	0.81%	0.98%	1.87%	1.84%
40. Uganda	1	21	62	154	252	252.00	0.07%	0.71%	1.26%	1.73%	1.84%

Country	Number of joint publications with a country					Growth of joint papers 2019 to 2000, times	Share of a country in all internationally collaborated publications (ICP s) of South Africa				
	2000	2005	2010	2015	2019		2000	2005	2010	2015	2019
42. Taiwan	7	16	35	183	249	35.57	0.46%	0.54%	0.71%	2.05%	1.81%
43. Pakistan	3	7	34	119	239	79.67	0.20%	0.24%	0.69%	1.34%	1.74%
44. Hong Kong	11	16	29	172	233	21.18	0.72%	0.54%	0.59%	1.93%	1.70%
45. Botswana	4	33	41	91	226	56.50	0.26%	1.11%	0.84%	1.02%	1.65%
46. Colombia	3	8	33	191	216	72.00	0.20%	0.27%	0.67%	2.14%	1.57%
47. Hungary	17	36	70	206	210	12.35	1.12%	1.21%	1.43%	2.31%	1.53%
48. Serbia	0	4	14	163	204	…	0.00%	0.13%	0.29%	1.83%	1.49%
49. Ethiopia	9	37	35	92	203	22.56	0.59%	1.24%	0.71%	1.03%	1.48%
49. Tanzania	1	19	48	123	203	203.00	0.07%	0.64%	0.98%	1.38%	1.48%

Note. Fifty most important partners of South Africa in research collaboration in Scopus in 2019 are shown in this table

Source: Authors' calculations from data taken from the Scopus database. The types of publications include articles, reviews and conference papers. Data is current as of May 2020

Table 7 Thematic structure of internationally collaborated publications (ICPs) of South Africa in years 2000, 2010 and 2017 by Scopus subject area

	Number of ICPs of South Africa in subject area, units		Share of Subject area in total number of ICPs of South Africa		Share of ICPs in the total number of publications in subject area	
Subject area	2005–2009	2015–2019	2005–2009	2015–2019	2005–2009	2015–2019
Agricultural and Biological Sciences	4231	9676	23.0%	17.3%	49.8%	60.2%
Arts and Humanities	520	2028	2.8%	3.6%	17.6%	26.8%
Biochemistry, Genetics and Molecular Biology	2272	6155	12.3%	11.0%	52.7%	63.5%
Business, Management and Accounting	286	1836	1.6%	3.3%	23.3%	29.8%
Chemical Engineering	514	1838	2.8%	3.3%	38.4%	49.9%
Chemistry	1247	3779	6.8%	6.8%	41.6%	53.7%
Computer Science	893	3262	4.9%	5.8%	30.2%	33.9%
Decision Sciences	153	738	0.8%	1.3%	30.7%	28.0%
Dentistry	51	80	0.3%	0.1%	35.9%	41.2%
Earth and Planetary Sciences	2145	5320	11.7%	9.5%	52.9%	64.6%
Economics, Econometrics and Finance	247	1263	1.3%	2.3%	27.6%	41.2%
Energy	317	1783	1.7%	3.2%	31.9%	43.5%
Engineering	1712	5894	9.3%	10.5%	30.8%	43.1%
Environmental Science	1542	5314	8.4%	9.5%	41.7%	56.7%
Health Professions	132	897	0.7%	1.6%	32.5%	52.2%
Immunology and Microbiology	1422	3000	7.7%	5.4%	56.3%	68.7%
Materials Science	1070	3903	5.8%	7.0%	39.8%	53.0%
Mathematics	1097	3240	6.0%	5.8%	45.3%	54.4%
Medicine	4540	14,569	24.7%	26.1%	47.2%	61.1%
Multidisciplinary	241	2201	1.3%	3.9%	79.0%	76.0%
Neuroscience	272	932	1.5%	1.7%	56.7%	69.7%

(continued)

Table 7 (continued)

	Number of ICPs of South Africa in subject area, units		Share of Subject area in total number of ICPs of South Africa		Share of ICPs in the total number of publications in subject area	
Subject area	2005–2009	2015–2019	2005–2009	2015–2019	2005–2009	2015–2019
Nursing	270	945	1.5%	1.7%	41.4%	51.9%
Pharmacology, Toxicology and Pharmaceutics	468	1597	2.5%	2.9%	45.8%	53.0%
Physics and Astronomy	2246	7687	12.2%	13.8%	57.0%	69.1%
Psychology	431	1614	2.3%	2.9%	36.7%	51.5%
Social Sciences	1558	6402	8.5%	11.5%	25.3%	32.8%
Veterinary	397	771	2.2%	1.4%	50.5%	62.4%
Total	18,407	55,873	100%	100%	41.3%	50.5%

Source: Authors' calculations from data taken from the Scopus database. The types of publications include articles, reviews and conference papers. Data is current as of May 2020

8 Data on Publication Activity of South Africa by S&T Domains in Microsoft Academic Graph Database

Tables 8, 9, 10, 11, 12, 13, 14, 15 provide the key results of bibliometric analysis of Microsoft Academic Graph (MAG)-indexed publications of South Africa in four domains of S&T development of a country (Health Innovation, Circular Economy, Education and High-tech Industry).

Table 8 shows the publication activity of South Africa in the Health Innovation domain. The country started producing more scientific publications particularly from 2015 onwards, contributing to 1% of the total publications produced in the field. The country's performance has increased in global rankings from the 26th in 2000 to 20th in 2019.

Regarding key citation indicators, South Africa's ranking in impact has been increasing. In 2019, South African publications received twice more citations. With this the country occupies the 20th position in the world. However, the impact of publications from South Africa has been stable and did not show much progress since then.

Table 9 illustrates the key citation indicators of South Africa and leading countries on "Health Innovation" domain in Microsoft Academic Graph database.

Table 10 shows the publication activity of South Africa in the Circular Economy domain. The country's quantity of publications has increased substantially since 2015 in this domain. With this, South Africa produces more than 2% of the scientific output in the World, which shows the increasing emphasis on the topic in scientific work. The country's position in global rankings is 16th.

Table 8 Key indicators of publication Activity of South Africa and leading countries in "Health Innovation" domain in Microsoft Academic Graph database

Country	2000	2005	2010	2015	2016	2017	2018	2019
Number of publication of a country in "Health Innovation" domain, thousands units								
World	397.8	677.9	1074.1	1380.6	1290.9	1184.2	1110.2	797.9
USA	123.2	191.2	264.7	342.8	333.6	307.6	281.3	226.9
China	11.0	52.9	119.5	119.6	103.7	107.0	109.9	97.1
UK	35.2	52.1	76.8	102.6	102.6	100.4	93.6	72.0
Germany	20.2	31.1	45.4	55.8	55.8	53.3	49.8	38.9
Australia	12.8	23.0	37.1	50.0	49.0	48.1	44.4	35.3
Canada	13.8	24.3	35.8	47.3	47.2	45.4	41.8	33.1
India	4.2	8.9	21.6	46.3	51.1	47.0	44.1	31.8
Italy	11.0	18.5	28.0	41.1	41.0	40.1	36.7	29.0
France	12.5	19.1	29.6	37.3	37.6	36.6	33.5	25.8
Japan	22.0	30.2	34.4	38.6	38.2	36.1	33.4	24.9
…	…	…	…	…	…	…	…	…
Belgium	4.0	6.5	10.8	14.6	14.1	13.4	12.9	9.9
Denmark	2.9	4.9	7.7	12.1	12.6	12.2	11.4	8.9
South Africa	2.1	4.1	7.7	12.4	12.6	12.6	11.9	8.9
Russia	3.1	4.1	5.9	11.5	12.6	12.6	13.4	8.7
Turkey	1.8	5.3	8.1	14.6	14.6	12.5	11.1	8.4
Share of country in the global number of publications in "Health Innovation" domain, %								
USA	30.97	28.20	24.64	24.83	25.84	25.97	25.34	28.44
China	2.77	7.81	11.13	8.66	8.03	9.04	9.90	12.17
UK	8.84	7.69	7.15	7.43	7.94	8.47	8.43	9.02
Germany	5.09	4.58	4.22	4.04	4.33	4.50	4.48	4.87
Australia	3.21	3.40	3.46	3.62	3.80	4.06	4.00	4.42
Canada	3.48	3.59	3.33	3.43	3.65	3.83	3.76	4.15
India	1.07	1.31	2.01	3.35	3.96	3.97	3.98	3.99
Italy	2.76	2.73	2.61	2.97	3.17	3.39	3.30	3.63
France	3.15	2.82	2.76	2.70	2.92	3.09	3.02	3.23
Japan	5.52	4.45	3.21	2.79	2.96	3.05	3.01	3.12
…	…	…	…	…	…	…	…	…
Belgium	0.99	0.96	1.00	1.06	1.09	1.14	1.16	1.24
Denmark	0.74	0.72	0.72	0.87	0.97	1.03	1.02	1.12
South Africa	0.52	0.61	0.71	0.90	0.97	1.06	1.07	1.12
Russia	0.78	0.61	0.55	0.83	0.97	1.06	1.20	1.09
Turkey	0.46	0.78	0.75	1.06	1.13	1.06	1.00	1.05
Position of a country in the global ranking by number of publications in "Health Innovation" domain (among top-50 countries by number of publications for 2000–2019)								
USA	1	1	1	1	1	1	1	1
China	8	2	2	2	2	2	2	2
UK	2	3	3	3	3	3	3	3

(continued)

Table 8 (continued)

Country	2000	2005	2010	2015	2016	2017	2018	2019
Germany	4	4	4	4	4	4	4	4
Australia	6	7	5	5	6	5	5	5
Canada	5	6	6	6	7	7	7	6
India	16	16	11	7	5	6	6	7
Italy	9	9	9	8	8	8	8	8
France	7	8	8	10	10	9	9	9
Japan	3	5	7	9	9	10	10	10
Belgium	17	17	17	18	19	18	19	18
Denmark	20	19	20	22	23	22	21	19
South Africa	26	24	21	21	21	19	20	20
Russia	19	25	26	23	22	20	18	21
Turkey	29	18	19	19	18	21	23	22

Source: Authors' calculations from data derived by iFORA from Microsoft Academic Graph database. All types of publications are included in the analysis. Data is current as of December 2019. Data for 2018 and 2019 is incomplete

Table 9 Key citation indicators of South Africa and leading countries on "Health Innovation" domain in Microsoft Academic Graph database

Country	2000	2005	2010	2015	2016	2017	2018	2019
Share of a country in the global number of citation received on publications in "Health Innovation" domain, %								
USA	55.99	54.15	51.20	47.38	46.86	44.51	43.95	42.40
China	1.96	3.81	7.14	12.05	13.13	15.14	16.99	17.75
UK	14.84	15.15	15.21	15.78	16.06	16.86	15.20	15.48
Germany	7.60	8.89	9.24	9.68	9.38	9.58	9.18	9.07
Australia	5.40	5.61	6.68	7.86	8.29	7.98	8.03	7.70
Canada	6.81	7.47	8.12	8.03	7.85	8.39	8.24	7.45
Italy	3.47	4.57	5.24	6.01	6.05	6.66	6.54	6.46
France	5.13	5.44	5.99	6.22	6.61	6.55	6.33	6.37
Netherlands	3.76	4.52	5.34	5.24	5.30	5.65	5.22	4.92
Spain	2.05	2.88	3.55	4.55	4.27	4.80	4.75	4.55
...	...	...	...	...	...	...	...	...
Brazil	0.94	1.45	2.05	2.69	2.64	2.87	2.85	2.51
Denmark	1.41	1.92	1.88	2.59	2.37	2.65	2.61	2.39
South Africa	0.84	1.26	1.56	2.05	1.73	2.24	2.08	1.74
Austria	1.00	1.45	1.58	1.85	1.61	1.75	1.57	1.69
Israel	1.19	1.27	1.20	1.59	1.45	1.53	1.52	1.59
Position of a country in the global ranking by number of citations received on publications in "Health Innovation" domain (among top-50 countries by number of publications for 2000–2019)								
USA	1	1	1	1	1	1	1	1
China	13	11	5	3	3	3	2	2
UK	2	2	2	2	2	2	3	3
Germany	3	3	3	4	4	4	4	4

(continued)

Table 9 (continued)

Country	2000	2005	2010	2015	2016	2017	2018	2019
Australia	5	5	6	6	5	6	6	5
Canada	4	4	4	5	6	5	5	6
Italy	9	8	9	8	8	7	7	7
France	6	6	7	7	7	8	8	8
Netherlands	8	9	8	9	9	9	9	9
Spain	12	13	12	10	10	10	10	10
…	…	…	…	…	…	…	…	…
Brazil	22	17	17	17	17	16	16	18
Denmark	15	15	18	18	18	17	18	19
South Africa	23	23	20	19	19	19	20	20
Austria	20	18	19	20	21	23	25	21
Israel	16	22	23	25	24	27	27	22
Citation impact of country in "Health Innovation" domain relative to world, points								
USA	1.81	1.92	2.08	1.91	1.81	1.71	1.73	1.49
China	0.71	0.49	0.64	1.39	1.63	1.67	1.72	1.46
UK	1.68	1.97	2.13	2.12	2.02	1.99	1.80	1.72
Germany	1.49	1.94	2.19	2.40	2.17	2.13	2.05	1.86
Australia	1.68	1.65	1.93	2.17	2.18	1.97	2.01	1.74
Canada	1.96	2.08	2.44	2.34	2.15	2.19	2.19	1.79
Italy	1.26	1.67	2.01	2.02	1.91	1.96	1.98	1.78
France	1.63	1.93	2.17	2.30	2.27	2.12	2.10	1.97
Netherlands	1.81	2.32	2.78	2.70	2.52	2.47	2.21	1.97
Spain	1.28	1.63	1.83	2.17	1.88	1.96	1.90	1.63
…	…	…	…	…	…	…	…	…
Brazil	0.84	1.06	1.10	1.19	1.12	1.22	1.16	0.97
Denmark	1.91	2.68	2.62	2.96	2.44	2.58	2.55	2.13
South Africa	1.62	2.07	2.19	2.28	1.78	2.10	1.95	1.56
Austria	1.50	2.19	2.58	2.89	2.43	2.52	2.31	2.20
Israel	1.38	1.84	2.21	3.06	2.50	2.58	2.67	2.34

Source: Authors' calculations from data derived by iFORA from Microsoft Academic Graph and Crossref database. All types of publications are included in the analysis. Data is current as of December 2019. Data for 2018 and 2019 is incomplete

Table 11 shows the key citation indicators of South Africa. Over the past few years, the country has received increased number of citations, but the year 2019 saw a decline. With this, the global ranking of the country is the 19th. The impact of the scientific work has remained pretty much stable since the year 2000.

When we look at the publication activity in the Education domain, similar to the other domains, South Africa has experienced a jump in 2015 (Table 12). The country's share in global scientific output has increased to 2.18%, which is significant. With these figures, the country occupies the 11th position in the world. This high ranking scientific position should be used as a driver for high-tech industrialization, transition to circular economy and overall development of socio-economic life.

Table 10 Key indicators of publication activity of South Africa and leading countries in "Circular economy" domain in Microsoft Academic Graph database

Country	2000	2005	2010	2015	2016	2017	2018	2019
Number of publication of a country in "Circular Economy" domain, thousands units								
World	125.7	227.7	448.7	634.8	598.8	575.8	563.3	410.8
USA	31.0	52.2	85.2	116.2	114.9	110.7	104.8	84.8
China	2.8	16.5	49.6	56.9	51.3	58.1	64.8	63.2
UK	9.0	14.4	25.1	37.2	38.3	39.5	37.9	30.3
India	1.8	3.7	11.2	25.7	28.8	28.5	28.4	22.1
Australia	4.0	7.9	15.9	22.8	22.4	22.5	21.3	17.9
Germany	4.4	7.8	15.1	22.3	23.4	23.1	22.5	17.6
Canada	3.7	6.7	11.9	17.4	17.9	17.8	17.0	14.0
Italy	2.1	4.0	8.3	15.4	15.5	16.6	15.9	12.8
Spain	1.7	3.8	8.1	13.2	13.7	14.4	14.2	11.9
France	2.8	5.3	10.3	15.2	15.7	15.5	14.6	11.3
...	...	...	...	...	...	...	...	...
Iran	0.1	0.8	3.3	8.2	8.6	8.7	9.0	8.6
Netherlands	2.2	4.3	7.4	10.4	10.9	11.0	10.7	8.4
South Africa	1.4	2.9	5.9	9.8	10.7	10.9	10.4	8.3
Sweden	1.5	2.8	5.3	8.4	7.9	7.7	7.7	6.0
Switzerland	1.4	2.6	5.3	7.7	7.8	7.9	7.2	5.5
Share of country in the global number of publications in "Circular Economy" domain, %								
USA	24.65	22.91	19.00	18.30	19.18	19.23	18.60	20.65
China	2.24	7.26	11.05	8.96	8.56	10.08	11.50	15.39
UK	7.20	6.33	5.58	5.86	6.40	6.85	6.72	7.39
India	1.47	1.62	2.49	4.04	4.80	4.95	5.05	5.37
Australia	3.19	3.49	3.55	3.60	3.74	3.91	3.79	4.36
Germany	3.51	3.44	3.37	3.52	3.90	4.02	3.99	4.27
Canada	2.98	2.95	2.66	2.74	2.99	3.08	3.02	3.40
Italy	1.68	1.78	1.84	2.43	2.60	2.88	2.82	3.10
Spain	1.34	1.65	1.81	2.07	2.28	2.49	2.51	2.90
France	2.26	2.31	2.30	2.39	2.62	2.70	2.60	2.75
...	...	...	...	...	...	...	...	...
Iran	0.10	0.36	0.75	1.29	1.44	1.51	1.59	2.08
Netherlands	1.77	1.87	1.66	1.64	1.82	1.92	1.90	2.04
South Africa	1.13	1.25	1.32	1.54	1.78	1.89	1.84	2.01
Sweden	1.17	1.24	1.17	1.32	1.31	1.33	1.38	1.45
Switzerland	1.13	1.16	1.18	1.21	1.31	1.38	1.27	1.35
Position of a country in the global ranking by number of publications in "Circular Economy" domain (among top-50 countries by number of publications for 2000–2019)								
USA	1	1	1	1	1	1	1	1
China	8	2	2	2	2	2	2	2
UK	2	3	3	3	3	3	3	3

(continued)

Table 10 (continued)

Country	2000	2005	2010	2015	2016	2017	2018	2019
India	11	12	7	4	4	4	4	4
Australia	5	4	4	5	6	6	6	5
Germany	4	6	5	6	5	5	5	6
Canada	6	7	6	7	7	7	7	7
Italy	10	10	10	8	9	8	8	8
Spain	12	11	11	11	11	10	10	9
France	7	8	9	9	8	9	9	10
…	…	…	…	…	…	…	…	…
Iran	46	34	20	18	17	17	16	14
Netherlands	9	9	13	14	14	13	13	15
South Africa	15	15	15	15	15	14	14	16
Sweden	13	16	17	17	18	19	19	17
Switzerland	14	17	16	19	19	18	21	18

Source: Authors' calculations from data derived by iFORA from Microsoft Academic Graph database. All types of publications are included in the analysis. Data is current as of December 2019. Data for 2018 and 2019 is incomplete

Table 11 Key citation indicators of South Africa and leading countries on "Circular Economy" domain in Microsoft Academic Graph database

Country	2000	2005	2010	2015	2016	2017	2018	2019
Share of a country in the global number of citation received on publications in "Circular Economy" domain, %								
USA	50.68	46.88	42.29	37.93	36.50	34.66	32.34	29.48
China	1.64	4.09	10.01	18.65	20.59	23.42	26.49	28.72
UK	13.66	15.45	13.19	13.69	13.39	13.22	12.09	11.47
Australia	6.74	6.18	7.49	8.33	7.96	8.14	7.73	8.19
Germany	6.24	7.43	7.79	8.32	7.87	7.81	7.45	6.68
Canada	6.75	7.31	7.51	6.78	6.28	6.77	5.81	5.77
India	1.97	2.50	4.48	5.14	5.22	5.58	5.78	5.61
Italy	2.28	3.20	3.80	5.31	4.77	5.37	4.96	4.79
Spain	2.03	3.46	4.13	4.78	4.43	4.84	4.97	4.59
France	4.20	5.80	5.36	5.26	5.17	5.32	4.55	4.45
…	…	…	…	…	…	…	…	…
Japan	3.15	3.67	3.36	3.83	3.48	3.53	3.20	2.47
Hong Kong	0.75	0.81	1.17	2.07	1.97	2.13	2.01	2.31
South Africa	1.39	1.87	2.08	2.26	2.03	2.49	2.35	1.89
Denmark	1.21	1.70	1.84	2.45	2.22	2.34	2.04	1.87
Pakistan	0.14	0.25	0.44	1.34	1.22	1.55	1.74	1.82
Position of a country in the global ranking by number of citations received on publications in "Circular Economy" domain (among top-50 countries by number of publications for 2000–2019)								

(continued)

Table 11 (continued)

Country	2000	2005	2010	2015	2016	2017	2018	2019
USA	1	1	1	1	1	1	1	1
China	14	8	3	2	2	2	2	2
UK	2	2	2	3	3	3	3	3
Australia	4	5	6	4	4	4	4	4
Germany	5	3	4	5	5	5	5	5
Canada	3	4	5	6	6	6	6	6
India	13	14	9	9	7	7	7	7
Italy	11	12	12	7	9	8	9	8
Spain	12	11	10	10	10	10	8	9
France	6	6	7	8	8	9	10	10
...	...	...	...	...	...	...	...	...
Japan	8	9	13	14	14	13	15	17
Hong Kong	26	25	25	23	22	21	20	18
South Africa	15	16	17	19	21	18	18	19
Denmark	18	17	19	18	19	19	19	20
Pakistan	43	41	38	31	30	30	26	21
Citation impact of country in "Circular Economy" domain relative to world, points								
USA	2.06	2.05	2.23	2.07	1.90	1.80	1.74	1.43
China	0.73	0.56	0.91	2.08	2.40	2.32	2.30	1.87
UK	1.90	2.44	2.36	2.34	2.09	1.93	1.80	1.55
Australia	2.11	1.77	2.11	2.31	2.13	2.08	2.04	1.88
Germany	1.78	2.16	2.31	2.37	2.02	1.94	1.87	1.56
Canada	2.27	2.47	2.82	2.47	2.10	2.19	1.93	1.70
India	1.34	1.54	1.80	1.27	1.09	1.13	1.15	1.04
Italy	1.36	1.80	2.07	2.18	1.84	1.87	1.76	1.54
Spain	1.52	2.09	2.28	2.30	1.94	1.94	1.98	1.58
France	1.86	2.51	2.33	2.20	1.98	1.97	1.75	1.62
...	...	...	...	...	...	...	...	...
Japan	0.78	1.06	1.40	1.81	1.52	1.50	1.41	1.04
Hong Kong	1.40	1.65	2.79	3.92	3.20	3.37	3.03	2.54
South Africa	1.23	1.49	1.58	1.47	1.14	1.32	1.27	0.94
Denmark	2.01	2.57	2.62	3.00	2.37	2.49	2.25	1.80
Pakistan	0.76	1.16	1.23	2.07	1.76	1.93	1.82	1.69

Source: Authors' calculations from data derived by iFORA from Microsoft Academic Graph and Crossref database. All types of publications are included in the analysis. Data is current as of December 2019. Data for 2018 and 2019 is incomplete

Table 12 Key indicators of publication Activity of South Africa and leading countries in "Education" domain in Microsoft Academic Graph database

Country	2000	2005	2010	2015	2016	2017	2018	2019
Number of publication of a country in "Education" domain, thousands units								
World	102.5	186.8	352.2	519.0	491.1	451.5	436.0	288.5
USA	29.2	46.4	70.3	96.3	93.5	85.9	82.5	64.5
UK	8.8	14.4	23.9	33.3	33.4	33.7	32.6	24.8
China	1.4	10.6	28.8	23.1	17.0	18.2	20.9	17.7
Australia	4.1	8.6	14.7	19.6	19.1	18.9	17.5	13.3
Canada	3.1	5.9	9.9	14.2	14.3	13.8	13.7	10.7
Germany	2.3	4.5	8.5	12.3	12.7	12.5	13.1	9.8
India	0.7	1.5	4.8	12.3	14.2	12.9	12.2	8.3
Spain	1.1	2.4	6.0	10.2	10.8	10.9	10.8	8.0
Italy	1.2	2.6	5.0	8.8	8.8	9.1	9.3	6.9
Brazil	0.8	2.1	5.8	10.1	10.6	9.0	9.5	6.6
South Africa	1.3	2.8	5.4	9.2	9.8	8.9	8.8	6.3
Netherlands	1.5	2.9	5.2	7.4	7.6	7.8	7.9	5.9
France	1.4	2.7	5.2	7.4	7.9	7.5	7.5	5.5
Japan	1.9	3.2	4.5	6.2	6.2	6.2	6.5	4.5
Sweden	1.0	2.0	3.4	5.5	5.0	4.9	4.9	3.7
Share of country in the global number of publications in "Education" domain, %								
World	100.0	100.0	100.0	100.0	100.0	100.0	100.0	100.0
USA	28.52	24.84	19.95	18.56	19.04	19.04	18.93	22.36
UK	8.57	7.70	6.78	6.42	6.79	7.46	7.49	8.59
China	1.33	5.66	8.18	4.46	3.46	4.03	4.80	6.12
Australia	3.98	4.61	4.17	3.78	3.89	4.18	4.00	4.61
Canada	2.98	3.15	2.82	2.74	2.91	3.05	3.14	3.69
Germany	2.25	2.41	2.40	2.37	2.58	2.76	3.01	3.40
India	0.64	0.80	1.36	2.37	2.89	2.85	2.81	2.87
Spain	1.04	1.31	1.69	1.97	2.20	2.41	2.48	2.76
Italy	1.16	1.39	1.42	1.70	1.80	2.01	2.13	2.40
Brazil	0.81	1.14	1.65	1.95	2.15	1.98	2.18	2.28
South Africa	1.29	1.52	1.55	1.77	1.99	1.97	2.01	2.18
Netherlands	1.46	1.53	1.47	1.42	1.55	1.73	1.82	2.03
France	1.33	1.44	1.49	1.42	1.61	1.67	1.72	1.91
Japan	1.82	1.69	1.28	1.19	1.26	1.36	1.50	1.55
Sweden	0.96	1.06	0.97	1.05	1.03	1.08	1.13	1.29
Position of a country in the global ranking by number of publications in "Education" domain (among top-50 countries by number of publications for 2000–2019)								
USA	1	1	1	1	1	1	1	1
UK	2	2	3	2	2	2	2	2
China	8	3	2	3	4	4	3	3
Australia	3	4	4	4	3	3	4	4

(continued)

Table 12 (continued)

Country	2000	2005	2010	2015	2016	2017	2018	2019
Canada	4	5	5	5	5	5	5	5
Germany	5	6	6	6	7	7	6	6
India	17	17	13	7	6	6	7	7
Spain	12	12	7	8	8	8	8	8
Italy	11	11	12	11	11	9	10	9
Brazil	14	13	8	9	9	10	9	10
South Africa	10	9	9	10	10	11	11	11
Netherlands	7	8	11	13	13	12	12	12
France	9	10	10	12	12	13	13	13
Japan	6	7	14	15	14	14	14	14
Sweden	13	14	16	17	19	18	17	15

Source: Authors' calculations from data derived by iFORA from Microsoft Academic Graph database. All types of publications are included in the analysis. Data is current as of December 2019. Data for 2018 and 2019 is incomplete

Table 13 Key citation indicators of South Africa and leading countries on "Education" domain in Microsoft Academic Graph database

Country	2000	2005	2010	2015	2016	2017	2018	2019
Share of a country in the global number of citation received on publications in "Education" domain, %								
USA	57.70	52.12	46.61	45.32	44.05	42.37	41.18	41.43
UK	13.58	14.85	14.77	16.63	16.76	18.39	16.79	15.88
China	0.83	2.03	4.67	8.25	9.29	11.56	13.49	12.28
Australia	8.22	6.73	8.13	9.39	9.94	9.61	8.85	7.91
Canada	6.68	6.71	7.50	7.74	8.15	8.73	7.11	6.53
Germany	3.95	5.30	5.98	7.16	7.29	8.76	7.57	6.24
Italy	1.75	2.32	2.95	4.68	4.78	5.44	5.16	4.57
Spain	1.38	1.93	2.82	4.28	4.39	5.10	4.95	4.22
Netherlands	2.20	3.73	4.59	5.06	4.95	5.69	5.15	3.96
France	2.06	2.79	3.21	4.07	4.94	5.11	4.41	3.81
India	0.76	0.88	2.01	3.18	3.48	4.10	3.86	3.31
Brazil	0.79	1.07	1.55	2.16	2.28	2.60	2.19	3.04
Switzerland	1.52	2.49	2.91	4.20	3.48	3.61	3.76	2.93
South Africa	1.29	2.40	2.67	3.14	3.02	3.24	2.90	2.25
Sweden	1.87	2.07	2.09	3.07	2.94	3.58	3.12	2.25
Position of a country in the global ranking by number of citations received on publications in "Education" domain (among top-50 countries by number of publications for 2000–2019)								
USA	1	1	1	1	1	1	1	1
UK	2	2	2	2	2	2	2	2
China	18	12	6	4	4	3	3	3
Australia	3	3	3	3	3	4	4	4

(continued)

Table 13 (continued)

Country	2000	2005	2010	2015	2016	2017	2018	2019
Canada	4	4	4	5	5	6	6	5
Germany	5	5	5	6	6	5	5	6
Italy	9	10	9	8	9	8	7	7
Spain	11	13	11	9	10	10	9	8
Netherlands	6	6	7	7	7	7	8	9
France	7	7	8	11	8	9	10	10
India	23	24	14	12	11	11	11	11
Brazil	20	22	19	19	18	18	18	12
Switzerland	10	8	10	10	12	12	12	13
South Africa	13	9	12	13	13	14	15	14
Sweden	8	11	13	14	14	13	13	15
Citation impact of country in "Education" domain relative to world, points								
World	1.00	1.00	1.00	1.00	1.00	1.00	1.00	1.00
USA	2.02	2.10	2.34	2.44	2.31	2.23	2.18	1.85
UK	1.58	1.93	2.18	2.59	2.47	2.46	2.24	1.85
China	0.63	0.36	0.57	1.85	2.69	2.87	2.81	2.01
Australia	2.06	1.46	1.95	2.49	2.56	2.30	2.21	1.72
Canada	2.25	2.13	2.66	2.83	2.80	2.86	2.26	1.77
Germany	1.76	2.20	2.49	3.02	2.83	3.17	2.52	1.84
Italy	1.51	1.67	2.08	2.76	2.65	2.71	2.43	1.90
Spain	1.33	1.48	1.67	2.17	2.00	2.12	1.99	1.53
Netherlands	1.51	2.45	3.12	3.56	3.20	3.29	2.83	1.94
France	1.55	1.95	2.15	2.86	3.08	3.06	2.56	1.99
India	1.18	1.10	1.48	1.34	1.21	1.44	1.37	1.15
Brazil	0.97	0.94	0.94	1.11	1.06	1.31	1.01	1.33
Switzerland	1.94	3.09	3.28	4.85	3.84	3.53	3.60	2.45
South Africa	1.00	1.58	1.73	1.77	1.52	1.64	1.44	1.03
Sweden	1.96	1.95	2.16	2.91	2.86	3.33	2.76	1.74

Source: Authors' calculations from data derived by iFORA from Microsoft Academic Graph and Crossref database. All types of publications are included in the analysis. Data is current as of December 2019. Data for 2018 and 2019 is incomplete

Table 13 shows the key citation indicators in South Africa. The country's share in global citations has been high since 2015. South Africa occupies the 14th position globally in terms of citations received. The global impact of the country has slightly increased in the past 20 years.

In the domain of the High-tech industry, South Africa's scientific output has continuously increased. A slight decrease in 2019 may be because of the incomplete publications in the Scopus database. South Africa's share in as percentage of global publications has continuously increased and reached to 1.65% in 2019.

Table 14 Key indicators of publication activity of South Africa and leading countries in "High-tech Industry" domain in Microsoft Academic Graph database

Country	2000	2005	2010	2015	2016	2017	2018	2019
Number of publication of a country in "High-tech Industry" domain, thousands units								
World	106.5	204.8	386.6	561.7	535.8	527.7	545.5	378.7
USA	26.6	45.4	69.3	97.3	97.6	99.2	103.6	80.6
China	2.8	18.1	50.2	50.5	46.5	53.8	65.0	55.6
UK	8.4	13.1	21.1	31.9	33.2	34.7	35.2	27.7
India	1.0	2.4	8.4	22.3	25.9	26.0	27.0	19.7
Germany	3.9	7.1	13.2	19.5	20.6	21.3	22.4	16.7
Australia	3.2	7.0	11.5	16.5	16.3	17.1	17.4	13.7
Canada	2.9	5.8	9.3	14.2	14.9	15.1	15.8	12.5
Italy	1.9	4.1	7.7	14.2	14.4	15.8	15.9	12.2
Spain	1.2	3.3	7.0	11.3	11.7	12.3	12.8	10.0
Korea Rep.	1.2	3.8	7.9	12.8	11.8	11.0	11.5	9.9
...	...	...	...	...	...	...	...	...
Netherlands	1.9	3.4	5.8	8.4	8.6	9.0	9.4	7.1
Iran	0.1	0.6	2.8	6.4	6.7	6.5	6.8	6.5
South Africa	1.2	2.4	4.7	7.9	8.4	8.3	8.0	6.2
Switzerland	1.3	2.3	4.3	6.5	6.5	6.7	6.7	5.2
Sweden	1.2	2.5	4.4	6.7	6.5	6.4	6.9	5.1
Share of country in the global number of publications in "High-tech Industry" domain, %								
World	100.0	100.0	100.0	100.0	100.0	100.0	100.0	100.0
USA	25.03	22.18	17.92	17.32	18.22	18.79	18.99	21.28
China	2.60	8.84	12.99	8.98	8.68	10.20	11.92	14.68
UK	7.85	6.39	5.45	5.69	6.20	6.58	6.46	7.31
India	0.91	1.16	2.18	3.97	4.83	4.93	4.95	5.20
Germany	3.67	3.47	3.40	3.47	3.84	4.03	4.11	4.40
Australia	2.98	3.41	2.98	2.93	3.05	3.24	3.20	3.62
Canada	2.69	2.84	2.42	2.53	2.78	2.87	2.91	3.31
Italy	1.82	1.99	1.98	2.53	2.69	2.99	2.91	3.23
Spain	1.16	1.60	1.81	2.00	2.19	2.32	2.35	2.65
Korea Rep.	1.15	1.86	2.05	2.27	2.21	2.09	2.11	2.62
...	...	...	...	...	...	...	...	...
Netherlands	1.74	1.67	1.50	1.50	1.61	1.71	1.73	1.89
Iran	0.08	0.30	0.73	1.14	1.24	1.24	1.24	1.71
South Africa	1.12	1.18	1.21	1.40	1.58	1.57	1.47	1.65
Switzerland	1.20	1.14	1.11	1.15	1.22	1.27	1.22	1.37
Sweden	1.10	1.20	1.13	1.20	1.21	1.22	1.26	1.34
Position of a country in the global ranking by number of publications in "High-tech Industry" domain (among top-50 countries by number of publications for 2000–2019)								
USA	1	1	1	1	1	1	1	1
China	7	2	2	2	2	2	2	2

(continued)

Table 14 (continued)

Country	2000	2005	2010	2015	2016	2017	2018	2019
UK	2	3	3	3	3	3	3	3
India	16	15	8	4	4	4	4	4
Germany	3	4	4	5	5	5	5	5
Australia	5	5	5	6	6	6	6	6
Canada	6	7	6	7	7	8	8	7
Italy	9	9	11	7	8	7	7	8
Spain	12	12	12	12	12	11	10	9
Korea Rep.	13	10	10	9	10	12	12	10
…	…	…	…	…	…	…	…	…
Netherlands	10	11	13	14	14	14	14	14
Iran	44	35	20	19	17	18	20	15
South Africa	14	14	15	15	15	15	15	16
Switzerland	11	16	17	18	18	17	21	17
Sweden	15	13	16	17	19	19	17	18

Source: Authors' calculations from data derived by iFORA from Microsoft Academic Graph database. All types of publications are included in the analysis. Data is current as of December 2019. Data for 2018 and 2019 is incomplete

Table 15 Key citation indicators of South Africa and leading countries on "High-tech Industry" domain in Microsoft Academic Graph database

Country	2000	2005	2010	2015	2016	2017	2018	2019
Share of a country in the global number of citation received on publications in "High-tech Industry" domain, %								
USA	51.50	47.10	43.94	41.20	40.05	37.18	35.95	33.40
China	1.89	5.15	10.92	18.96	20.61	22.17	25.47	26.57
UK	14.16	14.23	12.60	12.65	11.94	12.79	11.76	11.70
Australia	5.86	4.85	5.93	6.83	6.25	6.58	6.77	7.19
Germany	5.92	6.61	8.20	7.83	7.29	7.53	7.01	6.41
Canada	6.06	5.69	6.66	6.14	6.11	6.40	5.61	6.06
India	1.25	1.51	3.29	4.38	4.48	4.76	5.31	5.19
Italy	2.37	3.30	3.86	5.06	4.56	4.91	4.70	4.92
Spain	1.55	2.79	3.40	3.99	3.76	3.92	4.16	4.40
France	3.17	4.45	4.96	4.25	4.20	4.29	3.90	4.35
…	…	…	…	…	…	…	…	…
Singapore	0.81	1.02	1.64	2.45	2.49	2.36	2.32	2.24
Pakistan	0.12	0.21	0.40	1.05	0.99	1.26	1.24	1.65
South Africa	1.26	1.75	1.92	1.86	1.49	1.72	1.64	1.45
Malaysia	0.19	0.25	1.16	1.67	1.57	1.47	1.31	1.41
Belgium	1.45	1.85	2.04	1.91	1.68	1.57	1.71	1.41
Position of a country in the global ranking by number of citations received on publications in "High-tech Industry" domain (among top-50 countries by number of publications for 2000–2019)								

(continued)

Table 15 (continued)

Country	2000	2005	2010	2015	2016	2017	2018	2019
USA	1	1	1	1	1	1	1	1
China	11	5	3	2	2	2	2	2
UK	2	2	2	3	3	3	3	3
Australia	5	6	6	5	5	5	5	4
Germany	4	3	4	4	4	4	4	5
Canada	3	4	5	6	6	6	6	6
India	17	17	12	9	8	8	7	7
Italy	10	10	10	7	7	7	8	8
Spain	13	12	11	12	10	11	9	9
France	7	7	7	10	9	9	10	10
…	…	…	…	…	…	…	…	…
Singapore	22	24	19	16	15	17	17	19
Pakistan	40	39	36	32	31	27	26	20
South Africa	16	16	17	20	23	20	21	21
Malaysia	36	35	25	23	22	24	24	22
Belgium	15	15	16	19	20	22	20	23
Citation impact of country in "High-tech Industry" domain relative to world, points								
USA	2.06	2.12	2.45	2.38	2.20	1.98	1.89	1.57
China	0.73	0.58	0.84	2.11	2.37	2.17	2.14	1.81
UK	1.80	2.22	2.31	2.22	1.93	1.94	1.82	1.60
Australia	1.96	1.42	1.99	2.33	2.05	2.03	2.12	1.98
Germany	1.61	1.91	2.41	2.26	1.90	1.87	1.71	1.46
Canada	2.26	2.00	2.76	2.42	2.20	2.23	1.93	1.83
India	1.37	1.31	1.51	1.10	0.93	0.97	1.07	1.00
Italy	1.30	1.66	1.95	2.00	1.69	1.64	1.62	1.52
Spain	1.34	1.74	1.88	1.99	1.71	1.69	1.77	1.66
France	1.59	2.15	2.34	1.91	1.71	1.67	1.56	1.66
…	…	…	…	…	…	…	…	…
Singapore	1.52	1.65	3.22	4.05	3.62	3.05	2.88	2.29
Pakistan	0.86	0.96	1.15	1.90	1.61	1.68	1.37	1.60
South Africa	1.12	1.49	1.59	1.33	0.94	1.10	1.11	0.88
Malaysia	0.73	0.63	1.24	1.21	1.18	0.99	0.91	1.17
Belgium	1.87	2.29	2.63	2.39	2.03	1.93	2.13	1.59

Source: Authors' calculations from data derived by iFORA from Microsoft Academic Graph and Crossref database. All types of publications are included in the analysis. Data is current as of December 2019. Data for 2018 and 2019 is incomplete

The key indicators of publication activity of South Africa and leading countries in "High-tech Industry" domain in Microsoft Academic Graph database are shown in Table 14.

Table 15 shows the key citation indications in the High-tech industry domain. According to the table, South Africa's share in global citation figures has remained stable. In terms of citation-based global rankings, South Africa has dropped from the 16th position in 2000 to 21st in 2019. In parallel, the impact of went down. In order to meet targets regarding High-tech industrialization, the country needs to advance its scientific capacity.

9 Conclusions

Presenting the results of the analyses, the chapter began with a description of the South African research landscape with current trends in research outputs, thematic areas of specialization and scientific partners of South Africa across the world and in the African continent. Dynamic analysis of publications in research outputs indicated the trends in STI-related work in South Africa. Following the description of the research landscape and capacity, following chapters of the book take a detailed look at each of the STI domains identified by the South African National STI Foresight 2030 exercise. The semantic maps generated clearly illustrate the thrusts under each domain and identify emerging topics and issues related to them. This helps to set a development agenda and strategies under the STI domains. Further bibliometric analysis under each domain indicates the research capacity of South Africa, which helps to assess the country's ability to respond to the emerging trends and issues.

Acknowledgements The book chapter was prepared within the framework of the Basic Research Program at the HSE University.

Using Big Data for Foresight: Scientometric and Semantic Analysis for South Africa

Ozcan Saritas and Maxim Kotsemir

Abstract The South Africa 2030 Foresight study, utilizing big data analytics, aimed at advancing the national innovation system and enhancing knowledge generation capacity. This chapter presents outcomes from bibliometric and semantic analyses conducted by ISSEK HSE. Bibliometric analysis explored South Africa's research competences, scientific capacity and global collaborators, revealing mature and emerging research areas. Semantic analysis through iFORA delved into large datasets, offering insights into emerging issues within each priority domain and thrust identified in the National STI Foresight. The study identified 7 STI domains and 25 thrusts through rigorous criteria, providing valuable evidence for strategic decision-making. This evidence-based approach combines bibliometric and semantic analyses to inform South Africa's innovation and research objectives efficiently.

1 Introduction

The South Africa 2030 Foresight study aimed at contributing to the following STI objectives of the country, including advancing the capacity of the national innovation system to contribute to socio-economic development; and enhancing South Africa's capacity for generating knowledge to produce world class research outputs, as well as innovative products and processes thereafter. The foresight study reflected a fully-fledged application of the SFM methodology, with big data-based horizon scanning, scenario planning, priority setting and strategy formulation. Regarding big data analysis, bibliometric and semantic analysis methods were used as key methods. This chapter presents the results of the Big Data Analysis (BDA)

O. Saritas (✉)
Rochester Institute of Technology Dubai, Dubai Silicon Oasis, UAE
e-mail: oxscad1@rit.edu; osaritas@hse.ru

M. Kotsemir
National Research University Higher School of Economics, Moscow, Russia
e-mail: mkotsemir@hse.ru

O. Saritas et al. (eds.), *21st Century Foresight*, Science, Technology and Innovation Studies, https://doi.org/10.1007/978-3-031-69557-5_7

undertaken by the ISSEK HSE. Providing a new evidence-base, Big Data makes precious contributions for Foresight exercises.

Two types of analyses are involved in this study. First one is the bibliometric analysis of the scientific outputs produced by South Africa. The bibliometric analysis helped to understand South Africa's competences in research with key STI areas focused, scientific capacity within the country and key collaborators across the world and in Africa. The temporal analysis of scientific outputs indicates the mature and emerging areas in South African research landscape. The results of the bibliometric analysis are presented in the earlier chapter.

The second type of big data analysis work undertaken within the scope of the South African National STI Foresight exercise was the semantic analysis of large document sets through the "intelligent Foresight Analytics (iFORA)" system. iFORA is a proprietary tool developed by ISSEK HSE. It uses advanced semantic analysis, machine learning and AI algorithms to integrate diverse information sources including scientific articles, patents, news, grants and analytical reports among others, with the aim of exploring new knowledge on emerging and evolving trends. The semantic analysis helped to investigate each priority domain through large datasets consisting of scientific articles, patents, news, grants and analytical reports to describe emerging issues and topics. The iFORA system generated semantic maps for the domain areas and thrusts identified in the South African National STI Foresight exercise. Detailed analyses of maps are provided to describe emerging issues and topics in each domain and the thrusts under them. These are provided in the report with detailed interpretations.

2 Semantic Analysis of STI Domains and Thrusts

The National STI Foresight 2030 study identified seven STI domains and thrusts associated to them. STI domains are defined as broad but bounded areas, related to STI as well as to a societal need or issue. The Foresight study identified several thrusts for each of the STI domains. Thrusts are considered as STI-related priority areas, which indicate what South Africa wishes to achieve with the time horizon of 2030. Throughout the Foresight process, a number of STI areas were identified. These were prioritized using a set of criteria related to their (i) potentials for new impact, (ii) association to global and STI trends and (iii) contribution to socio-economic development. As a result of online and offline consultations with the stakeholders of the STI system in South Africa, first, eight STI domains were selected. The number was then reduced to seven as two of the domains were merged. Then thrusts were determined through the following stages of the Foresight study. First, a long list of thrusts was proposed through a Foresight workshop. Then, these were shortlisted using a new set of prioritization criteria by considering their importance and feasibility. The importance criteria included the (i) socio-economic impact; (ii) new impact; and (iii) strategic value of each thrust. The feasibility criteria included (i) availability of required knowledge and expertise; (ii) availability of institutional capacity; (iii) availability of infrastructure; (iv) required policy and regulatory environment in place; (v) social and ethical acceptability; (vi) amount of relevant funding currently allocated and (vii) ease of addressing barriers and obstacles. A total of 25 thrusts were identified under seven STI domains. These are:

CIRCULAR ECONOMY
CE1: Reduce, reuse and recycle waste
CE2: Ensuring sustainable water, energy and food (agriculture) security
CE3: Low-carbon and climate-resilient economy
CE4: Smart connectivity and mobility in communities
HIGH-TECH INDUSTRY
HT1: Enabling small business to adopt high tech
HT3: New thinking for new industries
HT4: New thinking for old industries
NUTRITION SECURITY FOR A HEALTHY POPULATION
NU1: Zero impact agriculture
NU2: Use and acceptance of modern biotechnology
NU3: Personalized information for healthy nutrition for all
NU4: Precision and big data in agri-businesses
THE OPPORTUNITIES, THREATS AND IMPACT OF ICTs, INCLUDING SMART SYSTEMS, ON SOCIETY
IT1: Checks and balances for a digital future
IT2: ICT infrastructure and Internet access
IT3: Big data, data analytics and decision support
IT4: Smart and sustainable municipal service delivery
HEALTH TECHNOLOGY TO PREVENT AND TREAT ILL-HEALTH, AND ADVANCE WELL-BEING FOR THOSE WHO ARE MARGINALIZED
HE1: Optimization of health systems
HE2: Improving the quality of healthcare
HE3: Digitization of health systems
SUSTAINABLE ENERGY TECHNOLOGIES FOR THE MARGINALIZED
EN1: Clean, affordable and sustainable energy for all
EN2: Renewable energy sources and technologies
EN3: Energy efficiency solutions for industry plus household use
EN4: Distributed energy generation and storage
EDUCATION FOR THE FUTURE
ED1: Skills for the fourth Industrial Revolution
ED2: Inclusive innovation and development
ED3: Curriculum development 2030

The following sections present the results of the semantic analysis with the iFORA system for each domain areas. Detailed analyses of maps are provided to describe emerging issues and topics in each domain and the thrusts under them.

2.1 *Circular Economy*

2.1.1 Overall Description of the Domain

The Circular Economy domain is concerned with the generation of products which are restorative and regenerative by design, and which circulate through the economy repeatedly, thereby minimizing waste. This includes the conversion of biological and nonbiological waste into new resources and materials as well as the restoration and protection of biodiversity with further implications for rural areas.

2.1.2 Overview of Research Potential of South Africa in "Circular Economy" Domain

Figure 1 shows that there is a growing scientific emphasis on the Circular Economy domain in South Africa. The number of publications has been growing steadily as well as the share of South Africa in the global scientific output in this domain.

Figure 2 shows the leading countries in the Circular Economy domain.

Among those countries, the key partners of South Africa are the United States, the United Kingdom, Australia, Germany and France (Table 1). Nigeria, Zimbabwe,

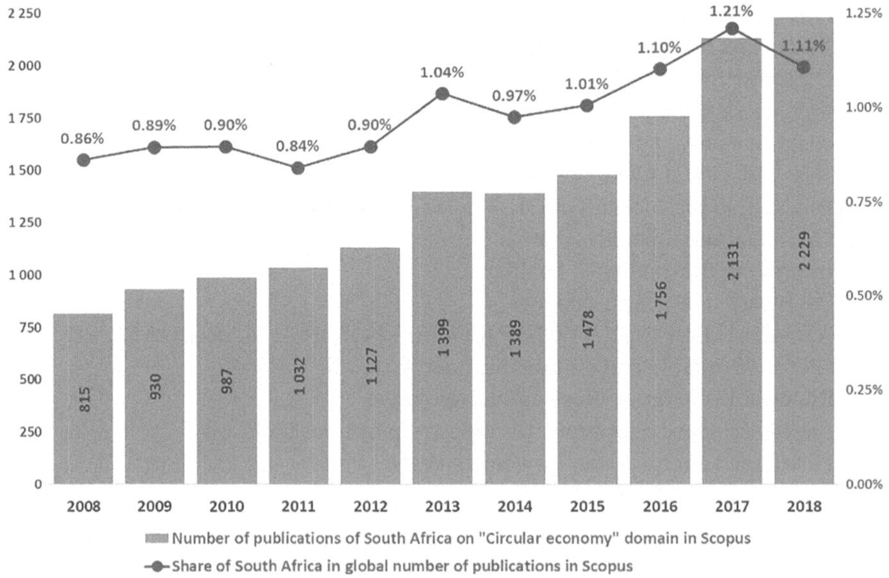

Fig. 1 Dynamics of publications of South Africa in "Circular economy" domain in Scopus in 2008–2018. Note: All types of documents, except of technical documents, indexed in Scopus are included in the analysis

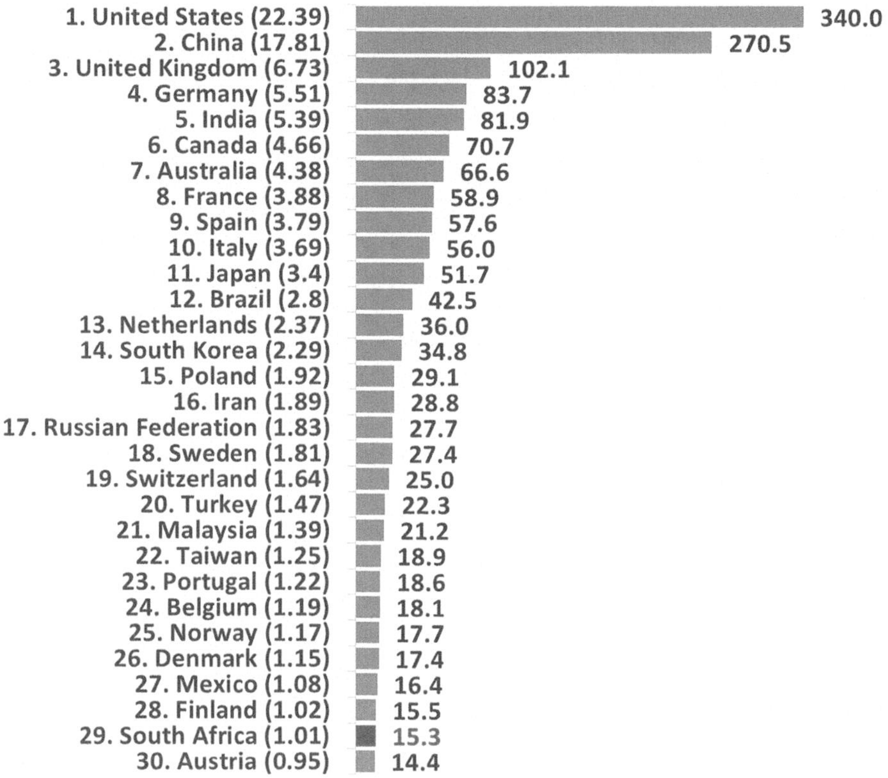

Fig. 2 Leading countries by number of publications ('000) in "Circular economy" domain in Scopus for 2008–2018. Note: 1. All types of documents, except of technical documents, indexed in Scopus are included in the analysis. 2. Share of a country in global number of publications is shown in parenthesis

Kenya, Namibia and Ethiopia are among the African scientific partners of South Africa.

Table 2 shows the leading organizations in South Africa publishing in the domain of Circular Economy with the total number of publications they produced in the last 10 years, as well as the percentage of their contribution in the total scientific output.

This section describes the "Circular Economy" STI domain and its thrusts based on the results of the semantic analysis produced by the iFORA system. First, the full domain is described with a domain map including all of the topics covered. Figure 3 shows the entire domain of the Circular Economy (CE) with its clusters and interrelations to each other. A detailed analysis of the "figure" provides an overview of CE thrusts and a number of existing and emerging topics for consideration under each of them. These are presented in Table 3. In the following sections, network maps are given to illustrate these thrusts and further issues emerging under each of them.

Table 1 Key partner countries of South Africa in internationally collaborated publications in "Circular economy" domain in Scopus for 2008–2018

Country	Number of joint papers with South Africa	Share of a country in all ICPs of South Africa
United States	2058	25.98%
United Kingdom	1730	21.84%
Australia	1126	14.22%
Germany	845	10.67%
France	753	9.51%
Netherlands	613	7.74%
Canada	548	6.92%
Sweden	476	6.01%
India	437	5.52%
Nigeria	433	5.47%
Spain	358	4.52%
China	352	4.44%
Italy	349	4.41%
Switzerland	329	4.15%
Norway	311	3.93%
Zimbabwe	292	3.69%
Belgium	281	3.55%
New Zealand	260	3.28%
Kenya	240	3.03%
Japan	237	2.99%
Brazil	200	2.53%
Denmark	199	2.51%
Austria	187	2.36%
Portugal	174	2.20%
Finland	162	2.05%
Namibia	157	1.98%
Ethiopia	139	1.76%
Czech Republic	132	1.67%
Botswana	127	1.60%
Tanzania	123	1.55%
Ghana	111	1.40%
Poland	111	1.40%
Argentina	102	1.29%
Mexico	95	1.20%
Iran	94	1.19%
Malaysia	92	1.16%
Uganda	91	1.15%
Chile	88	1.11%
Russian Federation	88	1.11%
Thailand	80	1.01%
Malawi	72	0.91%

Table 1 (continued)

Country	Number of joint papers with South Africa	Share of a country in all ICPs of South Africa
Zambia	71	0.90%
Cameroon	67	0.85%
Saudi Arabia	67	0.85%
Mozambique	65	0.82%
Israel	64	0.81%
Ireland	61	0.77%
Turkey	59	0.74%
Hungary	55	0.69%
Egypt	54	0.68%
Indonesia	52	0.66%
South Korea	50	0.63%

Note: 1. All types of documents, except of technical documents, indexed in Scopus are included in the analysis

Table 2 Leading organizations of South Africa by number of publications in "Circular economy" domain in Scopus for 2008–2018

Affiliation	Number of publications	Share in all South Africa publications
University of Cape Town	2127	8.89%
University of KwaZulu-Natal	1808	7.56%
Universiteit Stellenbosch	1775	7.42%
Universiteit van Pretoria	1731	7.24%
University of Witwatersrand	1271	5.31%
North-West University	910	3.80%
The Council for Scientific and Industrial Research	874	3.65%
University of Johannesburg	867	3.62%
Rhodes University	862	3.60%
University of the Free State	624	2.61%
Nelson Mandela Metropolitan University	536	2.24%
Tshwane University of Technology	495	2.07%
University of South Africa	414	1.73%
University of the Western Cape	403	1.68%
Agricultural Research Council, Pretoria	309	1.29%
University of Fort Hare	300	1.25%
University of Venda for Science and Technology	285	1.19%
South African National Biodiversity Institute	278	1.16%
Cape Peninsula University of Technology	258	1.08%
South African National Parks	219	0.92%
Marine and Coastal Management	214	0.89%

(continued)

Table 2 (continued)

Affiliation	Number of publications	Share in all South Africa publications
Durban University of Technology	207	0.87%
University of Limpopo	199	0.83%
South African Institute for Aquatic Biodiversity	181	0.76%
South African Environmental Observation Network	141	0.59%
South African Medical Research Council	135	0.56%
University of Zululand	116	0.48%
Vaal University of Technology	113	0.47%
International Water Management Institute, Pretoria	111	0.46%
Human Sciences Research Council of South Africa	95	0.40%
Plant Protection Research Institute, Pretoria	90	0.38%
Council for Geoscience	80	0.33%
University of Witwatersrand, School of Chemical and Metallurgical Engineering	76	0.32%
Kruger National Park	72	0.30%
Ezemvelo KZN Wildlife	68	0.28%
Endangered Wildlife Trust	64	0.27%
South African Water Research Commission	55	0.23%
National Research Foundation	54	0.23%
Central University of Technology, Free State	53	0.22%
South African Weather Service	53	0.22%

Note: 1. All types of documents, except of technical documents, indexed in Scopus are included in the analysis. 2. Organization names are presented as they are spelled in Scopus database. Semantic Analysis of "Circular Economy" Domain

2.1.3 CE1: Reduce, Reuse and Recycle Waste

This thrust is concerned with the development and deployment of performance improvements in "waste management." It is considered that development of efficient and effective waste management systems will make a significant contribution for strengthening the sustainable use of resources and thus reducing the environmental impact of waste. The analysis of the thrust through the iFORA system yielded two important action points. The first one is related to "waste reduction" and the second is "waste recycling and sustainable strategy." Issues around both action points are shown in the maps below (Figs. 4 and 5).

The first step in achieving breakthrough progress in CE is waste reduction. This covers the issues of garbage disposal, waste collection system, landfills and recycling solid waste. Points also arise regarding different types of waste such as food waste, industrial waste, human waste, organic waste, dry waste and hazardous

Fig. 3 "Circular economy" domain full semantic map for 2014–2018

Table 3 "Circular Economy" thrusts and topics

Circular economy thrusts	Topics emerging under each thrust
CE1: Reduce, reuse and recycle waste	Waste reduction (Fig. 2.1.4)
	Waste recycling and sustainable strategy (Fig. 2.1.5)
CE2: Ensuring sustainable water, energy and food (agriculture) security	Water resources (Fig. 2.1.6)
	"Water management" and "flood risk and extreme weather" (Fig. 2.1.7)
	Agricultural water use (Fig. 2.1.8)
	"Water security" and "Food security" (Fig. 2.1.9)
	Precision framing and farm efficiency (Fig. 2.1.10)
	Agricultural policy (Fig. 2.1.11)
CE3: Low-carbon and climate-resilient economy	Economic impact of sustainability (Fig. 2.1.12)
	"Energy efficiency" and "Renewable energy" (Fig. 2.1.13)
	Low carbon and low emissions (Fig. 2.1.14)
	Climate change (Fig. 2.1.15)
	"Biodiversity conservation" and "Land use" (Fig. 2.1.16)
	African environmental policy (Fig. 2.1.17)
CE4: Smart connectivity and mobility in communities, supporting the circular economy	Green transport (Fig. 2.1.18)

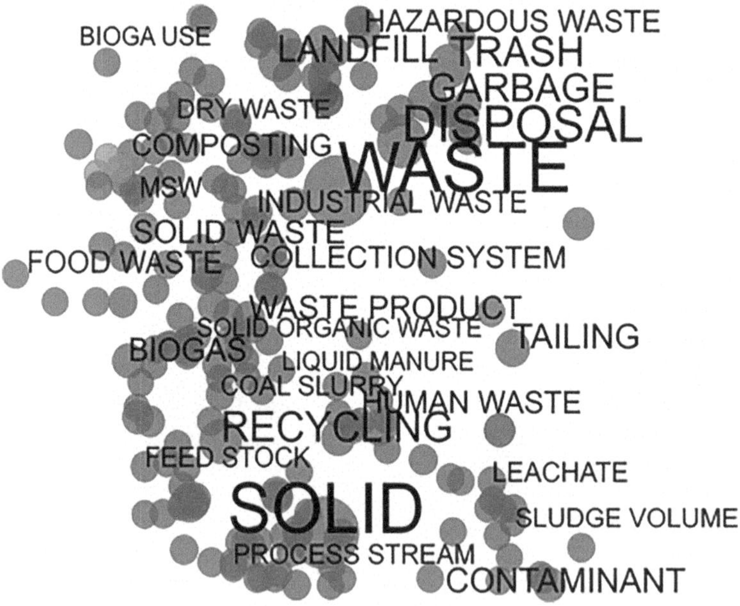

Fig. 4 "Waste reduction"

waste, all of which need to be treated and recycled in a certain way. Energy generation through biogas production appears to be a promising solution in terms of using waste.

Fig. 5 "Waste recycling and sustainable strategy"

Indeed, besides reducing waste and recycling and reusing them, it is also important to develop strategies for sustainable waste recycling.

Figure 3 shows the whole set of issues for the development of sustainable waste recycling strategies. There are a number of areas, which require action under this category. These include strategies for increasing efficiency such as making buildings greener and making them more sustainable, as well as collecting rainwater and using low flow toilets for increased water efficiency. Better management of different waste categories such as solid, plastic and hazardous, through waste reduction and recycling are also mentioned under this thrust.

2.1.4 CE2: Ensuring Sustainable Water, Energy and Food (Agriculture) Security

iFORA analysis yielded six important areas under this thrust:

1. Water resources
2. "Water management" and "Flood risk and extreme weather"
3. Agricultural water use
4. "Water security" and "Food security"
5. Precision framing and farm efficiency
6. Agricultural policy

The water resources category represents issues around water usage, operation of water resources and their management by municipalities, water pollution and water treatment through techniques like membranes and filtration (Fig. 6). Water harvesting and rainwater use also come as important areas under this cluster.

Management of water resources is strongly linked to the weather conditions and flood (Fig. 7). Extreme weather events, rainfalls and floods directly affect the availability of water. Unbalanced supply of water makes water management critical. There is a need to manage water when it is abundant as well as when it is not available. Therefore, besides introducing water flood management systems, there is a need to develop irrigation technologies for agricultural water use, as well as research and development on drought-resistant plants.

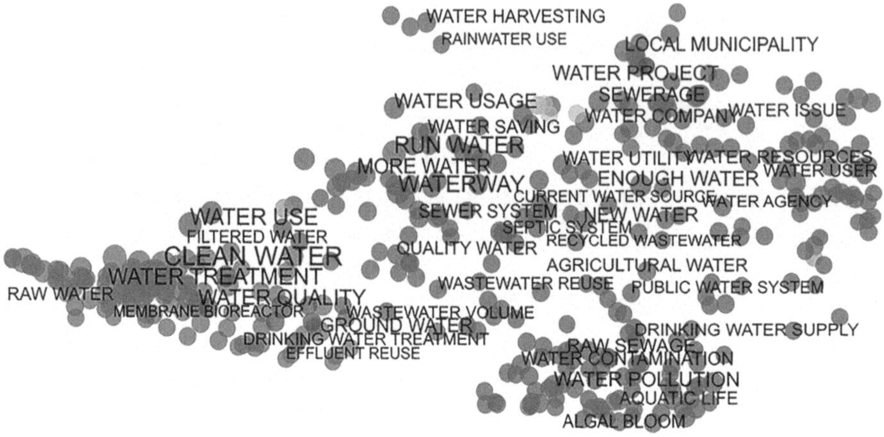

Fig. 6 “Water resources”

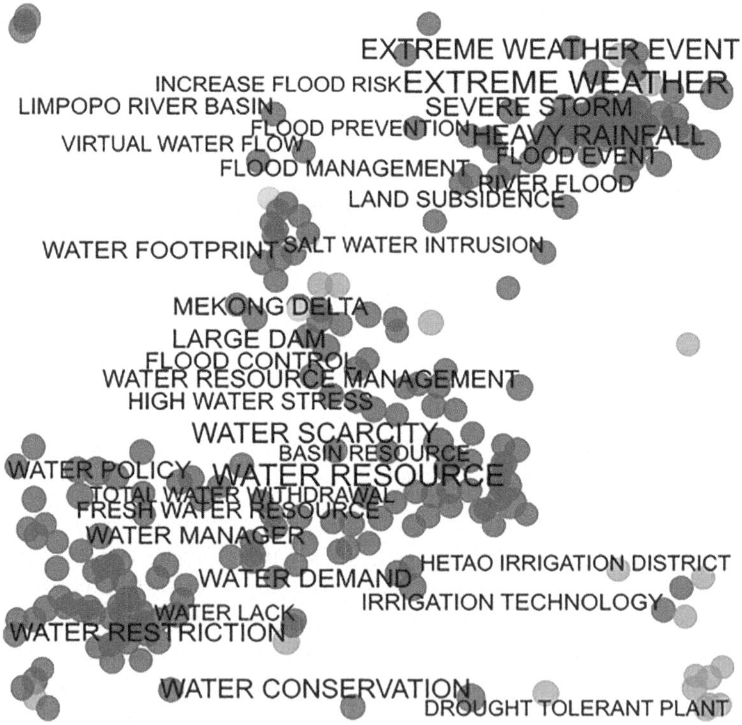

Fig. 7 “Water management” and “flood risk and extreme weather”

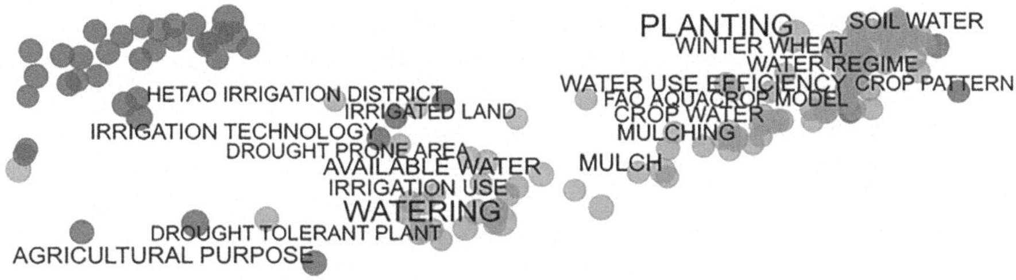

Fig. 8 "Agricultural water use"

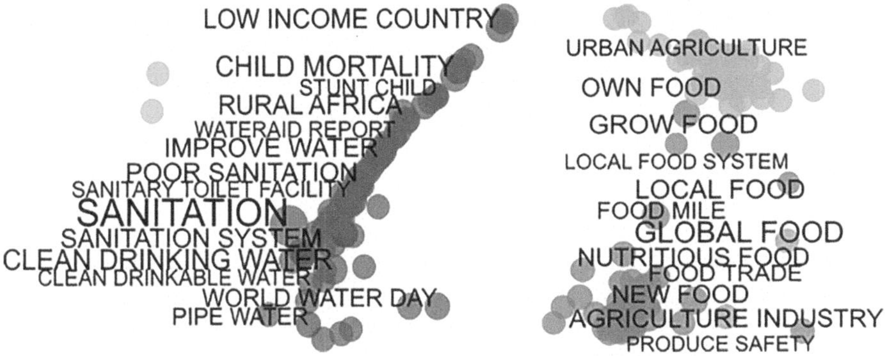

Fig. 9 "Water security" (left) and "food security" (right)

Under water security thrusts a particular emphasis emerges on agricultural water use. Here a number of issues arise regarding planting, crop water and water use efficiency as well as development of irrigation technologies and efficient watering (Fig. 8).

The water and food nexus emerge as an important issue. Under water security, the use of water for sanitation is highlighted with all other issues related that such as poor sanitation, sanitation systems and sanitary toilet facilities. Water security in terms of accessing clean drinkable water is also mentioned in this category and linked further to sanitation and low water access in rural Africa and associated child mortality (Fig. 9). Also in the right side of the figure, food security is mentioned in association to issues like growing food, local and urban food production systems as well as increasing nutrition and development of the agriculture industry with safe food production.

Among the technologies emerging with potential contributions to sustainable water, energy and food (agriculture) security is precision farming and other technologies for increasing farm efficiency. Smart farming and farm mechanization technologies also emerge as important areas for technological development. Increasing crop diversity, developing drought tolerant crops and further research in crop type and production appear to be areas for improving farm efficiency. Besides crops, soil health is also addressed as an important topic in this category (Fig. 10).

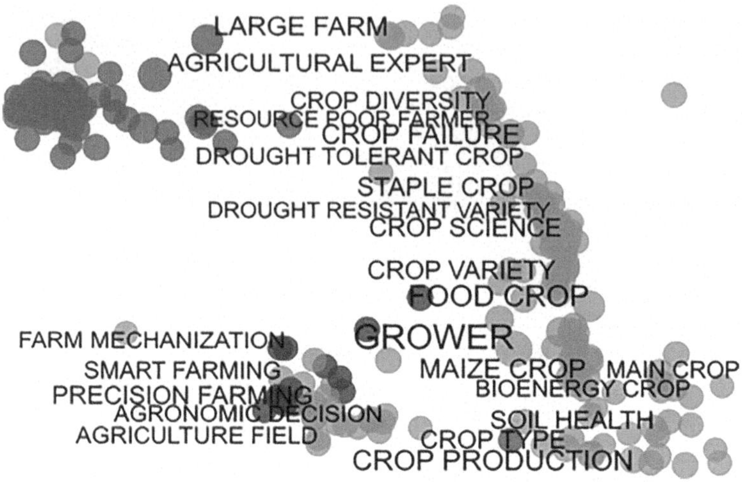

Fig. 10 "Precision farming and farm efficiency"

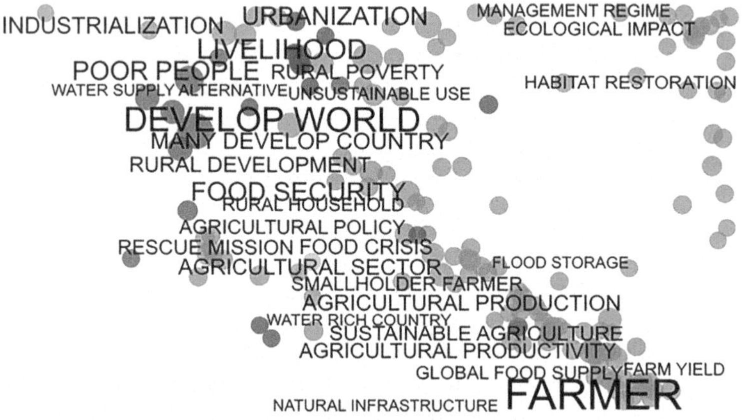

Fig. 11 "Agricultural policy"

Indeed, technologies are not solely enough for achieving sustainable water, energy and food security. There should be necessary agricultural policy in place for managing agricultural practices, providing food security and a more balanced development. Industrialization and urbanization also emerge as important and associate areas to the agricultural policy topic (Fig. 11).

2.1.5 CE3: Low-Carbon and Climate-Resilient Economy

The third thrust under CE domain is concerned with the low-carbon and climate-resilient economy. The iFORA system identified six important topics under this thrust, including:

1. Economic impact of sustainability
2. "Energy efficiency" and "Renewable energy"
3. Low carbon and low emissions
4. Climate change
5. "Biodiversity conservation" and "Land use"
6. African environmental policy

Economy plays a critical role for shifting to a low-carbon society and creating a more resilient economy under the effects of climate change. Reduction or carbon footprint, implementing greener initiatives and technologies, introduction of life cycle analysis and cleaner production are the key principles of this new economy. In order to achieve these, businesses need to be "greener" with more sustainable design of products, processes and services, and zero waste. Reusable bag emerges as a specific issue in this category, as high amount of plastic bag use appears to be an important problem (Fig. 12).

Efficient energy use and increasing use of renewables are also detected as important topics under low-carbon and climate-resistant economy. Under energy efficiency, the reduction of fossil fuel use and greenhouse gas emissions appear to be the key challenges for South Africa. Development of low energy input technologies, cogeneration, district heating and introduction of electric cars and necessary infrastructure for charging them appear to be important areas. Nuclear power is still on the agenda for South Africa together with coal power plants. The role and share of these in the energy mix should be determined toward the goals for higher energy efficiency and renewables (Fig. 13).

In addition to achieving energy efficiency, low carbon and low emissions are important concerns. All the issues related to these topics are indicated in Fig. 14 including setting targets, introducing necessary taxation systems, reducing fossil fuel use and associated emissions and taking policy measures to mitigate climate change while increasing the wealth of the country.

Climate change mitigation and environmental conservation also need to be considered not only at the local level, but also at the regional and global levels, as these

Fig. 12 "Economic impact of sustainability"

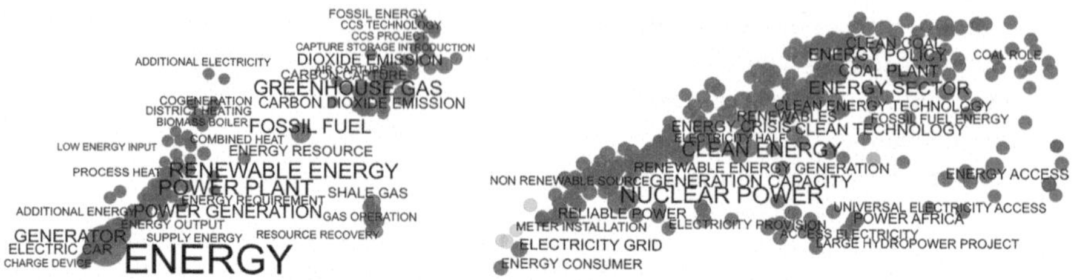

Fig. 13 "Energy efficiency" (left) and "renewable energy" (right)

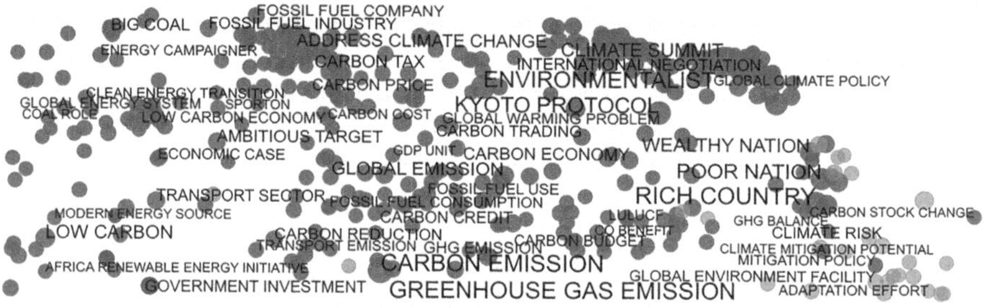

Fig. 14 "Low carbon and low emissions"

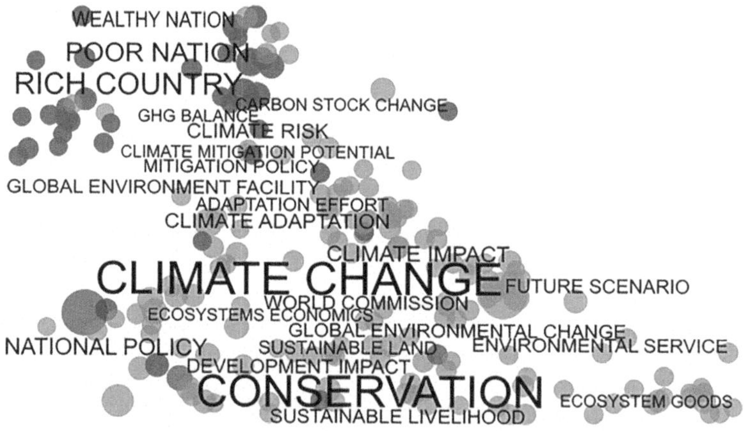

Fig. 15 "Climate change"

are among the key grand challenges for humanity. Synchronizing national policies, setting up carbon stock exchange systems, developing global systems for climate adaptation are among the important issues under this topic (Fig. 15).

Besides global, policy, economic and low-carbon aspects, the conservation of biodiversity and more sustainable use of land also emerge as important aspects leading to a low-carbon and climate-resilient economy. Issues arising under these topics

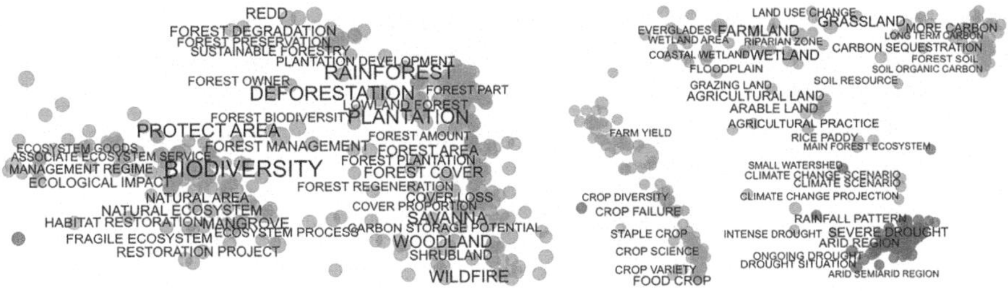

Fig. 16 "Biodiversity conservation" (left) and "land use" (right)

Fig. 17 "African environmental policy"

are indicated in Fig. 16. Regarding biodiversity conservation, a number of issues around forests emerge, including deforestation, plantation, the protection of woodland and rainforest. Habitat restoration, protection and management are important for biodiversity conservation. Indeed, the challenge of land use is closely associated to biodiversity conservation. At the intersection of both topics is the crop use with issues related to agricultural land use practices, and management of arable land, wetland, grazing land and farmland.

In order to achieve a low-carbon and climate-resistant economy, there are policy actions to be taken at the continental level in Africa. African governments and institutions need to develop joint and national policies for sustainable development, greener economy and cities with better infrastructures. These need to comply with the international and global sustainable development goals (Fig. 17).

2.1.6 CE4: Smart Connectivity and Mobility in Communities, Supporting the Circular Economy

For the final thrust in this category regarding smart connectivity and mobility, the iFORA system detected topics clustering around greener transport systems. These include use of greener and cleaner cars as well as setting up necessary urban infrastructures to develop greener cities and livelihoods. This also includes introduction of electric cars and improving electricity access (Fig. 18).

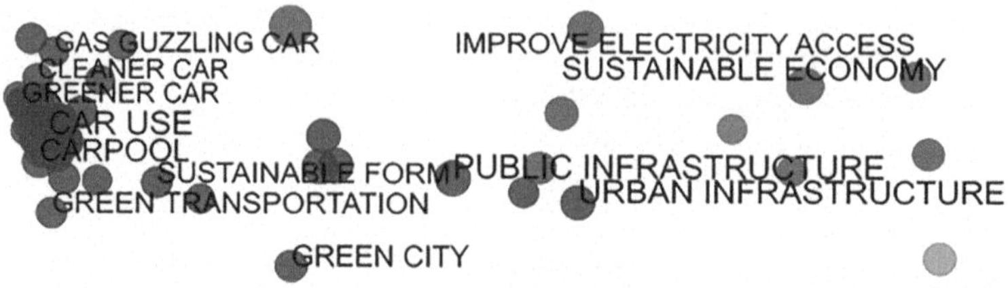

Fig. 18 "Green transport"

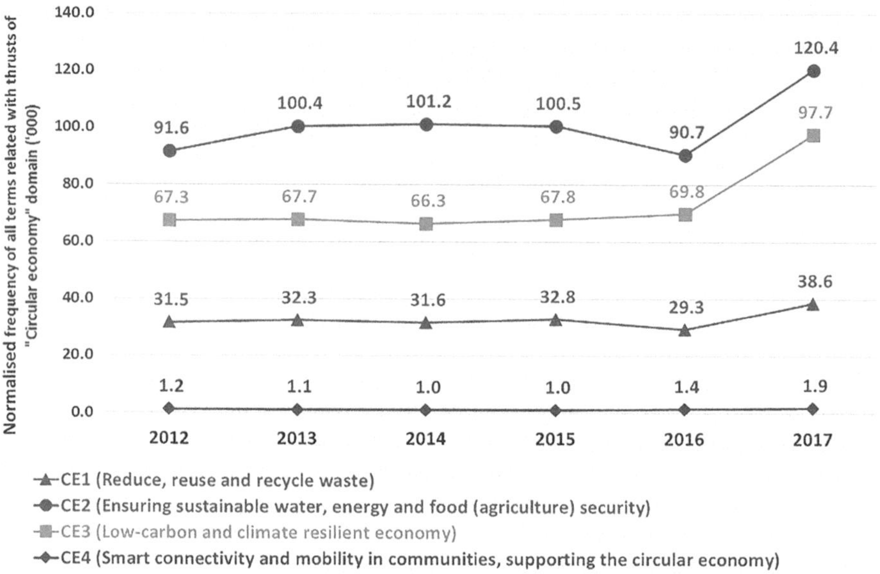

Fig. 19 Normalized frequency of all terms related with thrusts of "Circular Economy" domain ('000) in 2012–2017. Note: "Normalized frequency" for a specific thrust is a total normalized number of mentions (in thousands) of all terms (that are shown on a semantic map) related with this specific thrust in documents indexed in IFORA for a given year

2.1.7 Dynamics of Circular Economy Thrusts

In this section, the dynamics of the circular economy thrusts are shown based on their frequency of occurrence in the documents indexed in the iFORA database. Among the thrusts CE2 (Ensuring sustainable water, energy and food security) appears to be the highest on the agenda (Fig. 19). However, the progress with smart connectivity and mobility is extremely low (CE4). This thrust certainly needs more research for the achievement of the goals related to circular economy.

Furthermore, Table 4 illustrates to what extent the Circular Economy domain thrusts address the challenges. The table indicates that most of the thrusts in the CE domain are associated to population growth (except CE4). CE4 appears to be linked with job creation though at a lesser extent compared to CE1 and CE2.

Table 4 Impacts of all thrusts of "Circular Economy" domain on challenges

	Circular Economy			
Challenges	CE1	CE2	CE3	CE4
Affordable Education	1.0	0.0	0.0	0.0
Affordable Energy	3.0	15.5	4.2	0.0
Affordable Food	1.0	3.0	2.3	0.0
Carbon Emission Reduction	0.0	10.0	15.9	0.0
Clean Water Access	1.0	2.0	1.0	0.0
Crime Prevention	1.0	2.0	1.0	0.0
Crime Reduction	0.0	0.0	0.0	0.0
Economy Growth	0.0	4.0	1.3	0.0
Employment Growth	0.0	2.8	2.4	0.0
Export Growth	0.0	9.0	3.2	0.0
High Living Standard	2.0	1.0	1.0	0.0
High Quality Health Care Service	0.0	0.0	0.0	0.0
Income Growth	1.0	9.7	4.4	0.0
Job Creation	16.5	19.1	8.5	2.0
Low Greenhouse Gas Emission	2.3	3.3	2.0	0.0
Low Mortality	0.0	0.0	1.3	0.0
Population Growth	7.0	25.6	47.9	0.0
Poverty Alleviation	9.0	2.0	4.8	0.0
Reduce Water Demand	0.0	3.0	1.0	0.0
Renewable Energy Growth	0.0	3.0	2.0	0.0
Rural Community	7.3	14.4	3.9	0.0
Secure Water Supply	1.0	1.0	1.0	0.0
Skill Improvement	0.0	0.0	0.0	0.0
Universal Broadband Access	0.0	0.0	0.0	0.0

Notes: 1. Value in each specific cell of this table is a normalized number of co-occurrences of all terms (that are shown on a semantic map) related with a specific thrust (e.g., Thrust CE1) and a specific challenge (e.g., "Low Mortality") in documents indexed in iFORA database for 2012–2017. This value is treated as an impact of a specific thrust onto a specific challenge. 2. Cells are colored by bars as follows: within a column of a specific thrust the biggest green bar means the challenge for which this specific thrust has the highest impact

2.2 *High-Tech Industry*

2.2.1 Overall Description of the Domain

Industry 4.0 is expected to transform products, processes and services in a revolutionary way through the application of smart and connected systems. Following these lines, the High-tech Industry domain focuses on the implementation of advanced manufacturing in South Africa by utilizing the technologies like Robotics, Artificial Intelligence (AI), Internet of Things (IoT) and additive manufacturing among the others. Progress toward the High-tech Industry will transform old industries and will give rise to the emergence of new industries. Industrial actors including SMEs will need to equip themselves with necessary skills, infrastructure and capacity for a successful transition.

2.2.2 Overview of Research Potential of South Africa in "High-Tech Industry" Domain

Figure 20 shows that there is a fast growing scientific emphasis on the High-tech Industry domain in South Africa. The number of publications has been growing steadily as well as the share of South Africa in the global scientific output in this domain especially in last 3 years.

Figure 21 shows the leading countries in the high-tech industry domain. South Africa's scientific position in High-tech Industry is poor with the 41st place in the world. Top five key collaborators of the country in this domain are the United States, the United Kingdom, India, Germany and China. Other African countries rank also very low both in scientific capacity and level of collaboration (Table 5).

Table 6 shows the leading organizations in South Africa publishing in the domain of High-tech Industry with the total number of publications they produced in the last 10 years, as well as the percentage of their contribution in the total scientific output.

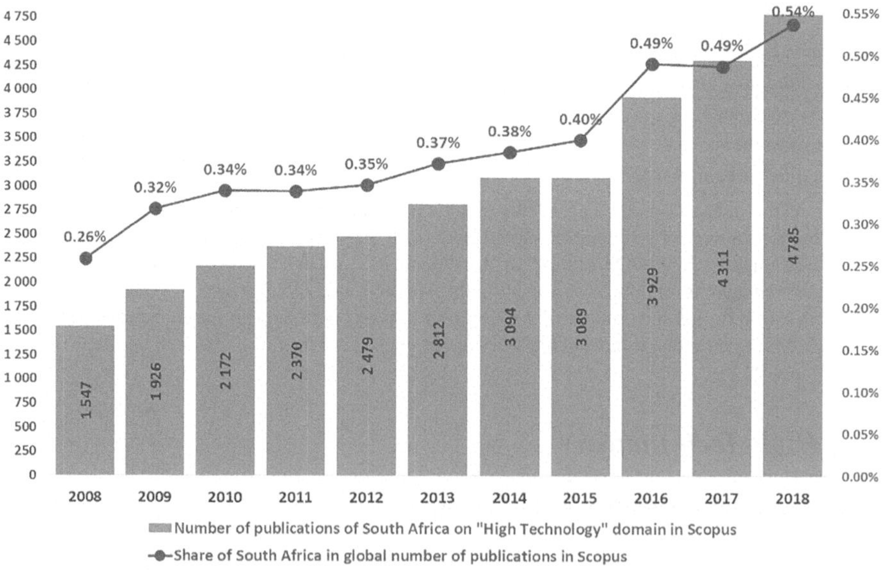

Fig. 20 Dynamics of publications of South Africa in "High-tech Industry" domain in Scopus in 2008–2018. Note: All types of documents, except of technical documents, indexed in Scopus are included in the analysis

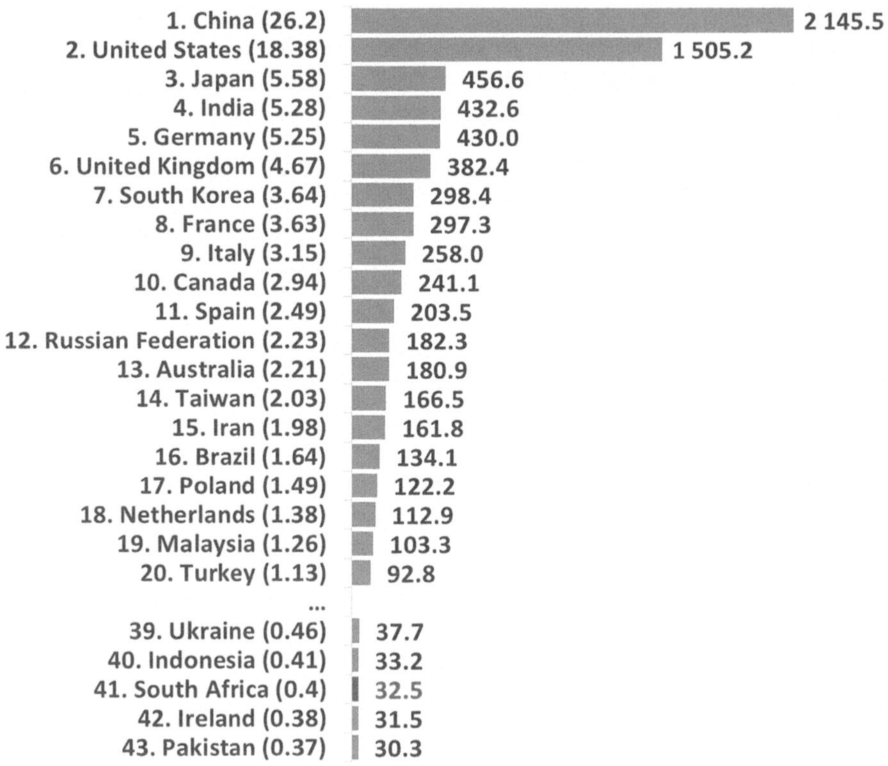

Fig. 21 Leading countries by number of publications in "High-tech Industry" domain in Scopus for 2008–2018. Note: 1. All types of documents, except of technical documents, indexed in Scopus are included in the analysis. 2. Share of a country in global number of publications is shown in parenthesis

Table 5 Key partner countries of South Africa in internationally collaborated publications in "High-tech Industry" domain in Scopus for 2008–2018

Country	Number of joint papers with South Africa	Share of a country in all ICPs of South Africa
United States	2665	19.10%
United Kingdom	1982	14.20%
India	1724	12.36%
Germany	1451	10.40%
China	1330	9.53%
France	1246	8.93%
Australia	1071	7.68%
Nigeria	932	6.68%
Canada	893	6.40%
Netherlands	879	6.30%
Italy	717	5.14%

(continued)

Table 5 (continued)

Country	Number of joint papers with South Africa	Share of a country in all ICPs of South Africa
Sweden	681	4.88%
Spain	611	4.38%
Switzerland	559	4.01%
Turkey	505	3.62%
Japan	503	3.60%
Saudi Arabia	503	3.60%
Brazil	467	3.35%
Poland	435	3.12%
Iran	421	3.02%
Russian Federation	414	2.97%
Norway	403	2.89%
Denmark	399	2.86%
Belgium	395	2.83%
Austria	368	2.64%
Czech Republic	341	2.44%
Greece	314	2.25%
Portugal	301	2.16%
Chile	300	2.15%
Taiwan	296	2.12%
Hungary	290	2.08%
Romania	290	2.08%
Israel	284	2.04%
Malaysia	282	2.02%
Argentina	278	1.99%
Slovakia	277	1.99%
South Korea	262	1.88%
Serbia	234	1.68%
Armenia	226	1.62%
Morocco	226	1.62%
Finland	225	1.61%
Slovenia	224	1.61%
Colombia	220	1.58%
Hong Kong	213	1.53%
Belarus	205	1.47%
Azerbaijan	204	1.46%
Kenya	201	1.44%
Zimbabwe	199	1.43%
New Zealand	195	1.40%
Pakistan	194	1.39%
Georgia	190	1.36%
Botswana	179	1.28%
Qatar	156	1.12%

Table 5 (continued)

Country	Number of joint papers with South Africa	Share of a country in all ICPs of South Africa
Egypt	152	1.09%
Mexico	143	1.02%
Algeria	141	1.01%
Namibia	131	0.94%
Ghana	103	0.74%
Ethiopia	100	0.72%
Thailand	100	0.72%

Table 6 Leading organizations of South Africa by number of publications in "High-tech Industry" domain in Scopus for 2008–2018

Affiliation	Number of publications	Share in all South Africa publications
Universiteit van Pretoria	4036	12.41%
University of Cape Town	3925	12.07%
Universiteit Stellenbosch	3591	11.04%
University of KwaZulu-Natal	3493	10.74%
University of Johannesburg	3436	10.57%
University of Witwatersrand	3166	9.74%
The Council for Scientific and Industrial Research	2226	6.85%
Tshwane University of Technology	1756	5.40%
North-West University	1293	3.98%
University of South Africa	1192	3.67%
University of the Free State	1063	3.27%
Rhodes University	872	2.68%
Nelson Mandela Metropolitan University	717	2.21%
University of the Western Cape	702	2.16%
Cape Peninsula University of Technology	685	2.11%
Durban University of Technology	609	1.87%
Vaal University of Technology	380	1.17%
University of Fort Hare	333	1.02%
University of Witwatersrand, School of Chemical and Metallurgical Engineering	299	0.92%
Central University of Technology, Free State	277	0.85%
South African Medical Research Council	275	0.85%
University of Zululand	249	0.77%
University of Venda for Science and Technology	222	0.68%
Meraka Institute	213	0.66%
National Research Foundation	210	0.65%
University of KwaZulu-Natal School of Chemical Engineering	203	0.62%

(continued)

Table 6 (continued)

Affiliation	Number of publications	Share in all South Africa publications
Mintek	201	0.62%
Sasol Technology Pty Ltd	194	0.60%
Agricultural Research Council, Pretoria	184	0.57%
University of Cape Town, Faculty of Health Sciences	160	0.49%
University of Limpopo	160	0.49%
Eskom	152	0.47%
Mangosuthu University of Technology	122	0.38%
National Health Laboratory Services	122	0.38%
NECSA	106	0.33%
Walter Sisulu University	105	0.32%

Note: 1. All types of documents, except of technical documents, indexed in Scopus are included in the analysis. 2. Organization names are presented as they are spelled in Scopus database

2.2.3 Semantic Analysis of "High-Tech Industry" Domain

This section describes the "High-Technology Industry" STI domain and its thrusts based on the results produced by the iFORA system. First, the full domain is described with a domain map including all of the topics covered. Then each of the thrusts is described by the emerging topics identified by iFORA. The next section presents topic clusters under each thrust with more detailed network maps with descriptions of emerging issues under them. Figure 22 shows the entire domain of the High-tech Industry (HT) with its clusters and interrelations to each other. A detailed analysis of Fig. 2.2.3 provides an overview of CE thrusts and a number of existing and emerging topics for consideration under each of them. These are presented in Table 7. In the following section, network maps are given to illustrate these maps and further issues emerging under each of them.

2.2.4 HT1: Enabling Small Business to Adopt High Tech

The iFORA system revealed two topics for the adoption of high technology in small businesses. The first one is related to mainstream IT adoption, which is necessary for all small businesses. Early adopters will gain competitive advantage to the others. Besides making existing small businesses more IT based, it is undoubtedly crucial to have high-tech start-ups for a more STI-based economy and society. Entrepreneurship and venture capital investment are two important drivers for setting up high-tech small businesses (Fig. 23).

IT adoption by small businesses is not only about a technological issue. Skills development for high-tech adoption is also an important aspect. Figure 24 demonstrates the issues related to skills development including job creation and markets, wage levels, skill requirements as well as job security and further career

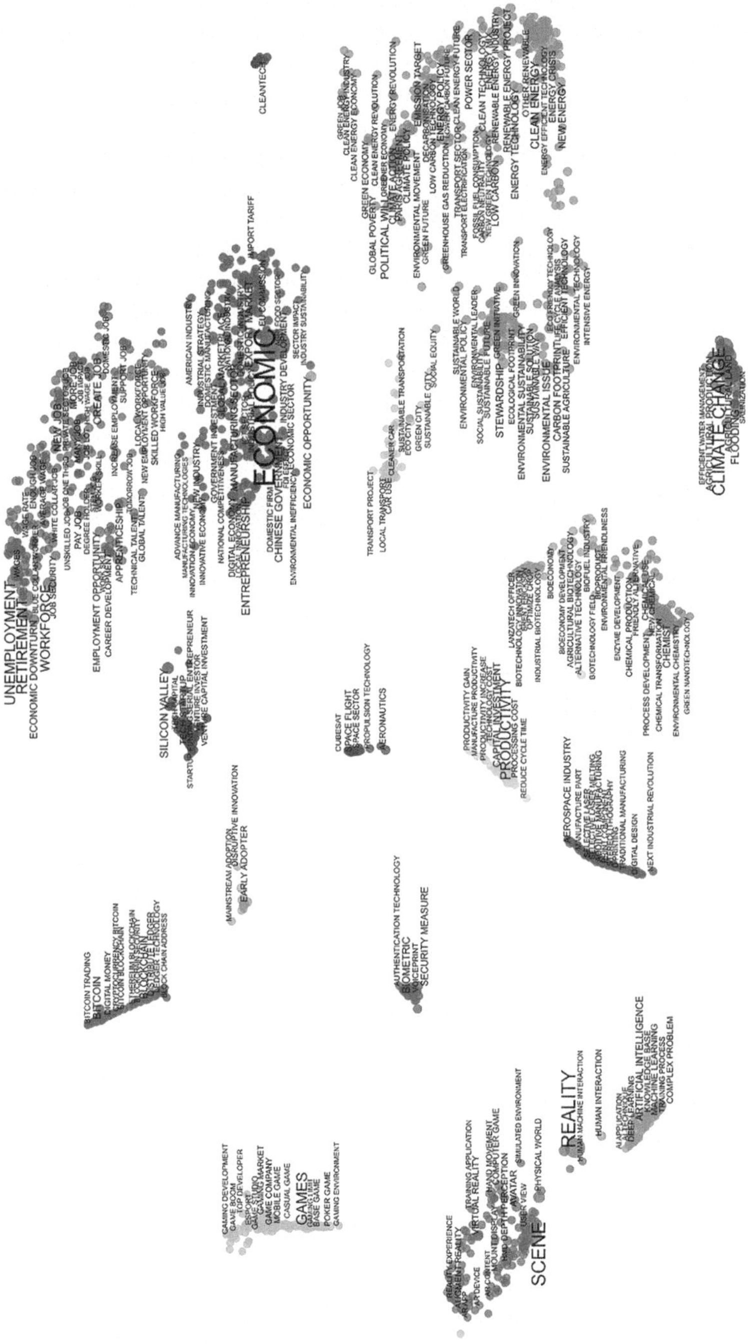

Fig. 22 "High-tech industry" domain full semantic map for 2014–2018 "High-Tech Industry" Thrusts

Table 7 "High-tech Industry (HT)" thrusts and topics

HT thrusts	Topics emerging under each thrust
HT1: Enabling small business to adopt high tech	"IT adoption" and "High-tech start-ups" (Fig. 2.2.4)
	"Skills development" (Fig. 2.2.5)
HT2: New thinking for new industries	"Artificial intelligence"; "Virtual reality" and "Gaming" (Fig. 2.2.6)
	"Cryptocurrency"; and "Biometry" (Fig. 2.2.7)
	"New technologies"; "Aerospace"; "Biotechnology" (Fig. 2.2.8)
HT3: New thinking for old industries	"Economy" and "Productivity" (Fig. 2.2.9)
	"Clean transport" and "Energy efficiency" (Fig. 2.2.10)
	"Sustainability" and "Environmental policy" (Fig. 2.2.11)
	"Climate change" and "Chemistry" (Fig. 2.2.12)

Fig. 23 "IT adoption" (left) and "High-tech start-ups" (right)

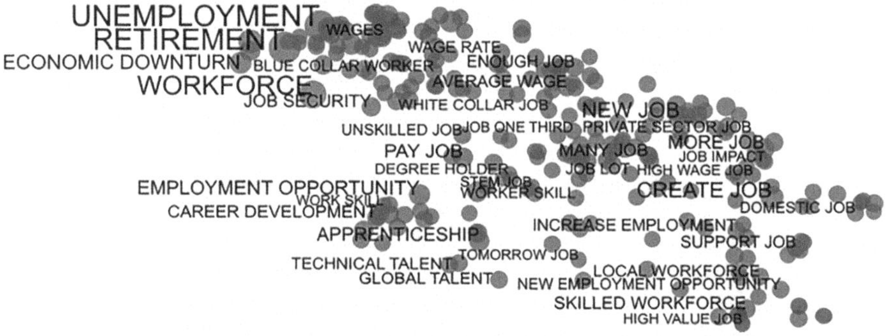

Fig. 24 "Skills development"

development. These factors combined with technical skills will have impacts on widespread IT adoption in small businesses.

2.2.5 HT2: New Thinking for New Industries

High-tech industry will certainly have impacts on new and old industries. Therefore, there is a need to shift in the thinking for new industries. The iFORA system identified three technological topics, which can be the drives for a breakthrough development of new industries, including Artificial Intelligence (AI), Virtual Reality (VR) and gaming. AI appears to be important for all sorts of industries and businesses. Necessary technologies and skills need to be developed for the implementation of AI systems, such as machine learning and deep learning systems. In conjunction with AI, VR systems provide larger application areas for new industries. Particularly human–machine interaction systems will transform the way products, production processes and provision of services. Gaming has also emerged as a new and growing market, not only for entertainment, but also for simulation and decision-making (Fig. 25).

The new high-technology business and industrial systems will also need to transform in terms of new economic and security systems. Cryptocurrencies are revolutionary in terms of creating more stable, secure and speedy financial transactions. Bitcoin is being experimented for several years now and is expected to be more widespread along with all other cryptocurrencies being continuously introduced into the financial market, such as Ethereum and others. In association of cryptocurrencies, the Blockchain concept is also developing fast. Actually, cryptocurrency is only one application for Blockchain as a distributed ledger system. It has potentials to transform all sorts of industries and businesses such as energy and agriculture, which are among the key drivers of South African social and economic systems. In parallel to the emergence of digital financial systems, increasing security issues need to be addressed. These include the development of biometric systems and authentication technologies with increasing security measures (Fig. 26).

Among the technologies for high-tech development are additive manufacturing, 3D printing and digital design. Aerospace and biotechnology emerge as two application areas for high-tech industries in South Africa. Particularly space technologies are considered to benefit from developments in this domain. Regarding biotechnology, the main areas for development are transformation into bioeconomy,

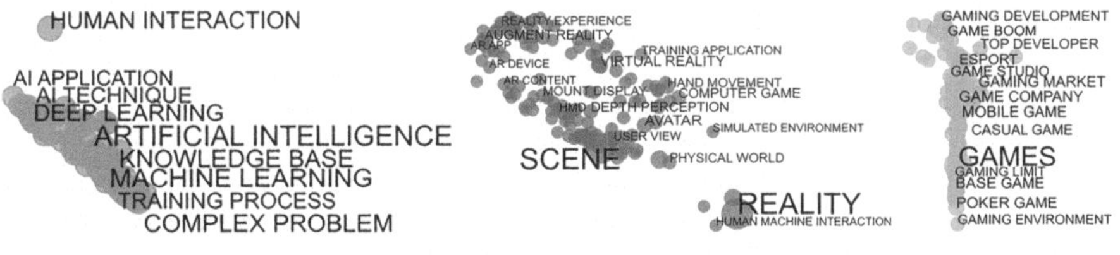

Fig. 25 "Artificial intelligence" (left); "Virtual reality" (center); "Gaming" (right)

Fig. 26 "Cryptocurrency" (left); "Biometry" (center); "Cleantech" (right)

Fig. 27 "New technologies" (left); "Aerospace" (center); "Biotechnology" (right)

Fig. 28 "Economy" (left) and "productivity" (right)

development of industrial biotechnology and biofuel industry. Without any doubt, environmental, social and ethical concerns need to be addressed regarding the implementation of biotechnologies (Fig. 27).

2.2.6 HT3: New Thinking for Old Industries

Besides setting up new industries with new technologies, it is also important to upgrade the old industries, which are the main backbones of development. Economic and productivity improvements are needed to make sure that old industries remain competitive. These include the development of the manufacturing industry at the national and domestic levels, supporting innovation, upgrading technologies and expanding export markets. Regarding productivity, technological and capital investments, as well as reducing cycle times and processing costs appear to be areas for action in all industries (Fig. 28).

Transport is one of the old, but rapidly developing sectors of the South African economy. Actions for the improvements of the transport systems begin with making the industry cleaner and more sustainable. Sustainable transport systems need to be developed with a balanced use of personal cars and local transport. Vehicles on the road should be clean and energy efficient. In order to increase energy efficiency, there is a need to introduce cleaner technologies for all sectors, including energy. Renewable and low-carbon energy technologies should be developed, fossil fuel use should be reduced and power sector should be upgraded (Fig. 29).

In line with energy efficiency, overall improvements in sustainability are needed. Green innovation and initiatives should be in place with reduced ecological footprint and decarbonization. Besides environmental sustainability, social sustainability is also important for a more sustainable future. This requires political will, leadership and stewardship with environmental policies in place (Fig. 30).

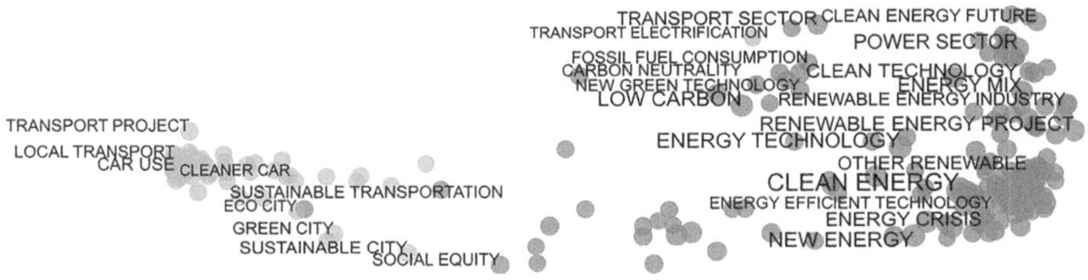

Fig. 29 "Clean transport" (left) and "energy efficiency" (right)

Fig. 30 "Sustainability" (left) and "environmental policy" (right)

Fig. 31 "Climate change" (left) and "chemistry" (right) fragments

Old industries like agriculture and chemistry should also be considered with new thinking. Climate change effects like flooding should be observed with mitigation policies and actions, such as through more efficient water management and agricultural production systems. Similarly, the chemistry industry needs to be development to make it greener and more environmentally friendly with advanced processes and production systems (Fig. 31).

2.2.7 Dynamics of "High-Tech Industry" Domain Thrusts

In this section, the dynamics of "high-tech industry" thrusts are shown based on their frequency of occurrence in the documents indexed in the iFORA database. Among the thrusts HT3 (New thinking for new industries) appears to the highest on the agenda with an increasing growth rate (Fig. 32). This is certainly promising as the research trends support the agenda of the fourth industrial revolution. However, research on small businesses to adopt high-tech is relatively low without a significant progress across years.

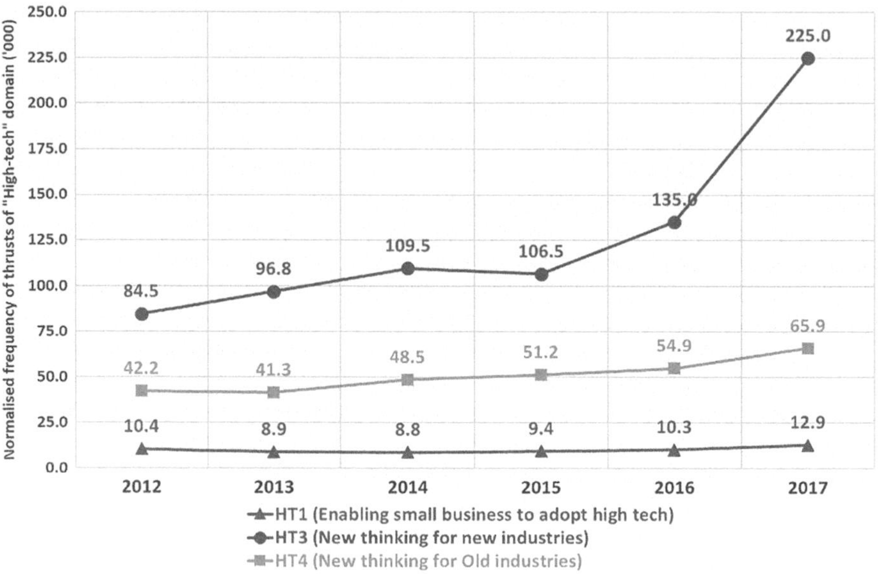

Fig. 32 Normalized frequency of all terms related with thrusts of "High-tech industry" domain ('000) in 2012–2017. Notes. 1. "Normalized frequency" for a specific thrust is a total normalized number of mentions (in thousands) of all terms (that are shown on a semantic map) related with this specific thrust in documents indexed in IFORA for a given year. 2. The HT2 thrust has been moved to "Education" domain and became ED1 "Skills for the 4th Industrial Revolution" thrust

Furthermore, Table 8 illustrates to what extent the "High-tech industry" domain thrusts address the challenges. HT4 is extremely linked to job creation. HT3 has also a strong linkage with the same challenge along with population growth. HT1 is the only thrust, which appears to be linked with income growth.

2.3 *Nutrition Security for a Healthy Population (Nutrition) Domain*

2.3.1 Overall Description of the Domain

Good nutrition is essential for a health population. There are currently serious issues around malnutrition and stunting in South Africa. Particularly women and children are in a disadvantaged position. There are economic (low income, unemployment etc.) and environmental (climate change) drivers of the health and nutrition related problems. Technologies like precision farming are considered to create opportunities for advancing farming in South Africa and making efficient use of arable land by reducing pressures due to climate change, waste and pollution. Hence, the nutrition security domain focuses on zero impact agriculture as well as the application of advanced technologies like biotechnologies, precision agriculture as well as big data.

Table 8 Impacts of all thrusts of "High-tech industry" domain on challenges

	High-tech		
Challenges	HT1	HT3	HT4
Affordable Education	0.0	5.0	1.5
Affordable Energy	0.0	0.0	9.0
Affordable Food	0.0	0.0	0.0
Carbon Emission Reduction	0.0	1.5	20
Clean Water Access	0.0	0.0	1.0
Come Prevention	0.0	1.0	1.0
Crime Reduction	0.0	2.0	0.0
Economy Growth	0.0	1.0	1.0
Employment Growth	0.0	3.0	23.0
Export Growth	0.0	1.0	3.0
High Living Standard	0.0	5.5	4.0
High Quality Health Care Service	0.0	0.0	0.0
Income Growth	1.0	1.3	6.0
Job Creation	0.0	36.6	133.1
Low Greenhouse Gas Emission	0.0	0.0	2.0
Low Mortality	0.0	0.0	0.0
Population Growth	0.0	53.8	24.0
Poverty Alleviation	0.0	2.0	11.8
Reduce Water Demand	0.0	0.0	1.0
Renewable Energy Growth	0.0	1.0	0.0
Rural Community	0.0	13.0	26.6
Secure Water Supply	0.0	0.0	0.0
Skill Improvement	0.0	0.0	0.0
Universal Broadband Access	0.0	1.0	1.5

Notes: 1. Value in each specific cell of this table is a normalized number of co-occurrences of all terms (that are shown on a semantic map) related with a specific thrust (e.g., Thrust HT1) and a specific challenge (e.g., "Low Mortality") in documents indexed in iFORA database for 2012–2017. This value is treated as an impact of a specific thrust onto a specific challenge. 2. Cells are colored by bars as follows: within a column of a specific thrust the biggest green bar means the challenge for which this specific thrust has the highest impact

2.3.2 Overview of Research Potential of South Africa in "Nutrition" Domain

Figure 33 shows that there is a growing scientific emphasis on the Nutrition security domain in South Africa. The number of publications has growing gradually as well as the share of South Africa in the global scientific output in this domain.

Figure 34 shows the leading countries in the Nutrition security domain. As the most advanced country in Africa, South Africa has the 25th position in the world in this domain.

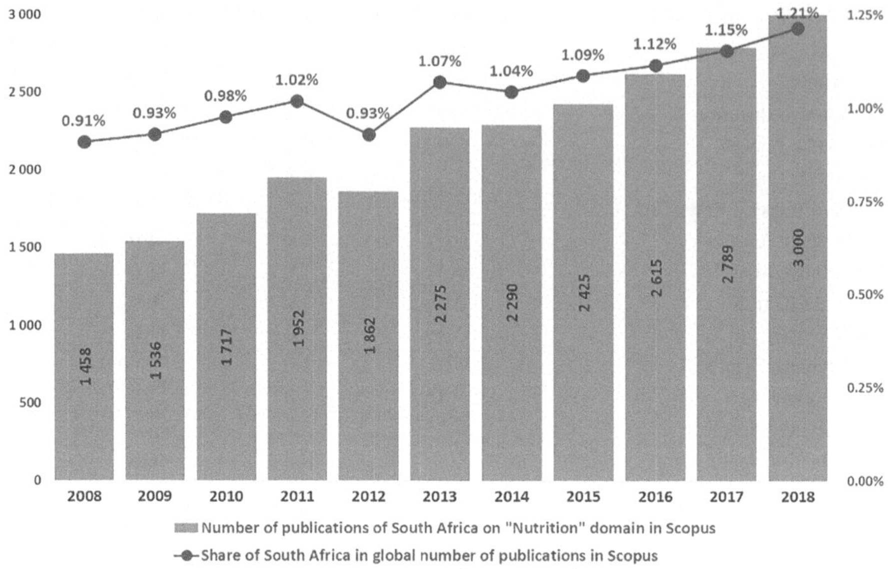

Fig. 33 Dynamics of publications of South Africa in "Nutrition" domain in Scopus in 2008–2018. Note: All types of documents, except of technical documents, indexed in Scopus are included in the analysis

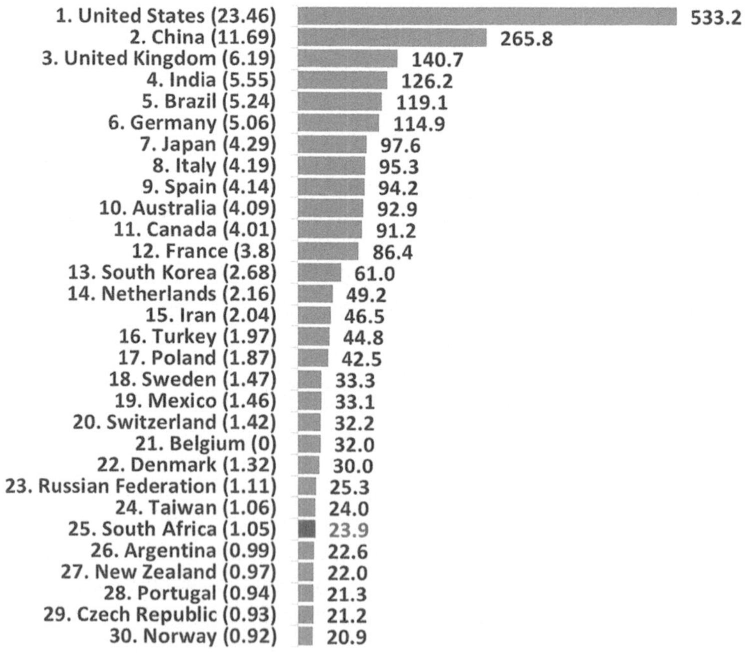

Fig. 34 Leading countries by number of publications ('000) in "Nutrition" domain in Scopus for 2008–2018. Note: 1. All types of documents, except of technical documents, indexed in Scopus are included in the analysis. 2. Share of a country in global number of publications is shown in parenthesis

Among those countries, the key partners of South Africa in the Nutrition security domain are the United States, the United Kingdom, Australia, Germany and France. African countries like Nigeria and Zimbabwe are among the top collaborators in the Africa continent (Table 9).

Table 10 shows the leading organizations in South Africa publishing in the domain of Nutrition security with the total number of publications they produced in the last 10 years. Top four universities publish more than half of research in this domain.

Table 9 Key partner countries of South Africa in internationally collaborated publications in "Nutrition" domain in Scopus for 2008–2018

Country	Number of joint papers with South Africa	Share of a country in all ICPs of South Africa
United States	3357	27.77%
United Kingdom	2244	18.56%
Australia	1457	12.05%
Germany	1106	9.15%
France	1060	8.77%
Netherlands	938	7.76%
Canada	829	6.86%
Nigeria	674	5.58%
Belgium	644	5.33%
Switzerland	594	4.91%
India	563	4.66%
Kenya	563	4.66%
Spain	556	4.60%
Italy	533	4.41%
China	532	4.40%
Sweden	501	4.14%
Brazil	457	3.78%
Zimbabwe	416	3.44%
Norway	383	3.17%
Portugal	373	3.09%
Denmark	356	2.95%
New Zealand	305	2.52%
Cameroon	293	2.42%
Japan	293	2.42%
Russian Federation	268	2.22%
Ethiopia	264	2.18%
Tanzania	262	2.17%
Czech Republic	239	1.98%
Argentina	237	1.96%
Finland	236	1.95%
Ghana	226	1.87%
Uganda	211	1.75%

Note: 1. All types of documents except of technical documents, indexed in Scopus are included in the analysis

Table 10 Leading organizations of South Africa by number of publications in "Nutrition" domain in Scopus for 2008–2018

Affiliation	Number of publications	Share in all South Africa publications
Universiteit van Pretoria	3969	16.59%
Universiteit Stellenbosch	3422	14.31%
University of KwaZulu-Natal	3167	13.24%
University of Cape Town	2968	12.41%
University of Witwatersrand	1679	7.02%
North-West University	1262	5.28%
Rhodes University	1202	5.03%
University of the Free State	1115	4.66%
Agricultural Research Council, Pretoria	1108	4.63%
South African Medical Research Council	1038	4.34%
University of Johannesburg	816	3.41%
South African National Biodiversity Institute	693	2.90%
University of the Western Cape	677	2.83%
University of Fort Hare	605	2.53%
Nelson Mandela Metropolitan University	570	2.38%
University of Limpopo	516	2.16%
Tshwane University of Technology	503	2.10%
The Council for Scientific and Industrial Research	396	1.66%
South African Institute for Aquatic Biodiversity	390	1.63%
Cape Peninsula University of Technology	351	1.47%
University of South Africa	347	1.45%
University of KwaZulu-Natal, Pietermaritzburg Campus	330	1.38%
Plant Protection Research Institute, Pretoria	328	1.37%
ARC Infruitec-Nietvoorbij	321	1.34%
University of Venda for Science and Technology	317	1.33%
National Health Laboratory Services	262	1.10%
Onderstepoort Veterinary Institute	243	1.02%
Groote Schuur Hospital	232	0.97%
South African Sugarcane Research Institute	220	0.92%
Compton Herbarium	210	0.88%
Human Sciences Research Council of South Africa	195	0.82%
Citrus Research International Pty Ltd, South Africa	192	0.80%
University of Zululand	174	0.73%
Marine and Coastal Management	174	0.73%
University of Cape Town, Faculty of Health Sciences	169	0.71%
National Institute for Communicable Diseases	163	0.68%
Durban University of Technology	146	0.61%
Walter Sisulu University	145	0.61%
The Nelson R. Mandela Medical School	141	0.59%
Sefako Makgatho Health Sciences University SMU	131	0.55%
South African Environmental Observation Network	121	0.51%
Tygerberg Hospital	102	0.43%

Note: 1. All types of documents, except of technical documents, indexed in Scopus are included in the analysis. 2. Organization names are presented as they are spelled in Scopus database

2.3.3 Semantic Analysis of "Nutrition" Domain

This section describes the "Nutrition Security for a Healthy Population" STI domain and its thrusts based on the results produced by the iFORA system. First, the full domain is described with a domain map including all of the topics covered. Then each of the thrusts is described by the emerging topics identified by iFORA. The next section presents topic clusters under each thrust with more detailed network maps with descriptions of emerging issues under them. Figure 2.3.3 shows the entire domain of the Nutrition Security for a Healthy Population (NU) with its clusters and interrelations to each other. A detailed analysis of Fig. 35 provides an overview of NU thrusts and a number of existing and emerging topics for consideration under each of them. These are presented in Table 11. In the following section, network maps are given to illustrate these maps and further issues emerging under each of them.

2.3.4 NU1: Zero Impact Agriculture

One of key steps for achieving zero impact agriculture is reducing waste. The iFORA analysis revealed that the biggest problem related to waste in Agriculture concerns water related issues. Besides polluted and wastewater, sewage and treatment, other types of waste such as food and kitchen waste and fish waste emerge as related issues. In order to reduce water waste and make use of limited water resources efficiently, managing agricultural water emerges as an important topic. There is a need to increase water productivity and develop water quality (Fig. 36).

Zero impact agriculture and providing healthy nutrition population require organic and ecological farming and food labeling in order to protect consumer health in terms of food safety and nutrition. Organic and ecological farming covers a range of issues from the quality and quantity of the farmland and crops to cultivation and eventual food yield. Food labeling is an important and direct way of communicating information about food and its ingredients to the consumer. Food labels include any written or pictorial information attached to a food package or a container. Labels include information on ingredients, quality and nutritional value of the food. Especially fat and sugar information and salt level of food are frequently looked at besides the other nutritional values. Food manufacturers/producers play an important role in indicating food labels accurately and in a visible and accessible way to consumers (Fig. 37).

Entrepreneurship and small agriculture emerge as two other issues under zero impact agriculture thrust. Entrepreneurship in this regard is concerned with startups and small businesses, which are crucial in employment and future economic growth in the country. Small agriculture also provides food for a large amount of population, especially for the ones, which have no economic means of accessing food markets and supermarkets. Besides these, small farming and growing own food is becoming popular in urban and peri-urban areas through urban farming and indoor farming, where food is produced nearby where it is consumed (Fig. 38).

Fig. 35 "Nutrition Security for a Healthy Population" domain full semantic map for 2014–2018 "Nutrition Security for a Healthy Population" Thrusts

Table 11 "Nutrition Security for a Healthy Population" thrusts and topics

Nutrition security thrusts	Topics emerging under each thrust
NU1: Zero Impact Agriculture	"Waste problems" and "Agricultural water management" (Fig. 2.3.4)
	"Organic and ecological farming" and "Food labeling" (Fig. 2.3.5)
	"Entrepreneurship" and "Small agriculture" (Fig. 2.3.6)
	"Agricultural policy" and "Sustainability" (Fig. 2.3.7)
NU2: Use and acceptance of modern biotechnology	"Crop GMO" and "Plant genomics" (Fig. 2.3.8)
	"Plant gene engineering" and "Pest-resistant crops" (Fig. 2.3.9)
	"Crop yield management" and "Crop science" (Fig. 2.3.10)
NU3: Personalized information for healthy nutrition for all	"Malnutrition" and "Nutrition diseases" (Fig. 2.3.11)
	"Eating behavior" and "Dieting problems" (Fig. 2.3.12)
	"Sugar food" and "Junk food" (Fig. 2.3.13)
	"Healthy eating behavior" and "Health food policy" (Fig. 2.3.14)
NU4: Precision & big data in agri-businesses	"Precision agriculture" (Fig. 2.3.15)

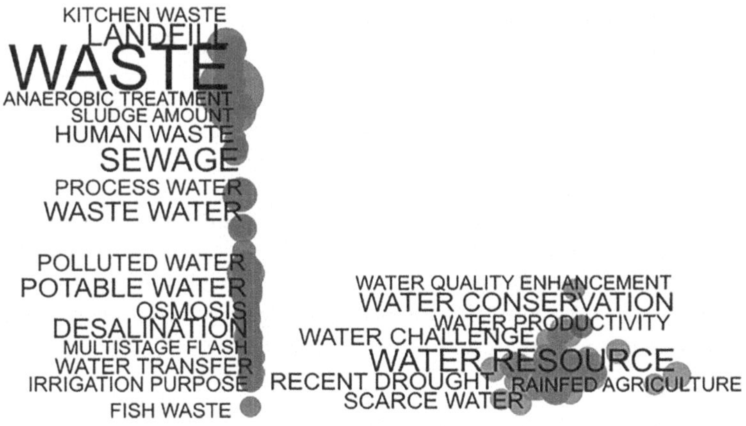

Fig. 36 "Waste problems" (left) and "agricultural water management" (right)

Zero impact agriculture and moving toward a society with healthy nutrition also require strong agricultural policies and measures for sustainability. There are a number of issues on the agenda of the agricultural policy, ranging from macro to micro level issues. These include policies for meeting sustainable/millennium development goals, world food supply and food security at the global level.

Fig. 37 "Organic and ecological farming" (left) and "food labeling" (right)

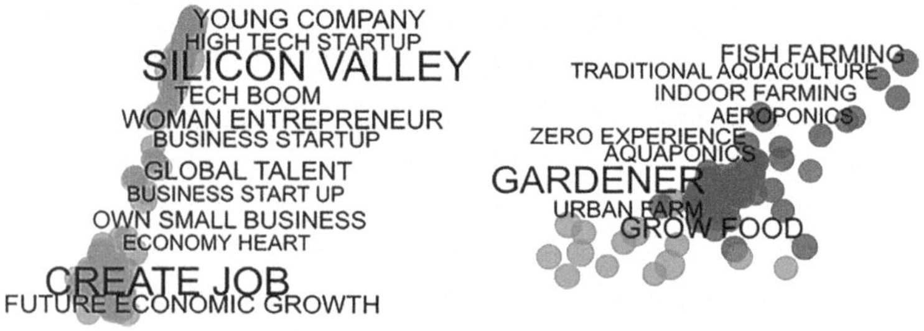

Fig. 38 "Entrepreneurship" (left) and "small agriculture" (right)

Fig. 39 "Agricultural policy" (left) and "sustainability" (right)

Developing the state of the African farmers, dealing with extreme poverty and high food prices are also on the agenda. Besides providing food for population, policy issues should also address the sustainability concerns, which cover achieving economic development and urbanization while mitigating the impacts of climate change, and making agriculture and food more sustainable (Fig. 39).

2.3.5 NU2: Use and Acceptance of Modern Biotechnology

Evolving and emerging technologies provide plenty of opportunities for tackling grand challenges. Similarly, biotechnologies are increasingly used for the agriculture and food sector. Genetically modified crops have been considered as a way to

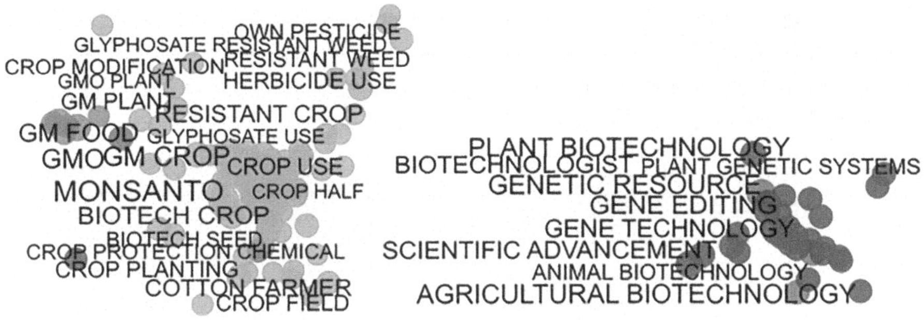

Fig. 40 "Crop GMO" (left) and "plant genomics" (right)

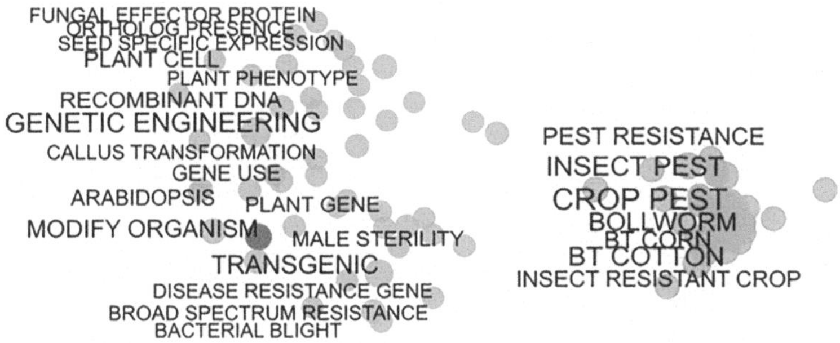

Fig. 41 "Plant gene engineering" (left) and "pest-resistant crops" (right)

produce resistant and nutritious crops through genetic modification. Although this appears beneficial, in principle, a number of concerns have been raised about the health and environmental impacts of GMOs. Issues around the GMO crops are related to crop modification, pesticide use and resistant weeds among the others. Regarding plant genomics, the key issues are clustered around gene technology, plant and animal biotechnology as well as gene editing, which are the scientific fields with rapid technological development in the world and in South Africa (Fig. 40).

Among the technologies used for plant gene engineering in South Africa, the iFORA analysis revealed several application areas in plant cells, phenotypes, callus transformation, transgenic and bacterial blights. Among the key aims of plant gene engineering are developing pest-resistant crops against insects and other pests (Fig. 41).

Modern technologies in agriculture are also used to improve crop yield to increase crop variety, improve herbicide tolerance as well as providing water use efficiency with lower environmental stress. Crop science related to breeding also appears as an important area of research in South Africa (Fig. 42).

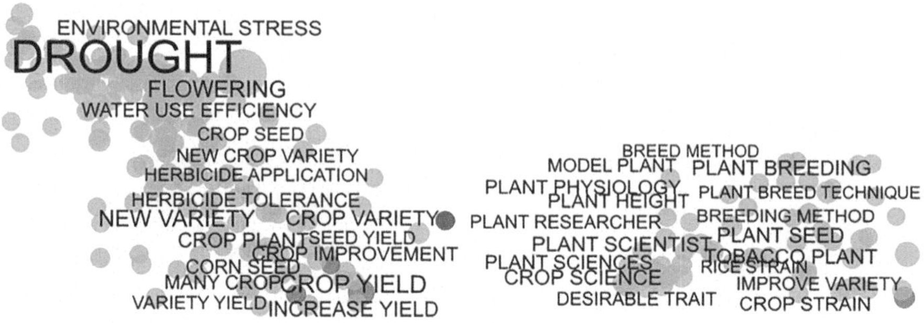

Fig. 42 "Crop yield management" (left) and "crop science" (right)

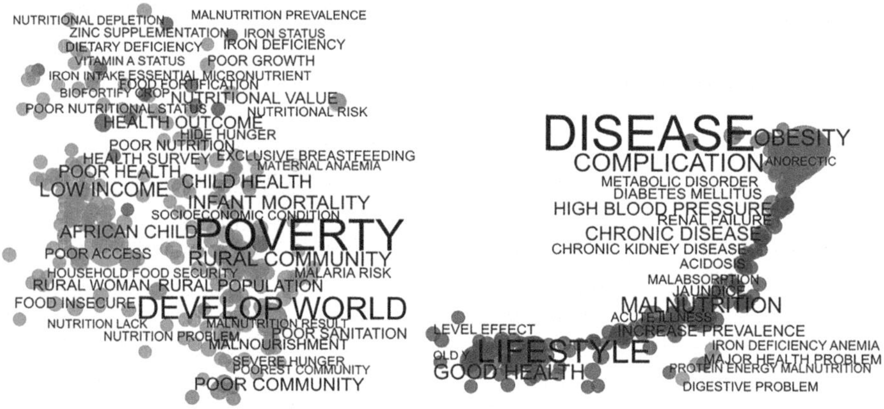

Fig. 43 "Malnutrition" (left) and "nutrition diseases" (right)

2.3.6 NU3: Personalized Information for Healthy Nutrition for All

Healthy nutrition for all is a challenge for multiple aspects including, agriculture, food, health, economic condition, environment, lifestyle as well as urban-rural issues. Malnutrition is a key problem in many developing countries in the world as well as in South Africa. Among the main drivers for malnutrition are low income and poverty, poor access to food, food insecurity and resultant poor health, infant mortality and other dietary deficiency related to the lack of iron, zinc and other nutrients. These give rise to a number of diseases and major health problems such as high blood pressure, metabolic disorders, diabetes, obesity, anemia, kidney disorders among the others, which are commonly seen in Africa (Fig. 43).

Eating behaviors are closely related to dietary problems. In today's world, large amount of populations in developed, developing or under-developed countries suffer from unhealthy diets. These are related to issues regarding daily calorie intake, balanced diet, vegetable and fruit consumption, and consumption of whole grain food. Failure in providing healthy diet results with hunger, thirst, high cholesterol, obesity and other metabolic dieting problems (Fig. 44).

Fig. 44 "Eating behavior" (left) and "dieting problems" (right)

Fig. 45 "Sugar food" (left) and "Junk food" (right)

In particular, sugar and junk food consumption emerge as two important problems in dietary problems. Sugar intake in the world has increased dramatically with sugary/sweetened drinks and food. Especially the use of unhealthy sugars in diet causes serious health problems. Taxes on fat soda are proposed in order to reduce the unhealthy consumption, which gives rise to higher levels of obesity. Junk food is one of the main causes of dieting problems due to the high level of sugar they contain. Changing lifestyles and food choices make people to consume more unhealthy food like processed meat, French fries and sugary snacks (Fig. 45).

Healthy eating behavior includes not only consuming healthy products and maintaining balanced meal. Ensuring food sustainability and reducing food print should also be among the main concerns. Proper health policies should be formulated for improving health, encouraging a healthy lifestyle, provide access to nutritious food by the larger groups of society while reducing bad diet and obesity (Fig. 46).

2.3.7 NU4: Precision and Big Data in Agri-Businesses

Besides bio- and crop technologies, other technologies are emerging to improve farming practice, and in turn producing more nutritious and sustainable food. Among these are precision farming, farm automation technologies and remote sensing technologies. Development and implementation of these technologies require

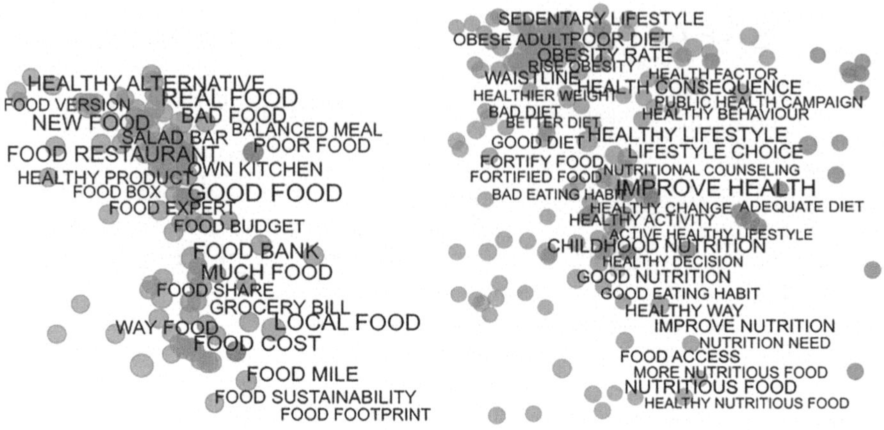

Fig. 46 "Healthy eating behavior" (left) and "health food policy" (right)

Fig. 47 "Precision agriculture"

close collaboration between scientists, industry and policymakers to make sure that these technologies are available and accessible by the farmers in daily agricultural practices (Fig. 47).

2.3.8 Dynamics of "Nutrition" Domain Thrusts

In this section, the dynamics of the "Nutrition" domain thrusts are shown based on their frequency of occurrence in the documents indexed in the iFORA database. Among the thrusts NU3 (Personalized information for healthy nutrition for all) appears to be the highest on the agenda (Fig. 48).

Table 12 illustrates to what extent the "Nutrition security" domain thrusts address the challenges. The use and acceptance of modern biotechnology (NU2) is strongly linked to affordable food. Personalized information for healthy nutrition for all (NU3) is instrumental in addressing low mortality and population growth. Finally zero impact agriculture (NU1) shows a considerable association to rural communities.

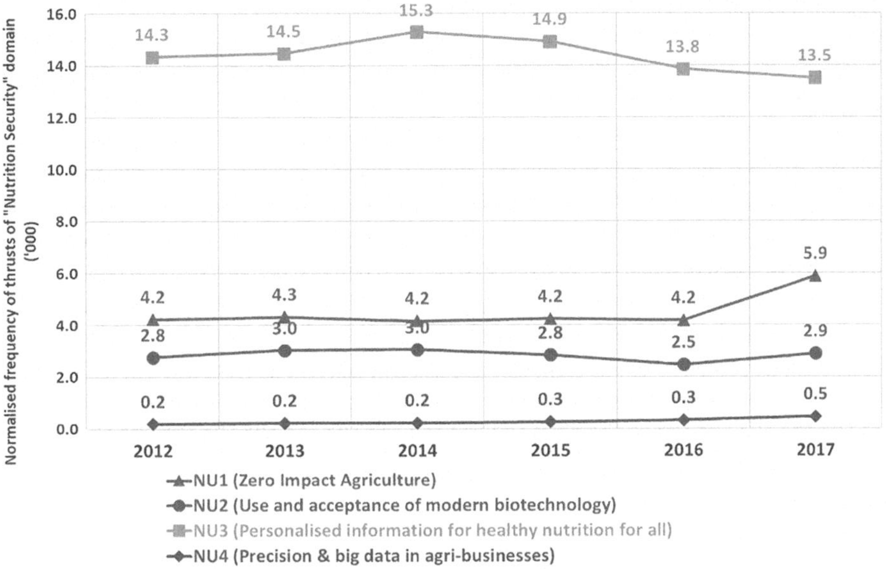

Fig. 48 Normalized frequency of all terms related with thrusts of "Nutrition" domain ('000) in 2012–2017. Note: "Normalized frequency …" for a specific thrust is a total normalized number of mentions (in thousands) of all terms (that are shown on a semantic map) related with this specific thrust in documents indexed in IFORA for a given year

Table 12 Impacts of all thrusts of "Nutrition" domain on challenges

	Nutrition Security			
Challenges	NU1	NU2	NU3	NU4
Affordable Education	0.0	0.0	1.0	0.0
Affordable Energy	5.5	0.0	4.3	0.0
Affordable Food	2.5	103.0	20.7	0.0
Carbon Emission Reduction	0.0	0.0	1.0	0.0
Clean Water Access	2.5	1.0	1.0	0.0
Crime Prevention	4.0	0.0	5.0	0.0
Crime Reduction	0.0	0.0	1.0	0.0
Economy Growth	0.0	0.0	0.0	0.0
Employment Growth	0.0	0.0	4.5	0.0
Export Growth	0.0	2.0	3.0	0.0
High Living Standard	3.0	3.0	2.0	0.0
High Quality Health Care Service	0.0	0.0	2.0	0.0
Income Growth	14.5	5.0	12.0	2.0
Job Creation	11.3	4.7	9.0	1.0
Low Greenhouse Gas Emission	2.0	2.0	2.7	0.0
Low Mortality	1.0	10.0	49.3	0.0

(continued)

Table 12 (continued)

Challenges	Nutrition Security			
	NU1	NU2	NU3	NU4
Population Growth	9.7	18.0	47.0	0.0
Poverty Alleviation	10.0	8.0	13.0	0.0
Reduce Water Demand	8.0	8.0	0.0	0.0
Renewable Energy Growth	0.0	0.0	1.0	0.0
Rural Community	28.0	9.8	21.3	5.0
Secure Water Supply	9.0	0.0	0.0	0.0
Skill Improvement	0.0	0.0	0.0	0.0
Universal Broadband Access	0.0	0.0	0.0	0.0

Notes: 1. Value in each specific cell of this table is a normalized number of co-occurrences of all terms (that are shown on a semantic map) related with a specific thrust (e.g., Thrust NU1) and a specific challenge (e.g., "Low Mortality") in documents indexed in iFORA database for 2012–2017. This value is treated as an impact of a specific thrust onto a specific challenge. 2. Cells are colored by bars as follows: within a column of a specific thrust the biggest green bar means the challenge for which this specific thrust has the highest impact

2.4 *Information and Communication Technologies: Opportunities, Threats and Impacts (ICT) Domain*

2.4.1 Overall Description of the Domain

Information and Communication Technologies (ICTs) are one of the key enablers of development in all domains from agriculture to health, industry to governance and service delivery. There are a number of technologies under the umbrella of ICTs. The Internet of Things (IoT) promises a hyper-connected and ultra-digitally responsive society that supports human, societal and environmental developments. Artificial intelligence offers unique opportunities to improve human lives and to address major societal challenges. Blockchain technology is expected to disrupt several markets by ensuring trustworthy transactions without the necessity of a third party. All these technologies bring opportunities and threats for socio-economic systems. Therefore, their development needs to be regulated by addressing the concerns regarding security, privacy, equity and integrity.

2.4.2 Overview of Research Potential of South Africa in "ICT" Domain

Figure 49 shows that there is a growing scientific emphasis on the ICT domain in South Africa. However, the number of publications and the share of South Africa in the global scientific output in this domain remain still very low. Clearly, more research in this domain is critical for the achievement of the 2030 goals and targets, as the ICTs are key drivers of development almost in all of the domains covered by the National Foresight 2030 study.

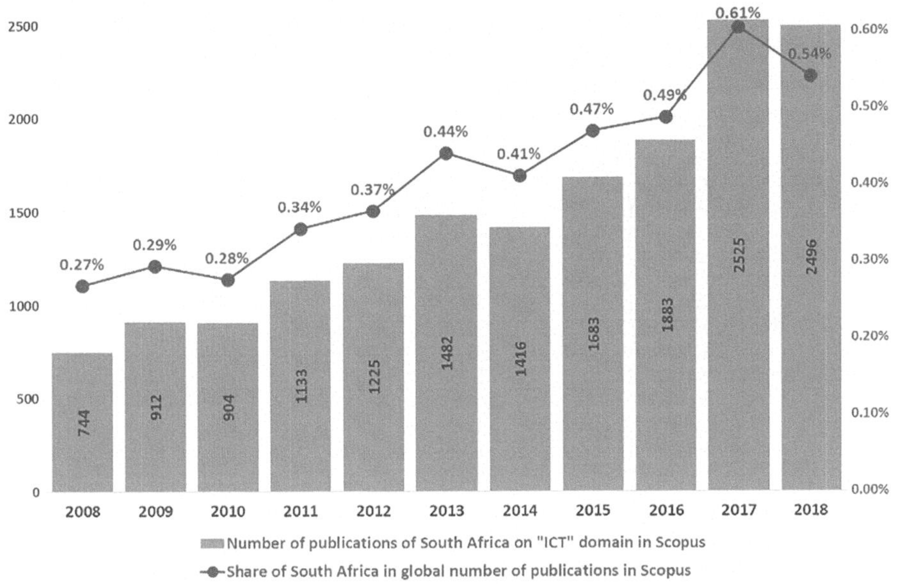

Fig. 49 Dynamics of publications of South Africa in "ICT" domain in Scopus in 2008–2018. Note: All types of documents, except of technical documents, indexed in Scopus are included in the analysis

Figure 50 shows the leading countries in the ICT domain. South Africa's position in ICTs is extremely low (45th place in the world). Top collaborators in this domain are similar to other domains (Table 13).

Table 14 shows the leading organizations in South Africa publishing in the domain of ICT with the total number of publications they produced in the last 10 years, as well as the percentage of their contribution in the total scientific output. The number of publications is considerably low considering the importance of the ICT domain.

2.4.3 Semantic Analysis of the "ICT" Domain

This section describes the "The Opportunities, Threats and Impact of ICTs (including Smart Systems, on Society)" STI domain its thrusts based on the results produced by the iFORA system. First, the full domain is described with a domain map including all the topics covered. Then each of the thrusts is described by the emerging topics identified by iFORA. The second section presents topic clusters under each thrust with more detailed network maps with descriptions of emerging issues under them. Figure 2.3.3 shows the entire domain of the Opportunities, Threats and Impact of ICTs, including Smart Systems, on Society (IT) with its clusters and interrelations to each other. A detailed analysis of Fig. 51 provides an overview of IT thrusts and a number of existing and emerging topics for consideration under

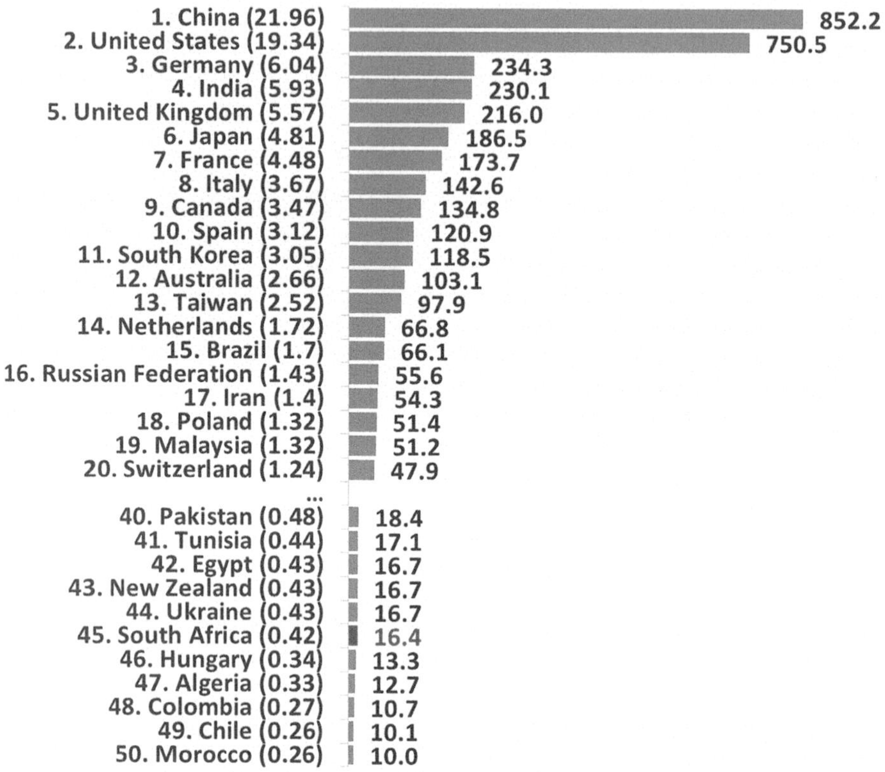

Fig. 50 Leading countries by number of publications on "ICT" domain in Scopus for 2008–2018. Note: 1. All types of documents, except of technical documents, indexed in Scopus are included in the analysis. 2. Share of a country in global number of publications is shown in parenthesis

Table 13 Key partner countries of South Africa in internationally collaborated publications in "ICT" domain in Scopus for 2008–2018

Country	Number of joint papers with SAR	Share of a country in all ICPs of South Africa
United States	1181	20.82%
United Kingdom	902	15.90%
Germany	627	11.05%
France	481	8.48%
China	433	7.63%
Australia	430	7.58%
Canada	392	6.91%
Netherlands	368	6.49%
India	341	6.01%
Italy	284	5.01%
Nigeria	273	4.81%
Sweden	206	3.63%

Table 13 (continued)

Country	Number of joint papers with SAR	Share of a country in all ICPs of South Africa
Belgium	155	2.73%
Switzerland	155	2.73%
Spain	152	2.68%
Finland	134	2.36%
Namibia	132	2.33%
Austria	124	2.19%
Brazil	115	2.03%
Zimbabwe	110	1.94%
Denmark	99	1.75%
Japan	99	1.75%
Norway	97	1.71%
Kenya	94	1.66%
Botswana	93	1.64%
Turkey	89	1.57%
New Zealand	87	1.53%
Saudi Arabia	84	1.48%
Poland	76	1.34%
South Korea	71	1.25%
Singapore	64	1.13%
Ghana	63	1.11%
Ireland	63	1.11%
Hong Kong	61	1.08%
Russian Federation	60	1.06%
Portugal	59	1.04%
Mexico	55	0.97%
Iran	53	0.93%
Malaysia	53	0.93%
Greece	51	0.90%
Argentina	50	0.88%

each of them. These are presented in Table 15. In the following section, network maps are given to illustrate these maps and further issues emerging under each of them.

2.4.4 IT1: Checks and Balances for a Digital Future

The first thrust of the IT domain is concerned with the societal impacts of the rapidly developing Information and Communication Technologies (ICTs). Many intended and unintended consequences are emerging with significant impacts. The analysis with the iFORA system revealed five important points, which are

Table 14 Leading organizations of South Africa by number of publications in "ICT" domain in Scopus for 2008–2018

Affiliation	Number of publications	Share in all South Africa publications
Universiteit van Pretoria	2349	14.32%
University of Cape Town	2224	13.56%
University of Johannesburg	1863	11.36%
University of KwaZulu-Natal	1567	9.55%
The Council for Scientific and Industrial Research	1551	9.46%
Universiteit Stellenbosch	1370	8.35%
University of Witwatersrand	1179	7.19%
University of South Africa	1053	6.42%
Tshwane University of Technology	914	5.57%
North-West University	627	3.82%
Meraka Institute	558	3.40%
Nelson Mandela Metropolitan University	456	2.78%
Rhodes University	410	2.50%
University of the Western Cape	337	2.05%
Cape Peninsula University of Technology	336	2.05%
Durban University of Technology	252	1.54%
University of the Free State	224	1.37%
University of Fort Hare	169	1.03%
Central University of Technology, Free State	160	0.98%
University of Zululand	154	0.94%
Vaal University of Technology	150	0.91%
University of Limpopo	106	0.65%
South African Medical Research Council	102	0.62%
South African Astronomical Observatory	93	0.57%
University of Venda for Science and Technology	79	0.48%
Eskom	57	0.35%
Walter Sisulu University	51	0.31%
African Institute for Mathematical Sciences	50	0.30%
Square Kilometre Array, South Africa	49	0.30%
Mangosuthu University of Technology	45	0.27%
Nelson Mandela University	37	0.23%
University of Witwatersrand, School of Chemical and Metallurgical Engineering	33	0.20%
South African National Space Agency	31	0.19%
Groote Schuur Hospital	27	0.16%
University of Cape Town, Faculty of Health Sciences	26	0.16%
Agricultural Research Council, Pretoria	25	0.15%
Stellenbosch Institute for Advanced Study	25	0.15%

Note: 1. All types of documents, except of technical documents, indexed in Scopus are included in the analysis. 2. Organization names are presented as they are spelled in Scopus database

Fig. 51 "The Opportunities, Threats and Impact of ICTs, including Smart Systems, on Society" domain full semantic map for 2014–2018 "The Opportunities, Threats and Impact of ICTs, including Smart Systems, on Society" Thrusts

Table 15 "The Opportunities, Threats and Impact of ICTs, including Smart Systems, on Society" thrusts and topics

IT thrusts	Topics emerging under each thrust
IT1: Checks and balances for a digital future	"Identity security" and "Cybersecurity" (Fig. 2.4.4)
	"Privacy issues" (Fig. 2.4.5)
	"Ethical issues of ICT" and "Media, news and trust" (Fig. 2.4.6)
IT2: Internet Access & the Network & Information Infrastructure	"Broadband access promotion" and "Cloud-based systems" (Fig. 2.4.7)
IT3: Big Data, Data Analytics and Decision Support	"Sensor networks and wireless infrastructure" and "Real time data" (Fig. 2.4.8)
	"Big data analytics" and "Decision support" (Fig. 2.4.9)
	"IT in education" (Fig. 2.4.10)
IT4: Smart and sustainable Municipal Service delivery	"Smart systems (city, energy, transport)" (Fig. 2.4.11)
	"Energy, electricity and utilities" (Fig. 2.4.12)
	"Smart governance" and "Cryptocurrency" (Fig. 2.4.13)

concerned with security and privacy as well as ethical and trust issues. These are presented respectively from Figs. 2.4.4 to 2.4.6. Identity theft and fraud are emerging as important concerns in parallel to the more widespread use of the ICTs. Personal information is collected through online systems by a wide variety of parties including the governmental agencies, internet operators, applications, online marketing and retail sites among the others. Therefore, strict protection and regulations are needed to make sure that this sensitive information does not go into wrong hands, which may result with ID theft. Today, the worldwide information security spending has almost reached 124 billion USD.[1] This large amount is an indicator of how crucial the issue is. Setting up a safe and secure digital infrastructure is the first step for cyber protection. Particularly governmental systems and networks may be vulnerable. However, private businesses are also under risk of cyber-attacks. There is a need to establish systems for identifying risk factors and developing incident response systems for increased security (Fig. 52).

In parallel to security issues, privacy issues emerge as important concerns in the widespread use of ICTs. Most of the topics related to privacy issues addresses policy measures, regulations and law in relation to privacy (Fig. 53).

Developments in ICTs create ethical dilemmas, and appropriate social and technological responses to be developed to tackle with them in terms of processes, policy initiatives and frameworks, data access and information sharing, technology solutions and legislation. Another aspect of the ethics concerns publication and sharing of information through media. With the widespread sharing of information through online and social media channels, it is getting more and more difficult to distinguish real news and stories from the fake ones. These create mixed social

[1] https://www.gartner.com/en/newsroom/press-releases/2018-08-15-gartner-forecasts-worldwide-information-security-spending-to-exceed-124-billion-in-2019

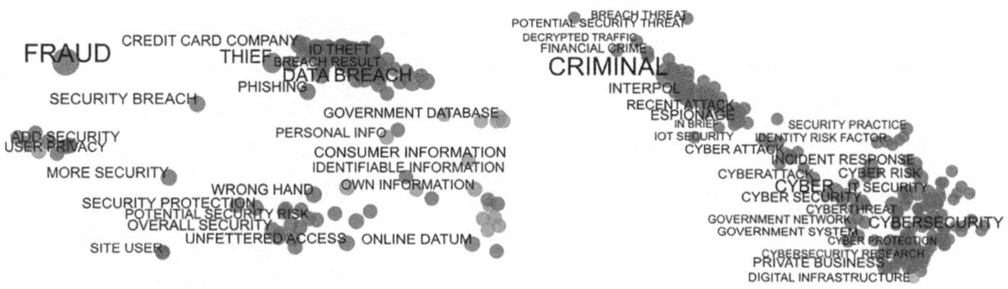

Fig. 52 "Identity security" (left) and "cybersecurity" (right)

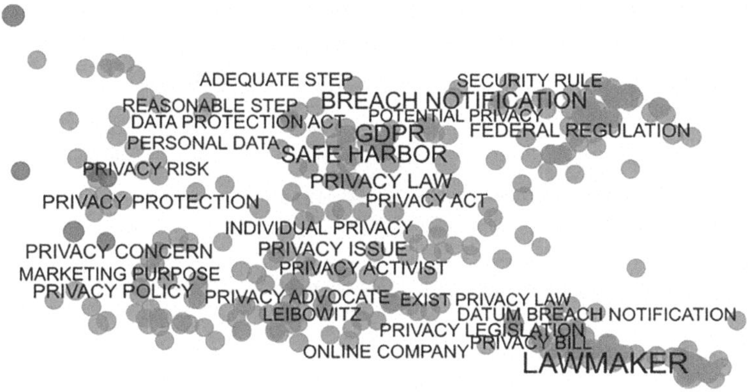

Fig. 53 "Privacy issues"

reactions to developments, which are shared by reliable as well as unreliable sources of information. Therefore, trust factor plays an important role in media and raises concerns about preventing the dissemination of fake news and information over ICT networks, for which there is no clear solution so far (Fig. 54).

2.4.5 IT2: Internet Access and the Network and Information Infrastructure

Accessing to the Internet and information infrastructure is crucial to exploit the opportunities offered by the digitalization process and digital economy. Two topics emerging related to this thrust are broadband access and cloud-based systems. Broadband access is important to make sure that large groups of the population both in urban and rural areas have access to the Internet in a fast, secure and affordable way. There is a need to develop necessary telecom infrastructure and provide broadband services for all. In addition to the Internet access, the storage of information is also an important issue. Cloud systems have been developed for making applications, services and resources available to large amount of users on demand via the Internet from a cloud computing providers' servers. Cloud services also require investments in infrastructures with servers, data centers and applications (Fig. 55).

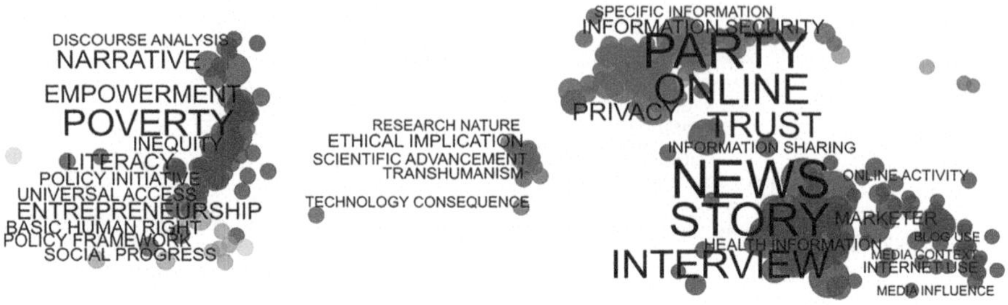

Fig. 54 "Ethical issues of ICT" (left) and "media, news and trust" (right)

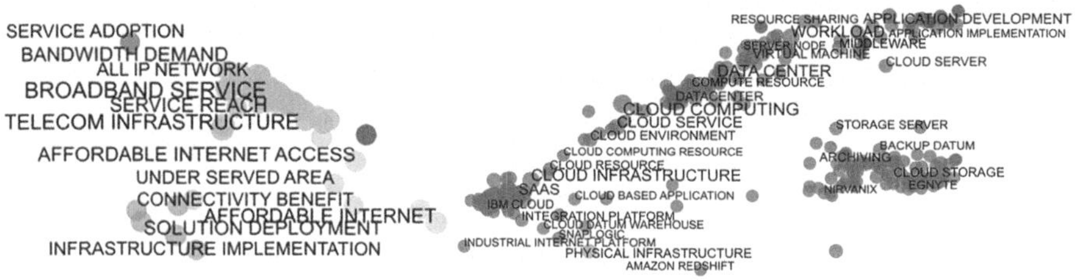

Fig. 55 "Broadband access promotion" (left) and "cloud-based systems" (right)

2.4.6 IT3: Big Data, Data Analytics and Decision Support

Increased amount of data and advancements in data analytics made the use of big data as a new source of evidence for decision-making. Two points related to this thrust are concerned with the collection of data. With the reduced cost of sensors and widespread implementation of wireless networks, large amount of information are collected related to climate, transportation, agriculture, health and all other sectors of the social and economic life. Besides collection of the historical data, there is now an increasing need and possibilities for collecting information real time to enable real time decision-making. A real time data collection system will require infrastructure for networks and communication systems, wireless networks as well as systems for processing data (Fig. 56).

Data collected and processed need to be analyzed through the big data analytics systems. This is certainly an area where South Africa, like many other countries in the world, should develop necessary skills, capacity and infrastructure. Systems to be developed include data collection, storage, processing, analytics, visualization and interpretation. A number of new skills will be required for the big data analytics systems including machine learning expert, data scientist, analyst, domain experts and programmers (Fig. 57).

In order to build necessary skills and capacity for a digital future, ICTs should be increasingly integrated into education systems. For instance, coding and big data analytics can be included in the curricula of tertiary schools, if not primary and secondary (Fig. 58).

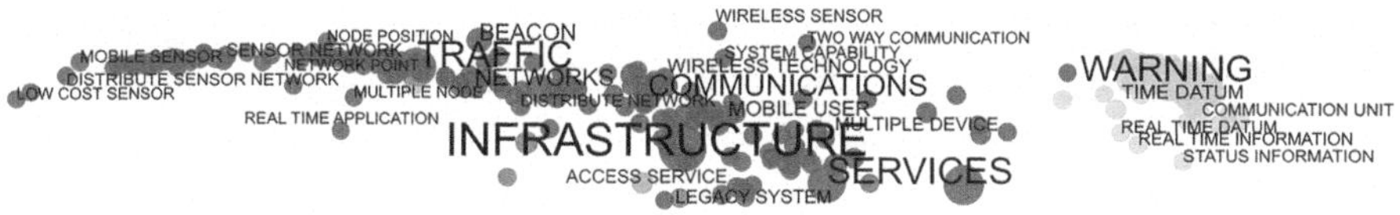

Fig. 56 "Sensor networks and wireless infrastructure" (left) and "real-time data" (right)

Fig. 57 "Big data analytics" (left) and "decision support" (right)

Fig. 58 "IT in education"

2.4.7 IT4: Smart and Sustainable Municipal Service Delivery

Information and Communication Technologies are also instrumental in terms of implementing smart cities, energy, transport and public services. A smart city refers to the use of ICTs to increase operational efficiency, information sharing and use, and service delivery to improve quality of life of citizens and provide sustainable systems with smart solutions. Smart energy use and transportation systems are two out of many other application areas for smart cities. Besides them, there are also technologies and applications for smart homes and buildings. The key enabling technology of these smart systems is the Internet of Things (IoT). It is expected that these technologies will find application areas not only in cities but also to improve quality of life in rural areas with better access to products, services and clean energy and efficient transport systems with smart and integrated solutions (Fig. 59).

Regarding smart energy systems, a number of topics arise around use of green energy sources, energy saving solutions, smart grid systems and energy information and management systems. Policy structures, energy infrastructure and utility companies need to be upgraded in order to implement smart energy solutions (Fig. 60).

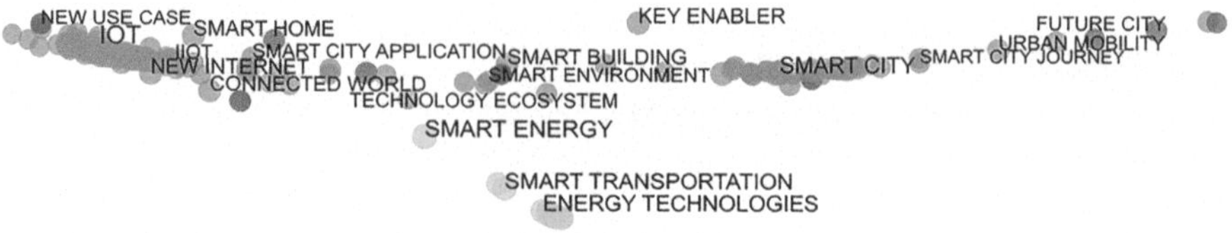

Fig. 59 "Smart systems (city, energy, transport)"

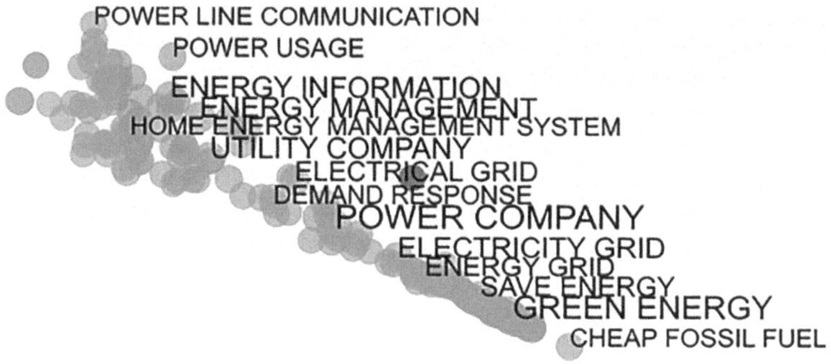

Fig. 60 "Energy, electricity and utilities"

Fig. 61 "Smart governance" (left) and "cryptocurrency" (right)

Implementing individual smart solutions certainly is not sufficient for setting up "smart systems." Smart governance is also needed for integrating and operating those components. Governance at the national, regional and local levels should be adapted to smart systems with the development of necessary public sector institutions, raising investments and involving enterprises in the process. The iFORA system also picked up the development of smart payment methods and digital currencies associated to the development of smart and sustainable municipal service delivery thrust (Fig. 61).

2.4.8 Dynamics of "ICT" Domain Thrusts

Dynamics of the "ICT" domain thrusts are shown based on their frequency of occurrence in the documents indexed in the iFORA database. Among the thrusts IT1 (Checks and balances for a digital future) appears to be the highest on the agenda (Fig. 62). The emphasis on big data, analytics and decision support (IT3) is growing; however, this growth is still insignificant. Use of ICTs for smart and sustainable municipal service delivery (IT4) is considerably low.

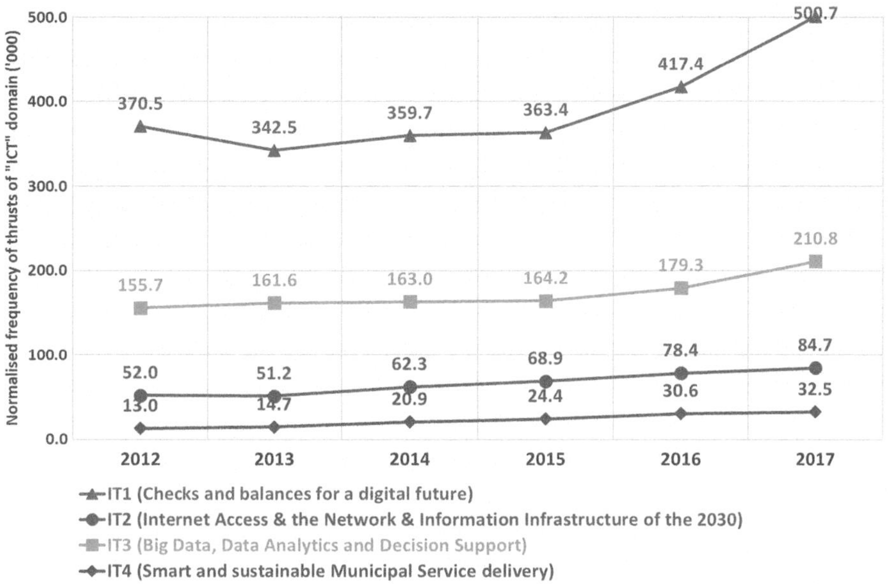

Fig. 62 Normalized frequency of all terms related with thrusts of "ICT" domain ('000) in 2012–2017. Note: "Normalized frequency" for a specific thrust is a total normalized number of mentions (in thousands) of all terms (that are shown on a semantic map) related with this specific thrust in documents indexed in IFORA for a given year

Furthermore, Table 16 illustrates to what extent the "ICT" domain thrusts address the challenges. Job creation, population growth and rural communities appear to be challenges linked to ICT thrusts.

2.5 *Health Technology To Prevent Ill-Health, And Advance Well-Being for Marginalized (Health) Domain*

2.5.1 Overall Description of the Domain

Within the scope of the South African National STI Foresight 2030, the Health domain refers to technologies and strategies to prevent and treat ill health while advancing well-being for marginalized. Overall, the health system needs to be optimized in order to deliver better diagnostic and treatment services and for the overall improvement of the healthcare system as well as drug development. An understanding needs to be developed in the society that prevention is cheaper than cure, so that people have better control over their own health. There is also needed to develop current health infrastructure and administration, which are inadequate, particularly in the rural areas of South Africa. Like in other domains, emerging technologies

Table 16 Impacts of all thrusts of "ICT" domain on challenges

	ICT			
Challenges	IT1	IT2	IT3	IT4
Affordable Education	0.0	0.0	1.0	0.0
Affordable Energy	1.0	0.0	0.0	0.0
Affordable Food	7.0	0.0	2.0	0.0
Carbon Emission Reduction	2.0	4.0	0.0	8.0
Clean Water Access	0.0	2.0	0.0	0.0
Crime Prevention	3.4	0.0	1.5	0.0
Crime Reduction	13.0	0.0	0.0	0.0
Economy Growth	0.0	0.0	0.0	1.0
Employment Growth	7.0	1.0	3.0	1.5
Export Growth	2.0	0.0	0.0	2.0
High Living Standard	1.0	0.0	0.0	1.0
High Quality Health Care Service	3.0	0.0	0.0	0.0
Income Growth	2.5	0.0	0.0	0.0
Job Creation	35.5	14.7	31.0	61.0
Low Greenhouse Gas Emission	0.0	0.0	0.0	0.0
Low Mortality	7.0	5.0	1.0	0.0
Population Growth	37.7	14.0	7.0	18.5
Poverty Alleviation	4.0	1.0	1.5	26.0
Reduce Water Demand	0.0	0.0	0.0	0.0
Renewable Energy Growth	0.0	0.0	0.0	2.0
Rural Community	16.0	28.7	9.0	21.0
Secure Water Supply	0.0	0.0	0.0	1.0
Skill Improvement	1.0	0.0	1.0	0.0
Universal Broadband Access	1.0	1.5	0.0	0.0

Notes: 1. Value in each specific cell of this table is a normalized number of co-occurrences of all terms (that are shown on a semantic map) related with a specific thrust (e.g., Thrust IT1) and a specific challenge (e.g., "Low Mortality") in documents indexed in iFORA database for 2012–2017. This value is treated as an impact of a specific thrust onto a specific challenge. 2. Cells are colored by bars as follows: within a column of a specific thrust the biggest green bar means the challenge for which this specific thrust has the highest impact

including mobile technologies, AI and Big Data will bring enormous opportunities for the development of the healthcare service delivery for all.

2.5.2 Overview of Research Potential of South Africa in "Health" Domain

Figure 63 shows that there is a growing scientific emphasis on the Health domain in South Africa. The number of publications in this domain the Health domain is considerably high compared to other STI domains identified in the National STI Foresight 2030 study. The number of publications has been growing continuously as well as the share of South Africa in the global scientific output in this domain.

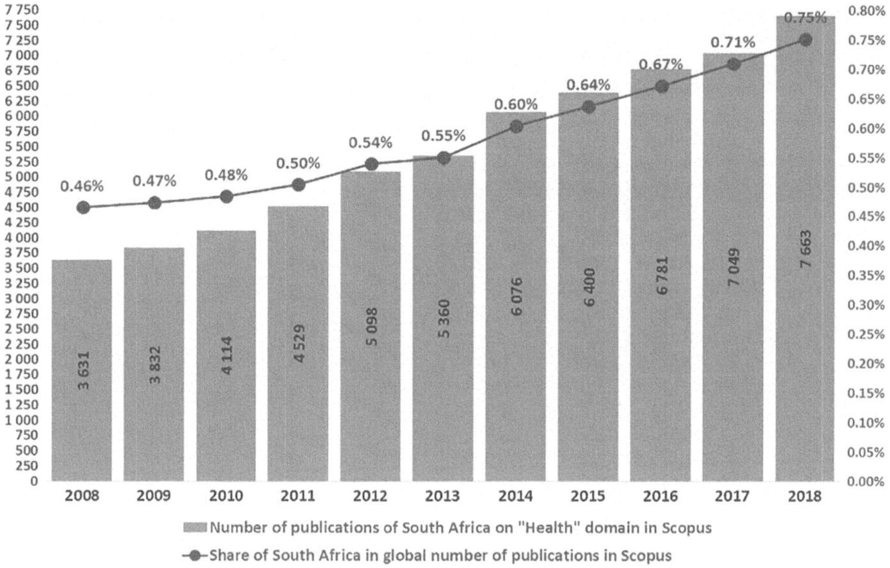

Fig. 63 Dynamics of publications of South Africa on "Health" domain in Scopus in 2008–2018. Note: All types of documents, except of technical documents, indexed in Scopus are included in the analysis

2.5.3 Profile of South Africa in "Health" Domain in Scopus

Figure 64 shows the leading countries in the Health domain. Although in terms of number of publications this domain appears to be one of the leading STI areas in South Africa, compared to the rest of the world, there is still a considerable progress the country needs to make. The country is currently positioned at the 31st place in the world.

Among those countries, the key partners of South Africa are the United States, the United Kingdom, Australia, the Netherlands and Canada (Table 17). Nigeria, Kenya, Uganda, Ghana and Zimbabwe are among the key African scientific collaborators.

Table 18 shows the leading organizations in South Africa publishing in the domain of Health with the total number of publications they produced in the last 10 years, as well as the percentage of their contribution in the total scientific output. University of Cape Town appears to have a significant capacity in publishing in the Health domain.

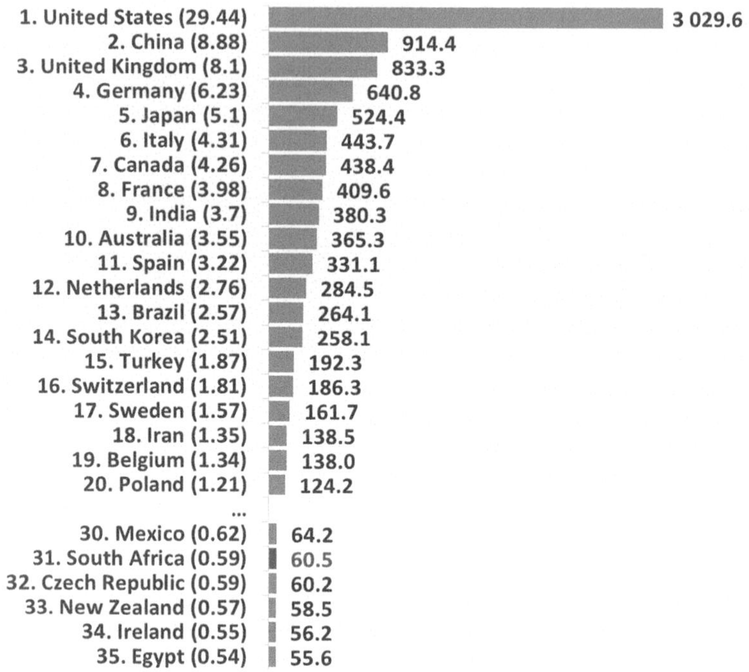

Fig. 64 Leading countries by number of publications in "Health" domain in Scopus for 2008–2018. Note: 1. All types of documents, except of technical documents, indexed in Scopus are included in the analysis. 2. Share of a country in global number of publications is shown in parenthesis

Table 17 Key partner countries of South Africa in internationally collaborated publications in "Health" domain in Scopus for 2008–2018

Country	Number of joint papers with South Africa	Share of a country in all ICPs of South Africa
United States	13,593	42.39%
United Kingdom	9168	28.59%
Australia	3895	12.15%
Netherlands	3343	10.42%
Canada	3156	9.84%
Germany	3022	9.42%
Switzerland	2750	8.58%
France	2475	7.72%
Belgium	2146	6.69%
India	1897	5.92%
Italy	1790	5.58%
Sweden	1780	5.55%
Nigeria	1681	5.24%
Spain	1617	5.04%

Table 17 (continued)

Country	Number of joint papers with South Africa	Share of a country in all ICPs of South Africa
Brazil	1436	4.48%
Kenya	1305	4.07%
China	1120	3.49%
Norway	1101	3.43%
Uganda	1041	3.25%
Denmark	1020	3.18%
Japan	879	2.74%
Ghana	846	2.64%
New Zealand	822	2.56%
Zimbabwe	777	2.42%
Thailand	766	2.39%
Cameroon	700	2.18%
Argentina	668	2.08%
Tanzania	660	2.06%
Malawi	641	2.00%
Austria	598	1.86%
Zambia	590	1.84%
Mexico	581	1.81%
Poland	577	1.80%
Ireland	571	1.78%
Finland	570	1.78%
Israel	559	1.74%
Saudi Arabia	522	1.63%
Portugal	501	1.56%

Note: 1. All types of documents except of technical documents, indexed in Scopus are included in the analysis

Table 18 Leading organizations of South Africa by number of publications in "Health" domain in Scopus for 2008–2018

Affiliation	Number of publications	Share in all South Africa publications
University of Cape Town	13,636	22.53%
University of Witwatersrand	9766	16.13%
Universiteit Stellenbosch	8345	13.79%
University of KwaZulu-Natal	7148	11.81%
Universiteit van Pretoria	6177	10.20%
South African Medical Research Council	4988	8.24%
North-West University	2952	4.88%
Groote Schuur Hospital	2427	4.01%
National Health Laboratory Services	2424	4.00%
University of the Western Cape	2081	3.44%

(continued)

Table 18 (continued)

Affiliation	Number of publications	Share in all South Africa publications
University of the Free State	1717	2.84%
University of Cape Town, Faculty of Health Sciences	1677	2.77%
The Nelson R. Mandela Medical School	1536	2.54%
University of Johannesburg	1529	2.53%
Tygerberg Hospital	1262	2.08%
National Institute for Communicable Diseases	1190	1.97%
Red Cross War Memorial Children's Hospital	1183	1.95%
University of South Africa	1153	1.90%
University of Limpopo	1019	1.68%
Human Sciences Research Council of South Africa	927	1.53%
Sefako Makgatho Health Sciences University SMU	861	1.42%
University of Fort Hare	843	1.39%
Rhodes University	826	1.36%
Baragwanath Hospital	715	1.18%
Tshwane University of Technology	644	1.06%
The Council for Scientific and Industrial Research	553	0.91%
University of the Witwatersrand, Faculty of Health Sciences, School of Pathology	551	0.91%
Nelson Mandela Metropolitan University	533	0.88%
Cape Peninsula University of Technology	523	0.86%
Centre for the AIDS Programme of Research in South Africa	521	0.86%
University of Venda for Science and Technology	464	0.77%
Walter Sisulu University	454	0.75%
Durban University of Technology	441	0.73%
National Research Foundation	424	0.70%
Africa Centre for Health and Population Studies	423	0.70%
Agricultural Research Council, Pretoria	400	0.66%
Inkosi Albert Luthuli Central Hospital	365	0.60%
Onderstepoort Veterinary Institute	279	0.46%
University of Cape Town Lung Institute	268	0.44%
University of the Witwatersrand, Faculty of Health Sciences, School of Clinical Medicine	248	0.41%
University of the Free State, School of Medicine	227	0.38%
University of Zululand	227	0.38%
University of the Witwatersrand, Faculty of Health Sciences	207	0.34%

Note: 1. All types of documents, except of technical documents, indexed in Scopus are included in the analysis. 2. Organization names are presented as they are spelled in Scopus database

2.5.4 Semantic Analysis of "Health" Domain

This section describes the "Health Technology to Prevent and Treat Ill-Health, and Advanced Well-Being, for Marginalized (HE)" domain and analysis of its thrusts based on the results produced by the iFORA system. First, the full domain is described with a domain map including all the topics covered. Figure 65 shows the

Fig. 65 "Health Technology to Prevent and Treat Ill-Health, and Advance Well-Being, for Marginalized" domain full semantic map for 2014–2018 "Health Technology to Prevent and Treat Ill-Health, and Advance Well-Being, for Marginalized" Thrusts

Table 19 "Health Technology to Prevent and Treat Ill-Health, and Advance Well-Being, for Marginalized (HE)" thrusts and topics

HE thrusts	Topics emerging under each thrust
HE1: Optimization of health systems	"Rapid diagnostic" and "Diagnostic tools" (Fig. 2.5.4)
	"Medical treatment" and "Patient treatment" (Fig. 2.5.5)
	"Healthcare system" and "Drug development" (Fig. 2.5.6)
HE2: Improving the quality of healthcare	"Birth care and child care" (Fig. 2.5.7)
	"Intensive care" (Fig. 2.5.8)
	"Healthcare organization management" (Fig. 2.5.9)
	"Healthcare staff issues" (Fig. 2.5.10)
	"Healthcare policy" (Fig. 2.5.11)
	"Healthcare economics" (Fig. 2.5.12)
	"Healthcare access" (Fig. 2.5.13)
	"Healthcare quality" and "Management of clinics" (Fig. 2.5.14)
	"Personalized medicine" and "Cancer treatment" (Fig. 2.5.15)
HE3: Digitization of health systems	"Hospital records" and "Patient records" (Fig. 2.5.16)
	"Heath information system" and "Health database" (Fig. 2.5.17)
	"Telemedicine," "Mobile medicine" and "Robotics" (Fig. 2.5.18)

Health domain with its clusters and interrelations to each other. A detailed analysis of the provides an overview of HE thrusts and a number of existing and emerging topics for consideration under each of them. These are presented in Table 19. In the following section, network maps are given to illustrate these maps and further issues emerging under each of them.

2.5.5 HE1: Optimization of Health Systems

The analysis of the "optimization of health systems" thrust revealed six topics regarding diagnostic, treatment and overall improvement of the healthcare system as well as drug development. In terms of diagnostics, rapid diagnostic systems are gaining importance. The cluster on this topic includes issues on different levels of diagnostics (clinical, medical or molecular), development of diagnostic platforms, sample collection methods and various technologies to be used for this purpose (including Bioelectronics and vapor nanobubble technologies (Fig. 66, left). In relation to that, Fig. 66 (right) presents various issues concerning diagnostics tools, and new methods and techniques for increasing accuracy.

Following the diagnosis, treatment comes as an important topic for the optimization of the health systems. The advancement of new technologies brings new methods and approaches for treatment. There is a need to improve patient examination

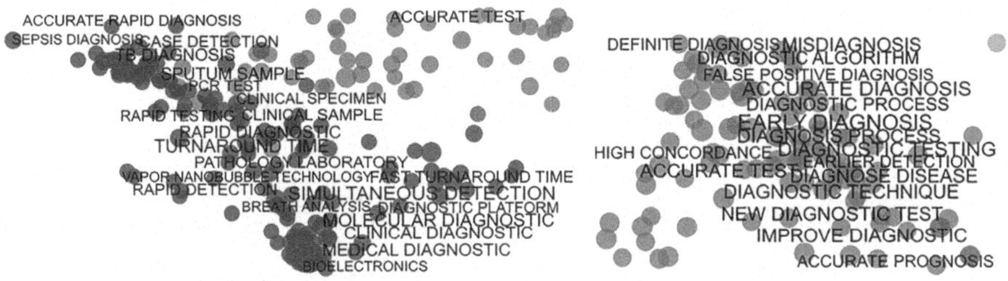

Fig. 66 "Rapid diagnostic" (left) and "diagnostic tools" (right)

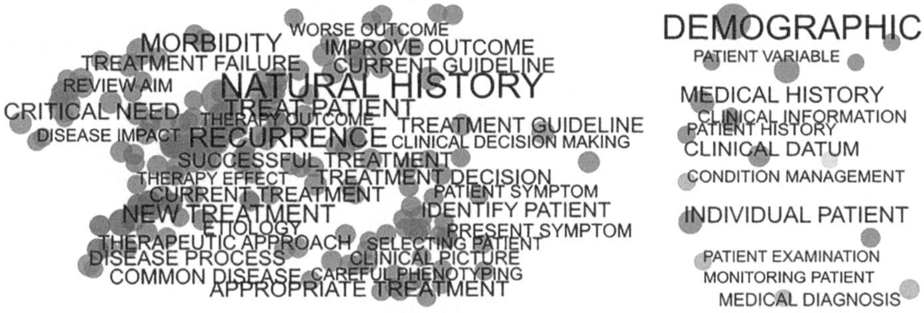

Fig. 67 "Medical treatment" (left) and "patient treatment" (right)

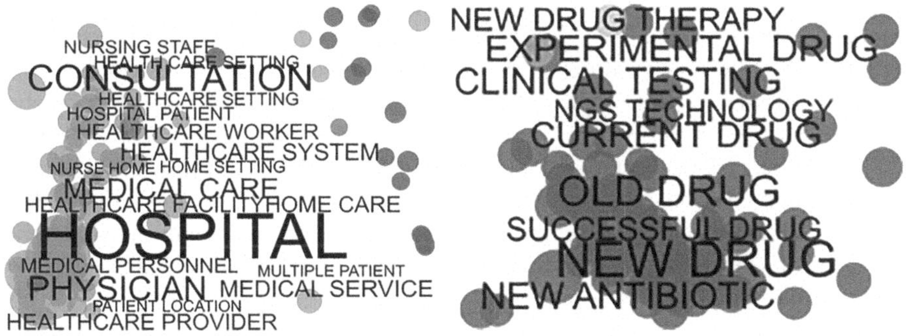

Fig. 68 "Healthcare system" (left) and "drug development" (right)

methods, keep clinical information and patient history, with the help of the ICT systems, and monitor patients' health conditions regularly (Fig. 67).

Besides improvements in diagnostics and treatment, the optimization of the health systems also requires its upgrading. These upgrades require substantial improvements in the hospitals and healthcare facilities, skills of the healthcare personnel and the quality of the medical services. Home care also appears to be an important point for the development as a part of the healthcare system (Fig. 68, left). Drugs are vital components of the healthcare. Development and use of new drugs will certainly have transformative effects on the healthcare system. The issues

Fig. 69 "Birth care and child care"

around the new drugs are populated around experiments and clinical testing of drugs and their successful use in therapy. The antibiotic use in the world is also a big debate in health circles. Development of new antibiotics also appears as a branch in new drug development activities (Fig. 68, right).

2.5.6 HE2: Improving the Quality of Healthcare

In addition to the optimization of the healthcare services, the second thrust of the HE domain is concerned with the improvement of the quality of the healthcare. A number of topics emerge regarding to this thrust ranging from the development of care from the birth to a comprehensive overhaul of the healthcare system from the way it is organized, staff employed, its economic aspects, quality, efficiency and access. These are described in the consecutive maps presented below. Children health is immensely important for a health population. Most of the diseases can be prevented already from the pregnancy stages of mothers. Reduced infant mortality is an important indicator of development. Therefore, necessary healthcare should be provided for women and parents from the pregnancy and maternity stage to early and late childhood (Fig. 69).

Another aspect of healthcare to be developed is intensive care. There are a range of issues to be considered starting from the admission to intervention, hospital stay and discharge (Fig. 70).

Besides the issues related to care, the improvement of healthcare requires transformations in the healthcare systems, which require action at multiple aspects. Beginning with the management of healthcare organizations, there are possibilities of using ICTs to implement and widespread use of electronic health records and patient data. These technologies, along with the application of personal health systems, are expected to make revolutionary changes in the way healthcare services are organized and delivered (Fig. 71).

Healthcare staff, including physicians, surgeons, cardiologists and nurses among the others, needs necessary skills and capabilities to keep up with the developments

Fig. 70 "Intensive care"

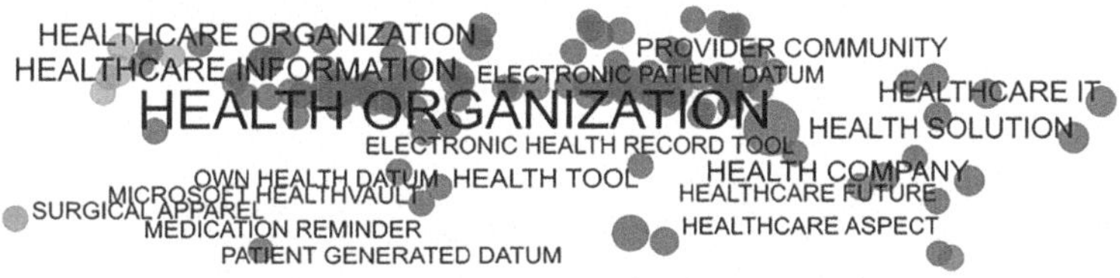

Fig. 71 "Healthcare organization management"

Fig. 72 "Healthcare staff issues"

in the healthcare system. Particularly skills related to the use of ICTs and health data as well as new technologies for diagnosis, treatment, surgery etc. seem to be important (Fig. 72).

Healthcare system also needs to be regulated with appropriate policy measures in order to be improved. Among the important aspects for regulation are access to the healthcare system and equity, management of healthcare costs, making healthcare more affordable for low-income populations, health insurance and community care among the others illustrated in Fig. 73.

As mentioned in the policy aspect, the availability and accessibility of healthcare services from the economic point of view come out as an important issue for the improvement of the healthcare systems. Reduction of the healthcare expenditures is

Fig. 73 "Healthcare policy"

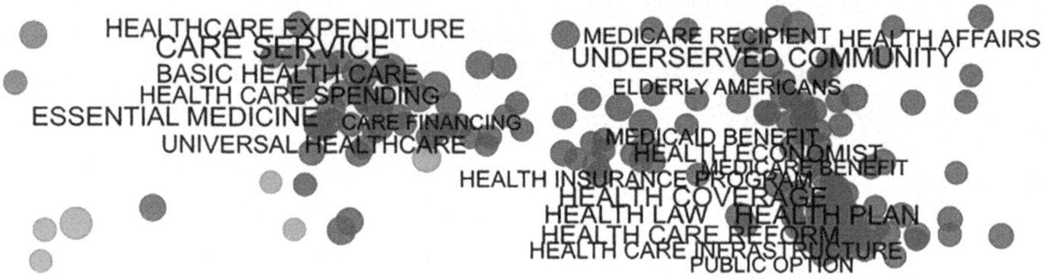

Fig. 74 "Healthcare economics"

Fig. 75 "Healthcare access"

a first step for accessing to basic healthcare services for the marginalized parts of the society. Meanwhile, it is important to reduce the burden of the healthcare spending on the economic system. This requires a more systemic approach, which relates to other STI domains, such as nutrition. With better nutrition of the population, the society will be healthier with less health problems. This is also related to education systems, economic systems and access to information and technology systems. Therefore, there is a need to have a systemic perspective when economic and policy aspects of healthcare are dealt with (Fig. 74).

Improved healthcare systems also mean accessible healthcare services for all layers of the society. There are a number of issues to be considered in terms of access, including income levels, medical costs, community care, disparity in healthcare and equity. Necessary insurance systems should be in place also for the marginalized to access healthcare services (Fig. 75).

While delivering healthcare services, besides their economy, the quality aspect should also be considered. Availability of medical personnel, improved infrastructure and services, patient experience are among the important determinants of the quality in healthcare. There is also a need for the efficient management of the clinics in order to maintain the quality of services (Fig. 76).

Improving healthcare quality is also directly related to the development of personalized medicine, where medical services, practices, interventions and products

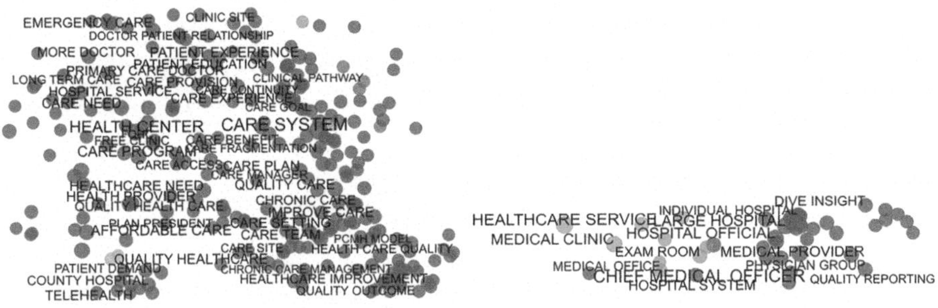

Fig. 76 "Healthcare quality" (left) and "management of clinics" (right)

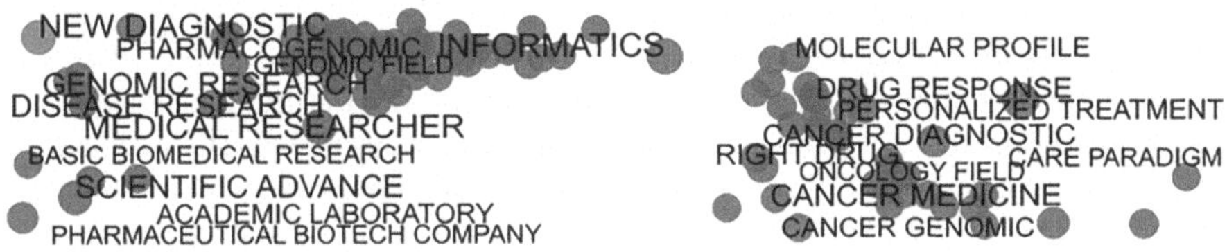

Fig. 77 "Personalized medicine" (left) and "cancer treatment" (right)

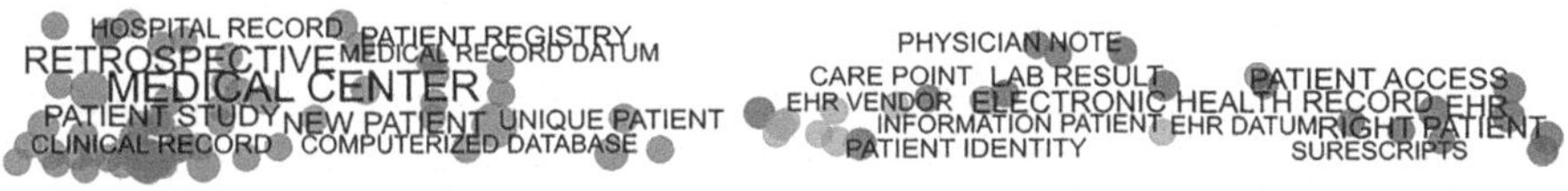

Fig. 78 "Hospital records" (left) and "patient records" (right)

are tailored to the individual patients based on their characteristics. Research in areas like genomics and pharmagenomics plays important role for personalized medicine. One of the areas where personalized medicine can play a significant role is cancer research, where personalized treatment is needed from the diagnostic stages to drug response (Fig. 77).

2.5.7 HE3: Digitization of Health Systems

The digitalization of health systems provides enormous opportunities for the optimization of healthcare services and increasing their quality. Electronic health records are one of the major application areas for the digitalization of health systems. Clinical records should be collected in a database in safe and secure ways by considering the privacy and ethical aspects. Necessary computer infrastructures need to be established with ICT skills in place (Figs. 78 and 79).

Recent years have also seen increasing possibilities for telemedicine, which allows caring for patients remotely when the healthcare provider and patient are not physically present with each other. Video consultations and virtual visits are

Fig. 79 "Heath information system" (left) and "health database" (right)

Fig. 80 "Telemedicine" (left), "mobile medicine" (center) and "robotics" (right)

technological alternatives as they enable remote care. Associated to the telemedicine concept is mobile medicine, which also serves for the remote delivery of healthcare services. The use of robotics in recent years moved telemedicine to a new dimension enlarging the services delivered from mere consultations to remote surgeries (Fig. 80).

2.5.8 Dynamics of "Health" Domain Thrusts

In this section, the dynamics of the "Health" domain thrusts are shown based on their frequency of occurrence in the documents indexed in the iFORA database. Among the thrusts HE1 (Optimization of health systems) and HE2 (Improving the quality of healthcare) appear to the highest on the agenda (Fig. 81). Digitization of health systems (HE3) is still not given sufficient attention in the Health domain.

Furthermore, Table 20 illustrates to what extent the "Health" domain thrusts address the challenges. In parallel to the other STI domains, the thrusts of the Health domain are strongly linked to rural community, job creation and population growth.

2.6 Sustainable Energy Technologies for the Marginalized (Energy) Domain

2.6.1 Overall Description of the Domain

With the advancement of societies, energy has become one of the basic human needs for a sustained and uninterrupted life. Although large cities in South Africa have a relatively stable energy supply, both rural and peri-urban marginalized communities suffer from accessing energy. Clean and affordable energy sources such as the harvesting of solar, wind and bio energy provide opportunities for marginalized communities to adopt these new technologies, as there is no overhead due to the

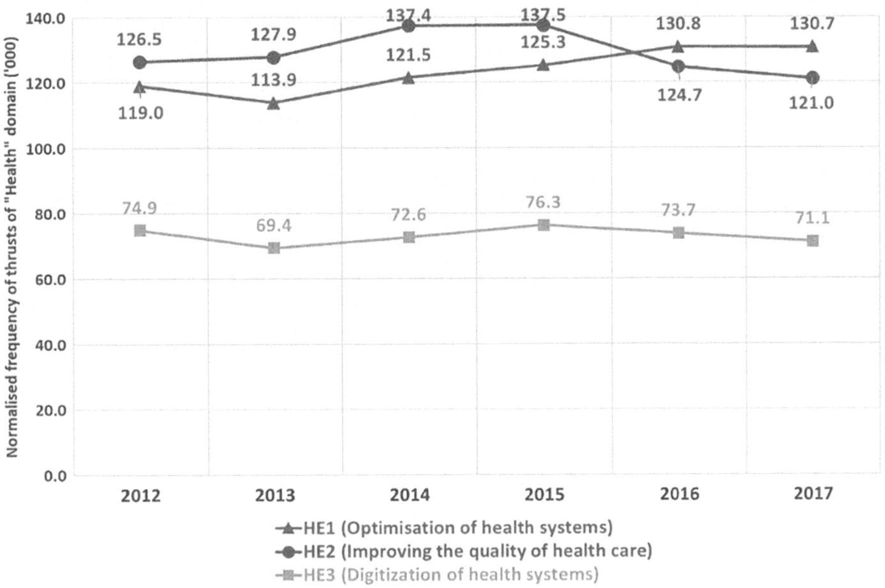

Fig. 81 Normalized frequency of all terms related with thrusts of "Health" domain ('000) in 2012–2017. Note: "Normalized frequency …" for a specific thrust is a total normalized number of mentions (in thousands) of all terms (that are shown on a semantic map) related with this specific thrust in documents indexed in IFORA for a given year

replacement of old technologies. Sustainable energy technologies can leapfrog old technologies and their many limitations. Possibilities for local sustainable energy production will give communities independence, self-sufficiency and cooperation toward a common good. Decentralized local energy sources will reduce dependence on the national grid and create economic opportunities for the marginalized across South Africa. Thus, the energy domain focuses on clean, affordable and renewable energy solutions, energy efficiency and distributed generation.

2.6.2 Overview of Research Potential of South Africa in "Energy" Domain

Figure 82 shows that there is a growing scientific emphasis on the Energy domain in South Africa. The number of publications has been growing rather faster compared to other STI domains. However, the share of South Africa's contribution in the global scientific output is still relatively low (under 1%).

Fig. 83 shows the leading countries in the Energy domain.

Among those countries, the key partners of South Africa are the United States, the United Kingdom, Germany, India and France. Collaboration with Nigeria is relatively high with the seventh position among the top collaborators (Table 21).

Table 20 Impacts of all thrusts of "Health" domain on challenges

	Health		
Challenges	HE1	HE2	HE3
Affordable Education	1.0	0.0	0.0
Affordable Energy	0.0	1.5	0.0
Affordable Food	1.0	4.5	0.0
Carbon Emission Reduction	0.0	0.0	0.0
Clean Water Access	1.0	1.0	0.0
Crime Prevention	1.0	2.0	1.0
Crime Reduction	2.5	1.0	0.0
Economy Growth	1.0	0.0	0.0
Employment Growth	4.5	1.0	1.0
Export Growth	0.0	1.0	1.0
High Living Standard	4.0	2.0	1.5
High Quality Health Care Service	3.5	2.0	1.0
Income Growth	0.0	2.0	0.0
Job Creation	14.2	15.2	26.0
Low Greenhouse Gas Emission	0.0	1.0	0.0
Low Mortality	3.3	9.8	6.0
Population Growth	17.8	28.7	1.3
Poverty Alleviation	2.0	1.0	3.0
Reduce Water Demand	0.0	0.0	0.0
Renewable Energy Growth	0.0	0.0	0.0
Rural Community	34.3	13.0	73.6
Secure Water Supply	0.0	0.0	0.0
Skill Improvement	0.0	0.0	0.0
Universal Broadband Access	0.0	0.0	1.0

Notes: 1. Value in each specific cell of this table is a normalized number of co-occurrences of all terms (that are shown on a semantic map) related with a specific thrust (e.g., Thrust HE1) and a specific challenge (e.g., "Low Mortality") in documents indexed in iFORA database for 2012–2017. This value is treated as an impact of a specific thrust onto a specific challenge. 2. Cells are colored by bars as follows: within a column of a specific thrust the biggest green bar means the challenge for which this specific thrust has the highest impact

Table 22 shows the leading organizations in South Africa publishing in the domain of Energy with the total number of publications they produced in the last 10 years, as well as the percentage of their contribution in the total scientific output.

2.6.3 Semantic Analysis of "Energy" Domain

This chapter describes the "Sustainable Energy Technologies for the Marginalized" domain and analysis of its thrusts based on the results produced by the iFORA system. First, the full domain is described with a domain map including all the topics covered. Then each of the thrusts is described by the emerging topics identified by

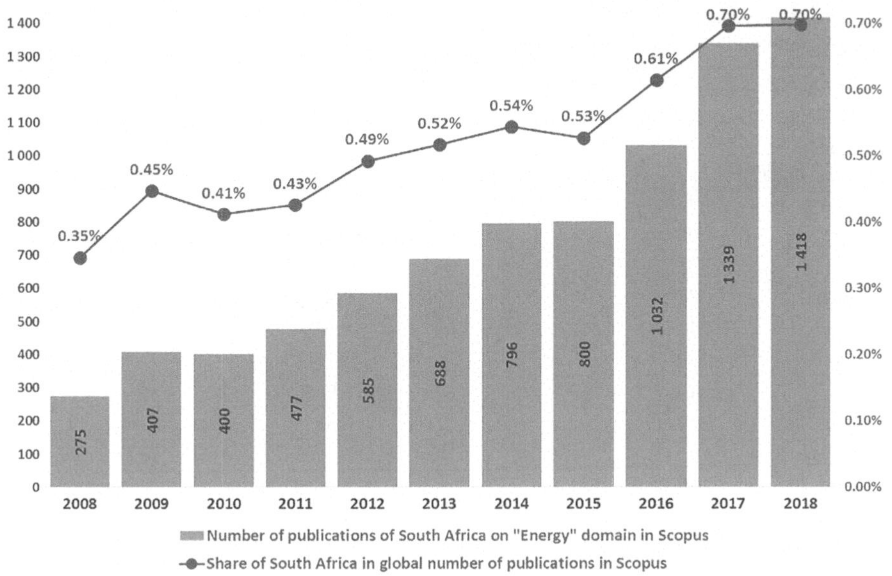

Fig. 82 Dynamics of publications of South Africa in "Energy" domain in Scopus in 2008–2018. Note: All types of documents, except of technical documents, indexed in Scopus are included in the analysis

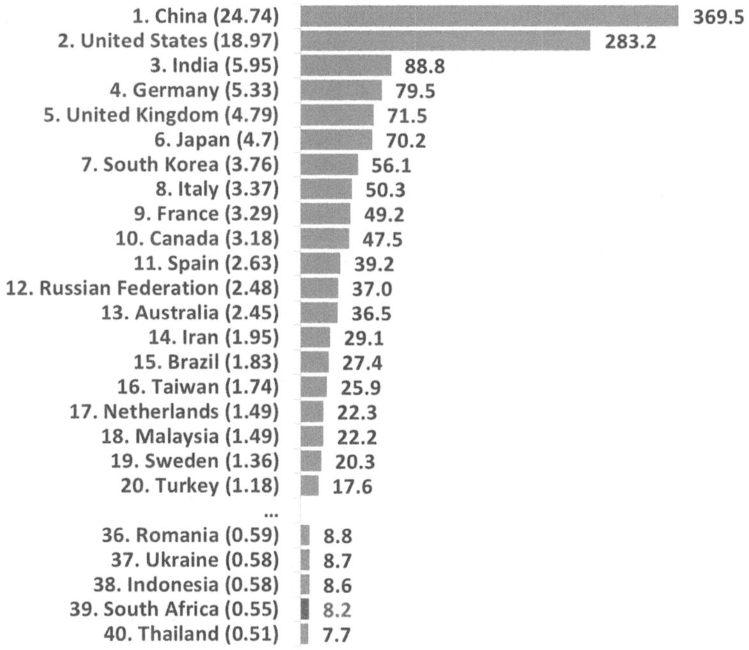

Fig. 83 Leading countries by number of publications ('000) in "Energy" domain in Scopus for 2008–2018. Note: 1. All types of documents, except of technical documents, indexed in Scopus are included in the analysis. 2. Share of a country in global number of publications is shown in parenthesis

Table 21 Key partner countries of South Africa in internationally collaborated publications in "Energy" domain in Scopus for 2008–2018

Country	Number of joint papers with South Africa	Share of a country in all ICPs of South Africa
United States	756	21.97%
United Kingdom	519	15.08%
Germany	397	11.54%
India	396	11.51%
France	328	9.53%
Australia	287	8.34%
Nigeria	277	8.05%
China	259	7.53%
Italy	240	6.97%
Canada	224	6.51%
Netherlands	195	5.67%
Iran	129	3.75%
Sweden	129	3.75%
Norway	107	3.11%
Spain	105	3.05%
Denmark	103	2.99%
Brazil	98	2.85%
Japan	97	2.82%
Poland	95	2.76%
Switzerland	85	2.47%
Austria	82	2.38%
Belgium	80	2.32%
Russian Federation	78	2.27%
Chile	75	2.18%
Finland	73	2.12%
New Zealand	68	1.98%
Saudi Arabia	63	1.83%
Malaysia	59	1.71%
Ireland	58	1.69%
Kenya	52	1.51%
Turkey	52	1.51%
Namibia	50	1.45%
Zimbabwe	50	1.45%

Note: 1. All types of documents except of technical documents, indexed in Scopus are included in the analysis

iFORA. The second section presents topic clusters under each thrust with more detailed network maps with descriptions of emerging issues under them. Figure 84 shows the entire domain of "Sustainable Energy Technologies for the Marginalized (EN)" with its clusters and interrelations to each other. A detailed analysis of Fig. 84

Table 22 Leading organizations of South Africa by number of publications in "Energy" domain in Scopus for 2008–2018

Affiliation	Number of publications	Share in all South Africa publications
University of Cape Town	954	11.61%
Universiteit van Pretoria	916	11.15%
University of KwaZulu-Natal	816	9.93%
University of Johannesburg	778	9.47%
North-West University	758	9.22%
Universiteit Stellenbosch	733	8.92%
University of Witwatersrand	553	6.73%
Tshwane University of Technology	463	5.63%
The Council for Scientific and Industrial Research	426	5.18%
University of the Western Cape	289	3.52%
University of South Africa	271	3.30%
Cape Peninsula University of Technology	251	3.05%
Durban University of Technology	161	1.96%
University of the Free State	161	1.96%
Eskom	149	1.81%
Nelson Mandela Metropolitan University	120	1.46%
Central University of Technology, Free State	108	1.31%
University of Fort Hare	107	1.30%
Vaal University of Technology	107	1.30%
Rhodes University	103	1.25%
University of Witwatersrand, School of Chemical and Metallurgical Engineering	98	1.19%
Sasol Technology Pty Ltd	87	1.06%
Pebble Bed Modular Reactor Pty Limited	74	0.90%
NECSA	63	0.77%
University of Venda for Science and Technology	58	0.71%
University of Limpopo	55	0.67%
Meraka Institute	54	0.66%
National Research Foundation	50	0.61%
Ithemba Laboratory for Accelerator-Based Sciences	47	0.57%
Mangosuthu University of Technology	41	0.50%
University of Zululand	38	0.46%
University of KwaZulu-Natal School of Chemical Engineering	36	0.44%
South African Astronomical Observatory	34	0.41%
Mintek	27	0.33%
Agricultural Research Council, Pretoria	26	0.32%

Note: 1. All types of documents, except of technical documents, indexed in Scopus are included in the analysis. 2. Organization names are presented as they are spelled in Scopus database

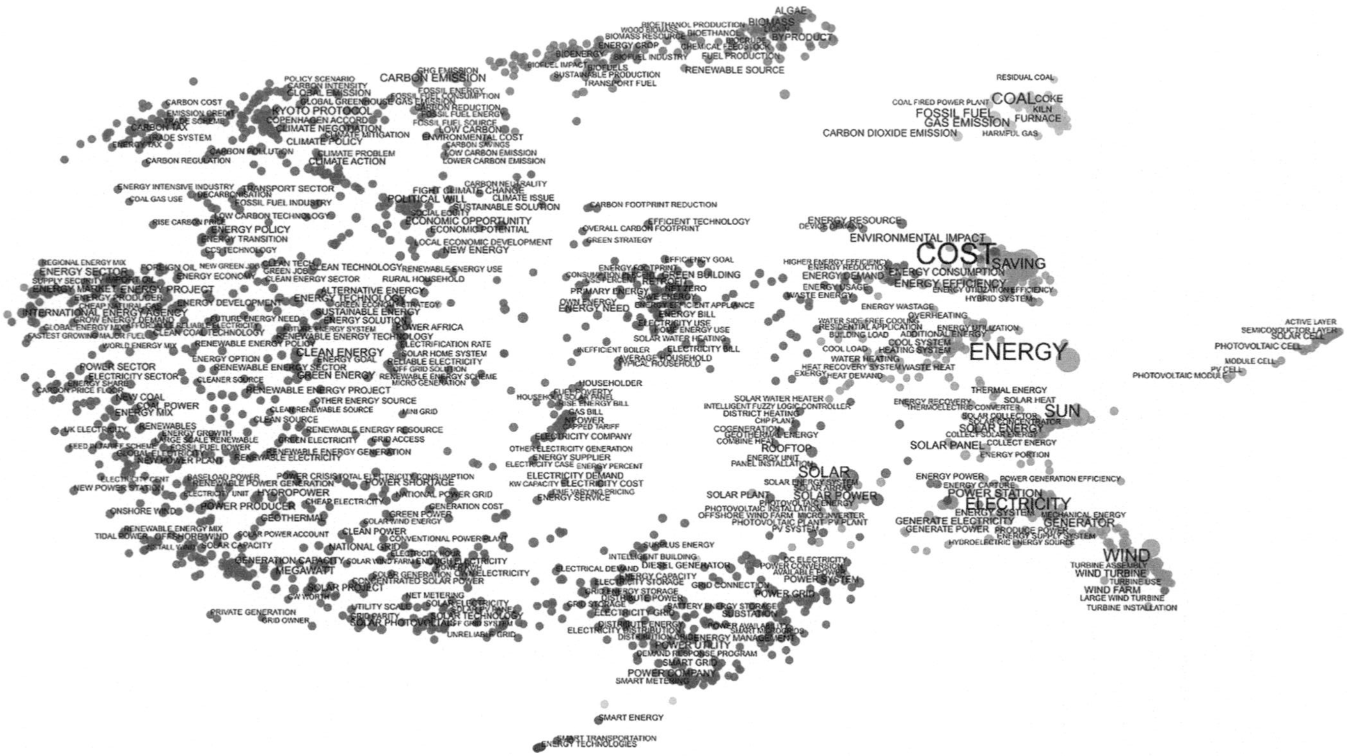

Fig. 84 "Sustainable Energy Technologies for the Marginalized" domain full semantic map for 2014–2018 "Sustainable Energy Technologies for the Marginalized" Thrusts

Table 23 "Sustainable Energy Technologies for the Marginalized" thrusts and topics

EN thrusts	Topics emerging under each thrust
EN1: Clean, affordable and sustainable energy for all	"Conventional energy sources" (Fig. 2.6.4)
	"Clean and renewable energy" (Fig. 2.6.5)
	"Bioenergy" (Fig. 2.6.6)
	"Energy policy" (Fig. 2.6.7)
	"Carbon emissions policy" (Fig. 2.6.8)
	"Energy costs and tariffs" (Fig. 2.6.9)
EN2: Renewable Energy sources and technologies	"Solar energy" (Fig. 2.6.10)
	"Wind energy" (Fig. 2.6.11)
EN3: Energy efficiency solutions for industry plus household use	"Energy efficiency" (Fig. 2.6.12)
	"Energy system" (Fig. 2.6.13)
EN4: Distributed energy generation and storage	"Distributed energy generation" (Fig. 2.6.14)
	"Grid energy generation" (Fig. 2.6.15)

Fig. 85 "Conventional energy sources"

provides an overview of EN thrusts and a number of existing and emerging topics for consideration under each of them. These are presented in Table 23. In the following section, network maps are given to illustrate these maps and further issues emerging under each of them.

2.6.4 EN1: Clean, Affordable and Sustainable Energy for All

Although renewable energy sources are used increasingly, it is expected that the conventional sources of energy will still have a considerable proportion in the energy mix in the next few decades to come. However, scientific and technological innovations should be introduced for the sustainable use of the conventional energy sources. South Africa is rich in coal reserves and the dependency of the country on fossil fuels is still high. In order to use these conventional sources in a more sustainable way, carbon dioxide emissions generated by them should be reduced, especially generated by the coal-fired power plants (Fig. 85).

Meanwhile, rich potentials of South Africa on clean and renewable energy sources also need to be exploited. Particularly, the potentials for solar energy are high in the country. There is a need to develop a more holistic and integrated approach for renewables in South Africa, including, the technologies for providing

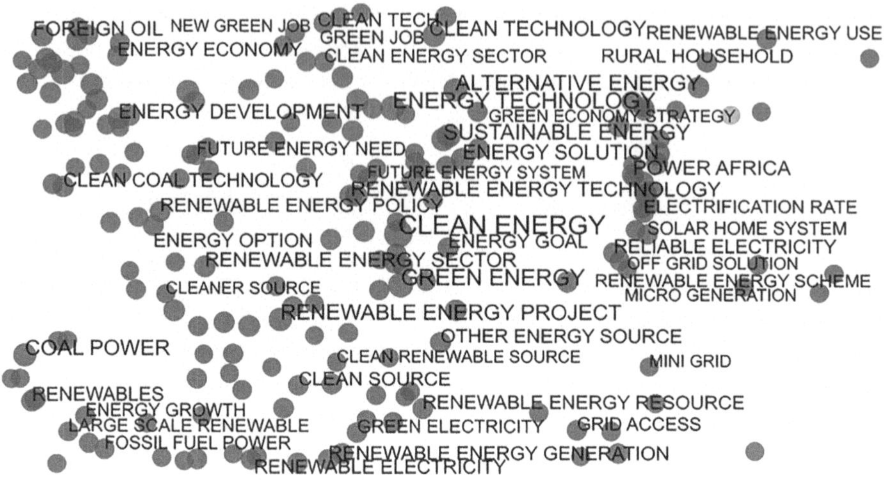

Fig. 86 "Clean and renewable energy"

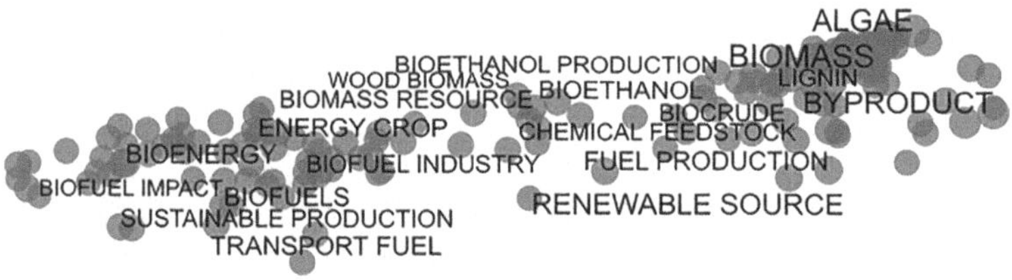

Fig. 87 "Bioenergy"

clean and renewable energy sources; solutions for urban and rural areas, such as smart grids in cities and micro energy generation in rural and remote areas; setting up energy storage systems; as well as building necessary technological and skills capacity in the country along with green jobs (Fig. 86).

Besides solar energy, South Africa also has great potentials on Bioenergy. Energy and fuel production from biomass, bioethanol and algae appear to be promising areas for development. A modern biofuel industry with sustainable production principles can be among the main suppliers of the transport fuel in the country (Fig. 87).

Providing clean, affordable and sustainable energy for all requires efficient energy policies. Green principles should be at the backbone for the energy policy with an overall aim of providing low-cost and low-carbon footprint energy for all in the county. Necessary political will should be in place in order to make sure that long term commitment for climate change mitigation, carbon neutrality and social equity targets can be achieved (Fig. 88).

Within energy policy, carbon emissions policy appears to have a special emphasis. Reduction of emissions from fossil fuels is the main global, regional and national priority. There are climate targets and commitments agreed in Kyoto protocol and in

Fig. 88 "Energy policy"

Fig. 89 "Carbon emissions policy"

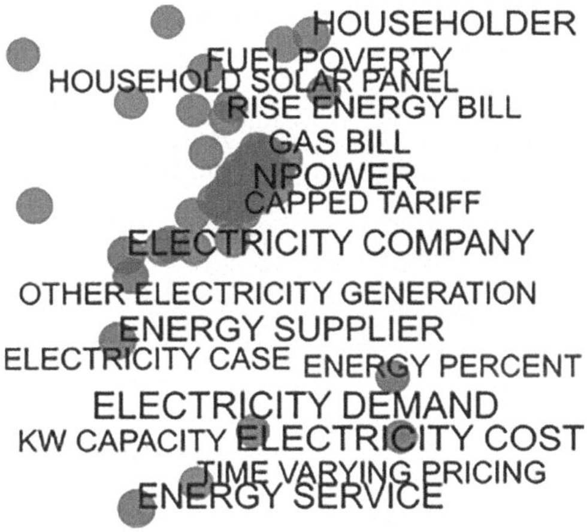

Fig. 90 "Energy costs and tariffs"

other international agreements. There is always a question on how to fund carbon reduction efforts. Carbon regulations and taxation schemes have been implemented to reduce emissions and generate income for scientific and technological innovations to meet emission targets. There is a need for widespread implementation of policy measures at all levels of governance from macro to micro levels—e.g., individual carbon credit/taxation systems (Fig. 89).

While innovating for new and renewable energy sources, and reducing carbon emission levels, there are expectations by the society for affordable energy. Costs and tariffs need to be balanced to make sure that society, particularly the marginalized, has access to sustainable energy supply. Among alternative ways of energy generation, household solar panel appears to be a technological solution. However, at present, this has a limited use. Necessary solutions should be developed for balancing energy demand and supply, tariffs and services (Fig. 90).

2.6.5 EN2: Renewable Energy Sources and Technologies

Regarding renewable energy sources and technologies, the analysis revealed that solar and wind energy should be high on the agenda in South Africa. As mentioned earlier, the country has considerable potentials for solar energy. This solution is particularly useful for the marginalized to meet their energy demand at an affordable price. Similarly, large-scale solar farms have been implemented in a number of countries for generating large amounts of energy. Widespread implementation of solar energy solutions requires overcoming several technological and technical challenges such as solar energy collection, conversion and combination (Fig. 91).

Besides solar energy, South Africa has considerable potentials for wind energy. Issues around the wind energy include the installation of wind turbines and setting up wind farms both on the land and offshore (Fig. 92).

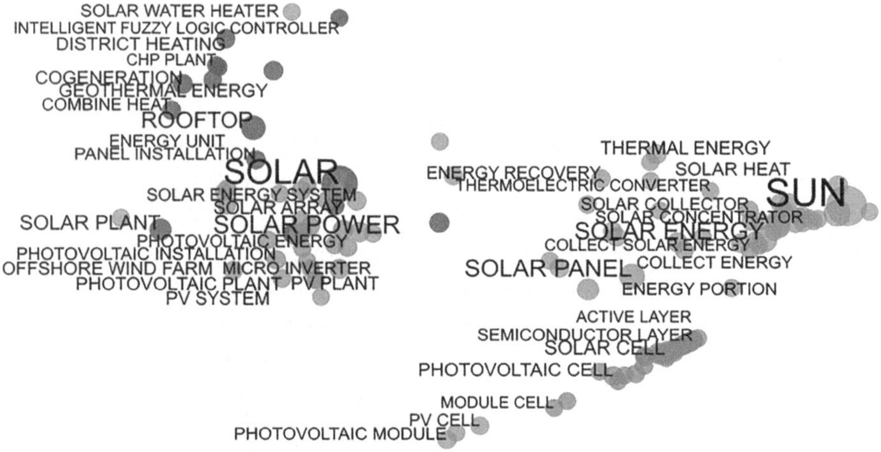

Fig. 91 "Solar energy"

Fig. 92 "Wind energy"

2.6.6 EN3: Energy Efficiency Solutions for Industry Plus Household Use

In addition to energy generation concerns, the consumption aspect should also be at the forefront. Energy costs and expenses can be reduced through the energy efficiency solutions both at the industrial and household levels. Heating and cooling systems should be efficient with lesser demand. Besides devices, building design and technologies should be developed for reduced energy consumption and energy waste (Fig. 93).

Recent years have seen advancements in developing smart energy systems with greater power generation efficiency, better energy capture and lesser grid loss. Besides smart energy, smart transport systems should have a higher priority on the South African development agenda (Fig. 94).

2.6.7 EN4: Distributed Energy Generation and Storage

Generating energy near the points of consumption is crucial for achieving greater efficiencies, lesser loses and reduced costs, particularly for the marginalized, who live in far and remote areas of the country. Energy generation, transfer and storage are among the areas for innovation in distributed energy. Furthermore, any use of

Fig. 93 "Energy efficiency"

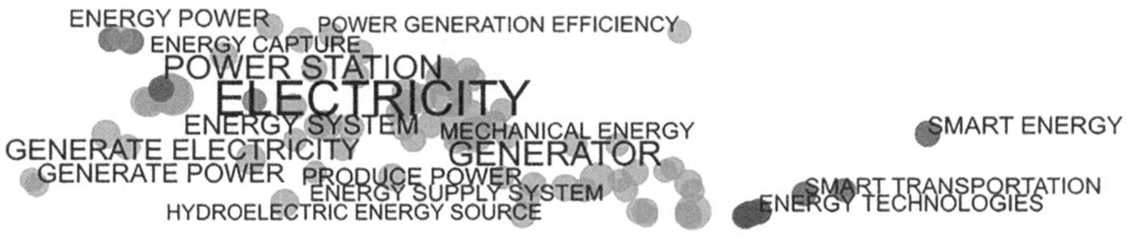

Fig. 94 "Energy system"

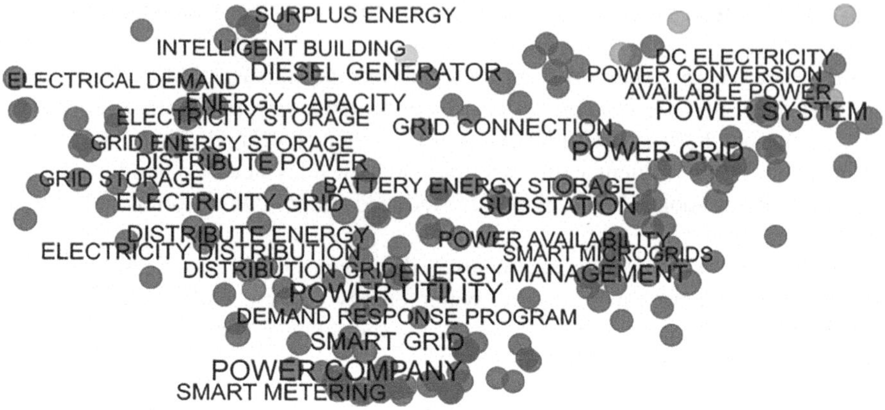

Fig. 95 "Distributed energy generation"

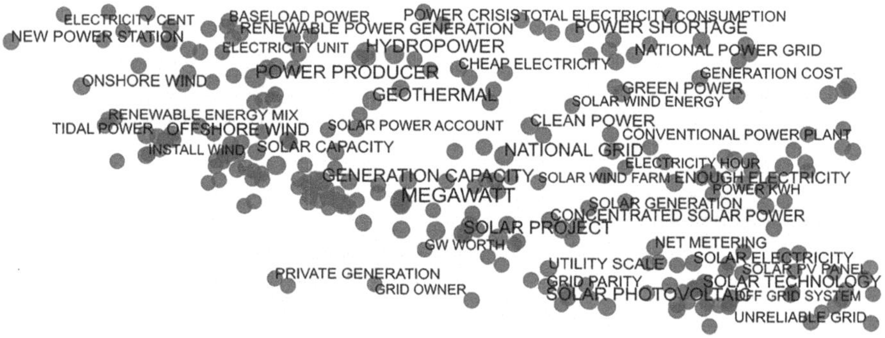

Fig. 96 Grid energy generation dynamics of "Energy" Domain Thrusts

advanced energy technologies will require setting up smart technologies like smart grids and metering systems (Fig. 95).

Figure 96 elaborates the issues around grid energy generation for aforementioned high-technology and renewable energy systems. Transition into new energy systems may require an initial investment; however, in the medium to long term, they will pay off with cheaper, cleaner and more sustainable solutions.

In this section, the dynamics of the "Energy" domain thrusts are shown based on their frequency of occurrence in the documents indexed in the iFORA database. Among the thrusts EN3 (Energy efficiency solutions for industry plus household use) appears to be the highest on the agenda (Fig. 97).

Table 24 illustrates to what extent the "Energy" domain thrusts address the challenges. Naturally, most of the thrusts in this domain are strongly linked to affordable energy (EN1, EN2 and EN3). EN4 appears to be highly linked with job creation. EN1 is also addressing the carbon emission reduction challenge.

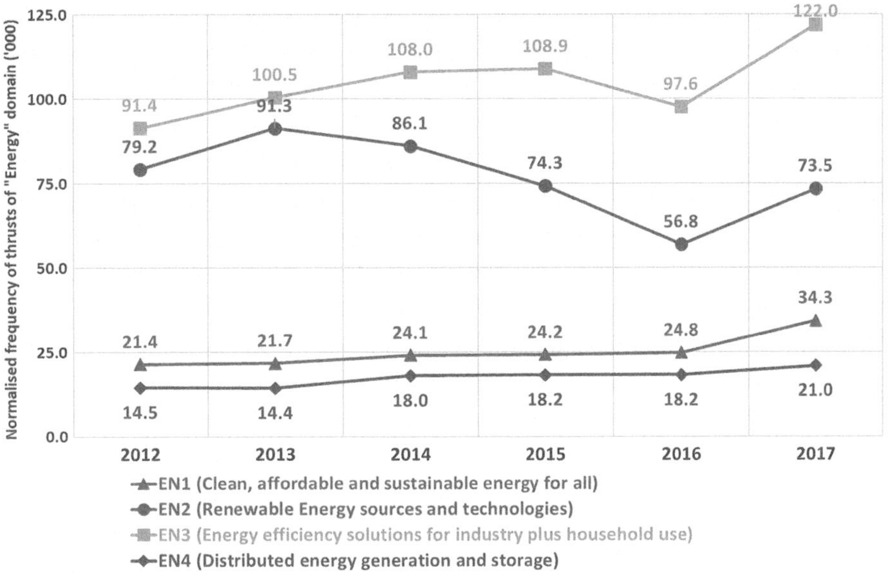

Fig. 97 Normalized frequency of all terms related with thrusts of "Sustainable Energy Technologies for the Marginalized" (Energy) domain ('000) in 2012–2017. Note: "Normalized frequency" for a specific thrust is a total normalized number of mentions (in thousands) of all terms (that are shown on a semantic map) related with this specific thrust in documents indexed in IFORA for a given year

2.7 *Education for the Future (Education) Domain*

2.7.1 Overall Description of the Domain

Education is a foundation of a fair society and successful economy. Toward the 2030s, South Africa wishes to provide all its citizens with quality education as a human right. Currently the public education system suffers from the problems of effectiveness of mathematics and science. Citizens of the country, particularly in rural areas, have expectations for accessing quality learning opportunities. Technology provides an ever-growing range of opportunities to provide this access. New and alternative learning technologies will not only give people basic education, but will also equip them with necessary skills for the future, while reducing divides in the society. Besides implementing new technologies, curriculum should also be developed in line to enable people to be more creative with skills for idea generation and problem solving.

Table 24 Impacts of all thrusts of "Energy" domain on challenges

Challenges	Energy			
	EN1	EN2	EN3	EN4
Affordable Education	0.0	0.0	0.0	0.0
Affordable Energy	553.7	103.8	117.0	3.0
Affordable Food	4.0	2.0	0.0	1.0
Carbon Emission Reduction	94.0	17.0	3.0	2.0
Clean Water Access	1.0	0.0	0.0	0.0
Crime Prevention	1.0	4.5	11.0	0.0
Crime Reduction	0.0	0.0	0.0	1.0
Economy Growth	5.0	1.0	0.0	1.0
Employment Growth	25.0	3.8	2.0	1.0
Export Growth	4.0	3.5	0.0	1.0
High Living Standard	1.0	2.0	0.0	3.0
High Quality Health Care Service	0.0	0.0	0.0	0.0
Income Growth	6.0	16.3	19.0	1.0
Job Creation	49.0	36.0	5.0	50.5
Low Greenhouse Gas Emission	12.3	35.0	64.0	1.0
Low Mortality	0.0	1.0	0.0	0.0
Population Growth	129.0	22.4	1.0	4.7
Poverty Alleviation	9.0	5.5	0.0	4.0
Reduce Water Demand	1.0	3.0	4.0	0.0
Renewable Energy Growth	44.0	7.3	0.0	0.0
Rural Community	10.5	22.3	2.0	26.0
Secure Water Supply	1.0	1.0	4.0	0.0
Skill Improvement	0.0	0.0	0.0	0.0
Universal Broadband Access	0.0	0.0	0.0	0.0

Notes: 1. Value in each specific cell of this table is a normalized number of co-occurrences of all terms (that are shown on a semantic map) related with a specific thrust (e.g., Thrust EN1) and a specific challenge (e.g., "Low Mortality") in documents indexed in iFORA database for 2012–2017. This value is treated as an impact of a specific thrust onto a specific challenge. 2. Cells are colored by bars as follows: within a column of a specific thrust the biggest green bar means the challenge for which this specific thrust has the highest impact

2.7.2 Overview of Research Potential of South Africa in "Education" Domain

Figure 98 shows that there is a growing scientific emphasis on the "Education" domain in South Africa. Although the number of publications is relatively low compared to other STI domains, some growth has been observed, particularly in year 2014. In recent years, share of South Africa's contribution to total scientific pool has accessed 1.22% with the high of 1.44% in 2014.

Figure 99 shows the leading countries in the Education domain. Compared to other domains South Africa is positioned high in this domain with the 16th place.

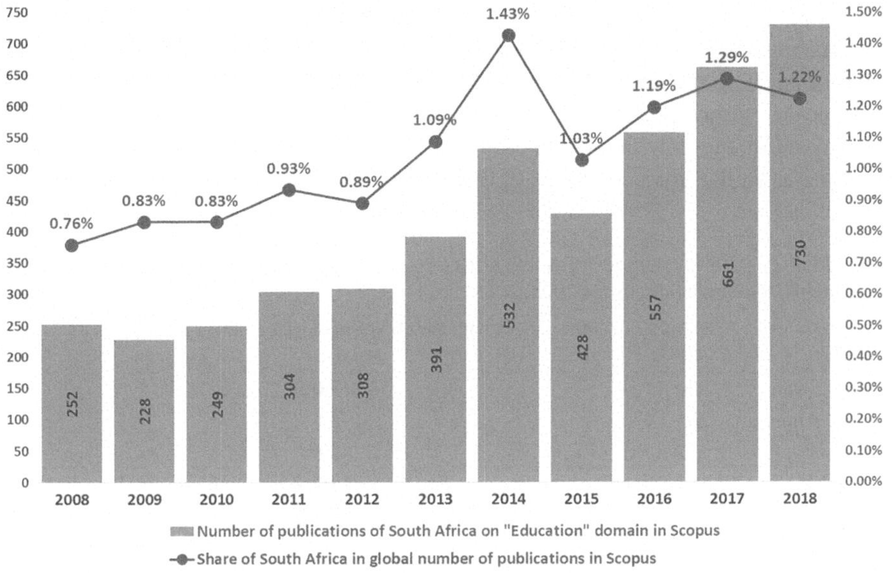

Fig. 98 Dynamics of publications of South Africa in "Education" domain in Scopus in 2008–2018. Note: All types of documents, except of technical documents, indexed in Scopus are included in the analysis

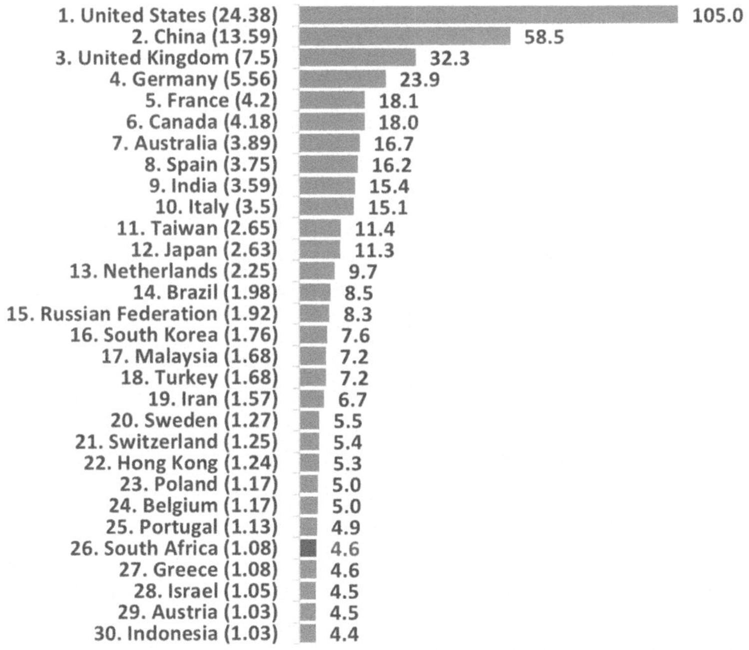

Fig. 99 Leading countries by number of publications ('000) on "Education" domain in Scopus for 2008–2018. Note: 1. All types of documents, except of technical documents, indexed in Scopus are included in the analysis. 2. Share of a country in global number of publications is shown in parenthesis

Among the key partner countries of South Africa are the United States, the United Kingdom, Germany, Australia and Canada. Nigeria comes as the first African collaborator with the ninth place among the top collaborators (Table 25).

Table 26 shows the leading organizations in South Africa publishing in the domain of Education with the total number of publications they produced in the last 10 years, as well as the percentage of their contribution in the total scientific output.

Table 25 Key partner countries of South Africa in internationally collaborated publications in "Education" domain in Scopus for 2008–2018

Country	Number of joint papers with South Africa	Share of a country in all ICPs of South Africa
United States	300	21.60%
United Kingdom	265	19.08%
Germany	122	8.78%
Australia	111	7.99%
Canada	108	7.78%
Netherlands	84	6.05%
China	81	5.83%
Sweden	63	4.54%
Nigeria	56	4.03%
India	55	3.96%
France	53	3.82%
Belgium	51	3.67%
Zimbabwe	48	3.46%
Italy	40	2.88%
Spain	38	2.74%
Switzerland	38	2.74%
Greece	34	2.45%
Russian Federation	32	2.30%
South Korea	31	2.23%
Finland	30	2.16%
Norway	30	2.16%
Pakistan	29	2.09%
Ghana	24	1.73%
Austria	23	1.66%
Botswana	22	1.58%
Kenya	22	1.58%
Denmark	21	1.51%
Turkey	21	1.51%
Brazil	20	1.44%
Czech Republic	20	1.44%

Note: 1. All types of documents, except of technical documents, indexed in Scopus are included in the analysis

Table 26 Leading organizations of South Africa by number of publications in "Education" domain in Scopus for 2008–2018

Affiliation	Number of publications	Share in all South Africa publications
University of South Africa	536	11.55%
Universiteit van Pretoria	518	11.16%
University of Witwatersrand	463	9.98%
University of Cape Town	455	9.81%
University of KwaZulu-Natal	418	9.01%
University of Johannesburg	387	8.34%
North-West University	358	7.72%
Universiteit Stellenbosch	327	7.05%
University of the Free State	195	4.20%
University of the Western Cape	170	3.66%
Tshwane University of Technology	139	3.00%
Cape Peninsula University of Technology	125	2.69%
Rhodes University	117	2.52%
The Council for Scientific and Industrial Research	89	1.92%
Nelson Mandela Metropolitan University	83	1.79%
South African Medical Research Council	65	1.40%
Durban University of Technology	64	1.38%
Central University of Technology, Free State	64	1.38%
University of Fort Hare	56	1.21%
University of Limpopo	55	1.19%
Human Sciences Research Council of South Africa	54	1.16%
Meraka Institute	47	1.01%
Vaal University of Technology	39	0.84%
Walter Sisulu University	38	0.82%
University of Zululand	37	0.80%
University of Venda for Science and Technology	37	0.80%
Mangosuthu University of Technology	15	0.32%
University of Cape Town, Faculty of Health Sciences	11	0.24%
The Nelson R. Mandela Medical School	10	0.22%

Note: 1. All types of documents, except of technical documents, indexed in Scopus are included in the analysis. 2. Organization names are presented as they are spelled in Scopus database

2.7.3 Semantic Analysis of "Education" Domain

This section describes the "Education for the Future" domain and analysis of its thrusts based on the results produced by the iFORA system. First, the full domain is described with a domain map including all of the topics covered. Then each of the thrusts is described by the emerging topics identified by iFORA. The next section presents topic clusters under each thrust with more detailed network maps with descriptions of emerging issues under them. Figure 100 shows the entire domain of "Education for the Future (ED)" with its clusters and interrelations to each other. A detailed analysis of Fig. 100 provides an overview of ED thrusts and a number of existing emerging topics for consideration under each of them. These are presented in Table 27. In the following section, network maps are given to illustrate these maps and further issues emerging under each of them.

2.7.4 ED1: Skills for the 4th Industrial Revolution

Innovations and transformations always require new skills and capabilities. The fourth Industrial Revolution is not an exception. Two of the main drivers of the fourth Industrial Revolution are technology development and digitalization. There are a number of technologies, which are likely to bring disruptions. Among them are robotics, Artificial Intelligence and new methods for production. The digitalization trend itself transforms economy, energy, health, communication and all other industries. In the age of digitalization, it is important for society to be a part of the process. At the very basic level, most of the students, if not all, need to have Internet access with higher broadband adoption rates. Failing in providing Internet access to the society and marginalized may increase the digital divide in society (Fig. 101).

New technological developments and new jobs to be created will require development of new skills and capabilities. These include innovation and entrepreneurship skills. Particularly innovative start-ups are likely to make social, economic and technological contributions, while generating highly qualified and skilled employment. In order to start such initiatives, it is important to create a culture of entrepreneurial thinking, skills and programs for supporting young entrepreneurs. Furthermore, new industrial revolution will also require skills like creativity, problem solving and idea generation, which are the keys for innovation. Design thinking, teamwork and critical thinking appear to be other important skills, which will make a difference for humans in the age of smart technologies and machines (Fig. 102).

2.7.5 ED2: Inclusive Innovation and Development

Inclusive innovation and development require empowerment of poor people and disadvantaged groups into the process of innovation. Education and schooling are certainly the first steps in generating awareness and achieving greater inclusivity. Education should be supported with a continuous empowerment through the generation of new and highly qualified jobs for younger people. Regular and higher income

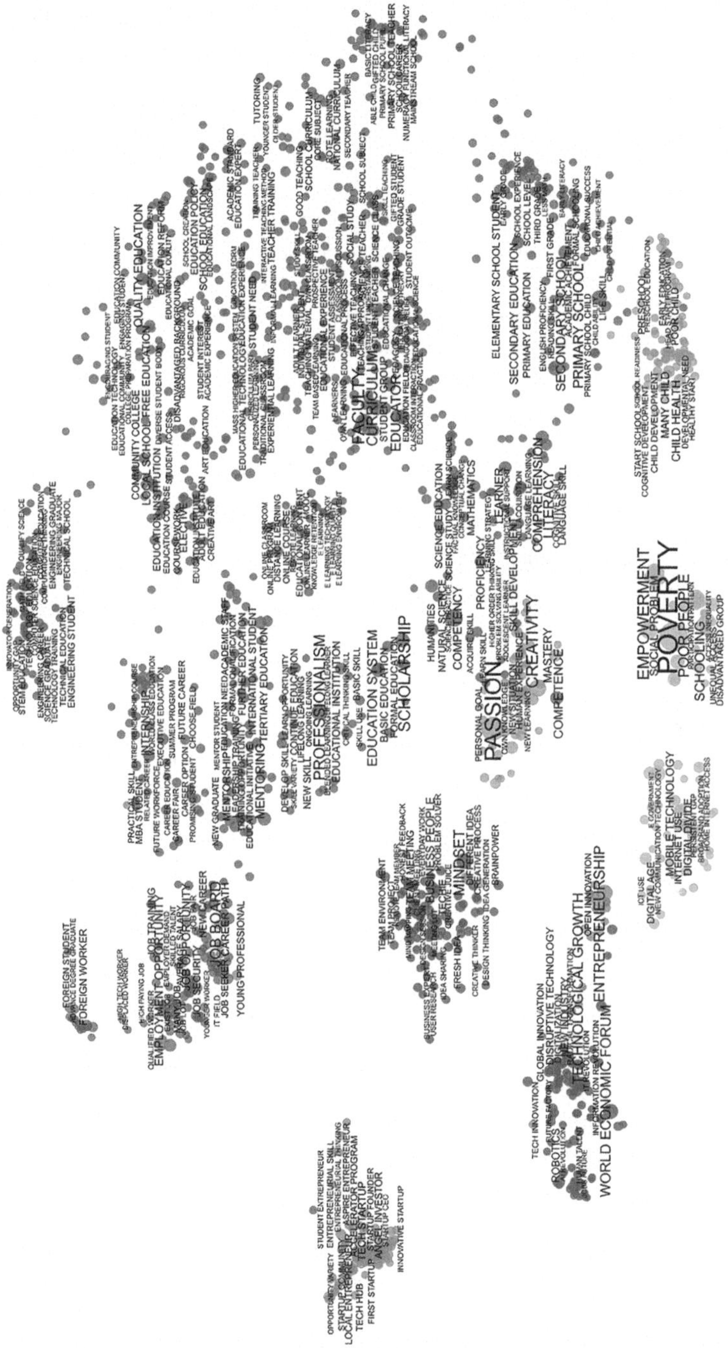

Fig. 100 "Education for the Future" domain full semantic map for 2014–2018"Education for the Future" Thrusts

Table 27 "Education for the Future" thrusts and topics

ED thrusts	Topics emerging under each thrust
ED1: Skills for the fourth Industrial Revolution	"Technology development" and "Digitalization" (Fig. 2.7.4)
	"Innovation and entrepreneurship" and "Creativity, problem solving and idea generation" (Fig. 2.7.5)
ED2: Inclusive innovation & development	"Inclusive development" and "Employment and job search" (Fig. 2.7.6)
	"Student career development" and "preschool learning" (Fig. 2.7.7)
ED3: Curriculum development 2030	"Curriculum development" (Fig. 2.7.8)
	"Skill development" and "Basic skills" (Fig. 2.7.9)
	"Primary and secondary school" (Fig. 2.7.10)
	"Higher education system" (Fig. 2.7.11)
	"Engineering education" (Fig. 2.7.12)
	"Online education" (Fig. 2.7.13)
	"Education policy" (Fig. 2.7.14)

Fig. 101 "Technology development" (left) and "digitalization" (right)

Fig. 102 "Innovation and entrepreneurship" (left) and "creativity, problem solving and idea generation" (right)

jobs with clear career paths will help to create a healthier society with a more sustainable economic system and greater possibilities for the marginalized (Fig. 103).

Career development should be started from the earlier levels of education even starting at the preschool level. At the school level, students should be provided necessary mentorship and guidance. Future career options should be shown to them. Promising students should be directed toward future careers where they may indicate higher potentials in tertiary and higher education. The process can begin at the preschool stage. Child development, readiness for school and cognitive development are among the most important factors for a healthy start to school and future careers. Poor children also need to be considered at this stage for their successful integration into inclusive innovation and development processes (Fig. 104).

Fig. 103 "Inclusive development" (left) and "employment and job search" (right)

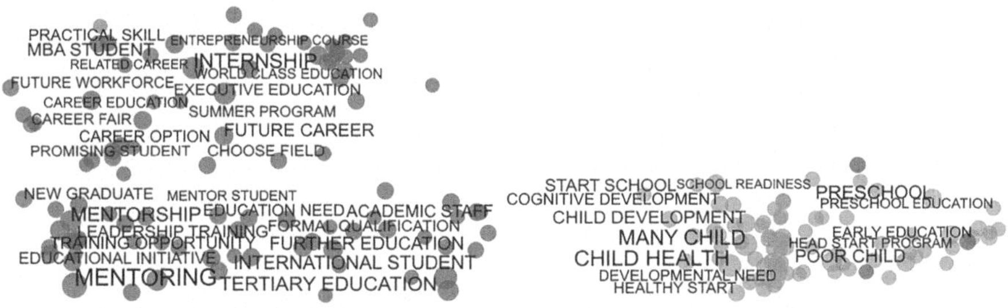

Fig. 104 "Student career development" (left) and "preschool learning" (right)

2.7.6 ED3: Curriculum Development 2030

Continuous curriculum development is a must for the education system to prepare students for the future. This holds true for all levels of education from pre–/primary school to higher education. New curriculum to be developed should allow up-to-date teaching materials and effective teaching methods and should allow an interactive process with students actively involved in classroom discussions and working in teams where possible. Good teaching practices should be developed and should be made widespread across schools. Continuous education of educators is also a must for adapting teachers into new teaching methods with relevant teaching materials (Fig. 105).

Curriculum development should help to develop skills for the new industrial developments discussed earlier. As mentioned earlier, new curriculum to be developed should contribute to cognitive and critical thinking skills, as well as creativity and problem-solving abilities. For sure, the knowledge society enabled by new technologies requires a solid science education, mathematics and language learning as basic skills (Fig. 106).

Starting from the primary school, reading and language seem to be important skills to be gained. Understanding children's potentials from the early ages is important for future educational success (Fig. 107).

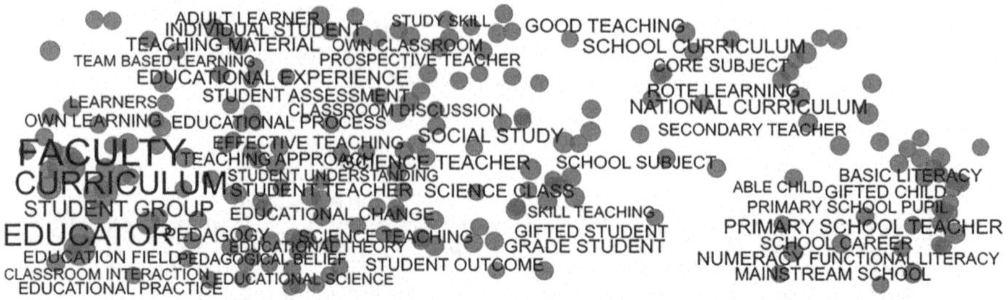

Fig. 105 "Curriculum development"

Fig. 106 "Skill development" (left) and "basic skills" (right)

Fig. 107 "Primary and secondary school"

Similarly, at the level of higher education, it is important to increase academic and educational standards. Newest educational technologies should be exploited to enable experiential and personalized learning with interactive teaching methods (Fig. 108).

In relation to the new industrial revolution and curriculum development to address the emerging scientific and technological needs, an overhaul in the engineering education is crucial. Science, computer, mathematics and technical education will remain at the core of the engineering curriculum (Fig. 109).

The increasing use of Information and Communication Technologies will also have implications on the way education is delivered. There is an increasing trend for delivering courses online. This will enable possibilities for distant learning, which

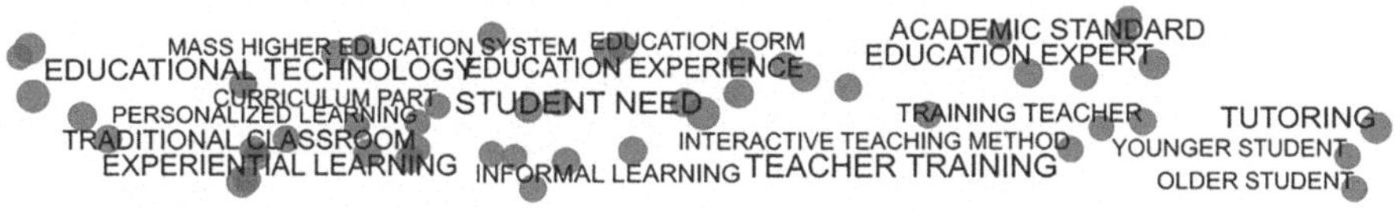

Fig. 108 "Higher education system"

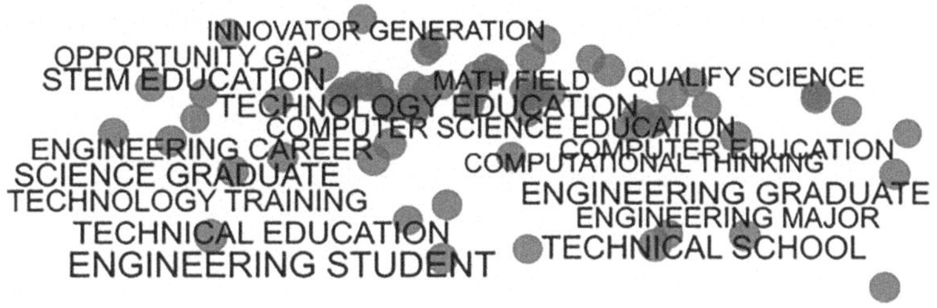

Fig. 109 "Engineering education"

Fig. 110 "Online education"

would also help the marginalized and remote layers of the society to access the education systems. For instance, Massive Online Open Courses (MOOC) courses enable accessing online courses not at the national level, but also at the global level from all centers of excellence in the world. Necessary e-learning technologies, infrastructure and ecosystem need to be set up to enable online education (Fig. 110).

In order to implement the aforementioned transformations in the education system, there is a need for efficient education policies. There are a wide variety of issues to be addressed. Overall, the quality of the education should be increased at all levels. Particularly for the marginalized and disadvantaged parts of the society education should be delivered for free or at least at minimal costs for increased access. Besides student education, adult education should also be given emphasis so that there will not be educational, social or technological divides in the society. Scientific and technical skills are important; however, necessary education for creativity and arts should also be delivered in a balanced way. Community colleges and local schools should be empowered (Fig. 111).

Fig. 111 "Education policy"

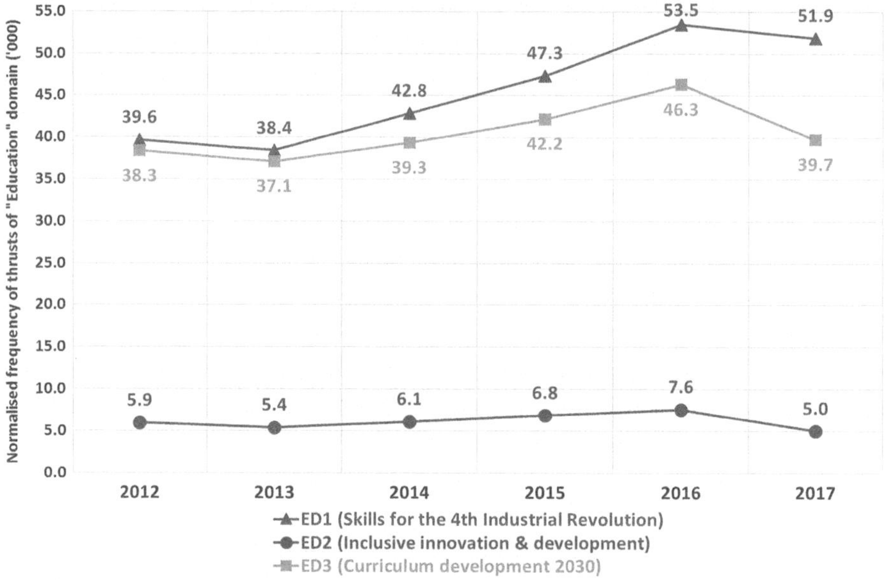

Fig. 112 Normalized frequency of all terms related with thrusts of "Education" domain ('000) in 2012–2017. Note: "Normalized frequency" for a specific thrust is a total normalized number of mentions (in thousands) of all terms (that are shown on a semantic map) related with this specific thrust in documents indexed in IFORA for a given year

2.7.7 Dynamics of "Education" Domain Thrusts

In this section, the dynamics of the "Education" domain thrusts are shown based on their frequency of occurrence in the documents indexed in the iFORA database. Among the thrusts, ED1 (Skills for the fourth Industrial Revolution) and ED3 (Curriculum development 2030) appear to be the highest on the agenda. ED2 remains largely insignificant (Fig. 112).

Furthermore, Table 28 illustrates to what extent the "Education" domain thrusts address the challenges. Education thrusts are linked to job creation and employment growth with contributions in addressing rural community, population growth and skill improvement.

Table 28 Impacts of all thrusts of "Education" domain on challenges

	Education		
Challenges	ED1	ED2	ED3
Affordable Education	0.0	0.0	14.0
Affordable Energy	0.0	0.0	3.3
Affordable Food	0.0	0.0	0.0
Carbon Emission Reduction	1.5	0.0	0.0
Clean Water Access	0.0	0.0	0.0
Crime Prevention	1.0	0.0	4.0
Crime Reduction	1.0	0.0	1.0
Economy Growth	0.0	0.0	0.0
Employment Growth	6.0	2.0	2.5
Export Growth	3.0	0.0	2.0
High Living Standard	1.0	0.0	9.0
High Quality Health Care Service	0.0	0.0	0.0
Income Growth	0.0	0.0	9.5
Job Creation	7.0	1.0	71.3
Low Greenhouse Gas Emission	0.0	0.0	1.0
Low Mortality	0.0	0.0	0.0
Population Growth	4.3	00	3.3
Poverty Alleviation	1.0	0.0	6.0
Reduce Water Demand	0.0	0.0	0.0
Renewable Energy Growth	0.0	0.0	0.0
Rural Community	3.7	0.0	32.0
Secure Water Supply	0.0	0.0	0.0
Skill Improvement	3.5	0.0	0.0
Universal Broadband Access	0.0	0.0	0.0

Notes: 1. Value in each specific cell of this table is a normalized number of co-occurrences of all terms (that are shown on a semantic map) related with a specific thrust (e.g., Thrust ED1) and a specific challenge (e.g., "Low Mortality") in documents indexed in iFORA database for 2012–2017. This value is treated as an impact of a specific thrust onto a specific challenge. 2. Cells are colored by bars as follows: within a column of a specific thrust the biggest green bar means the challenge for which this specific thrust has the highest impact

3 Conclusions

Development of internationally competitive science and technology is able to contribute to the creation of national wealth and societal progress is among the key prerequisite of success for any big nation including South Africa. Selection of promising fields of research and relevant areas of practical applications of scientific findings is one of major issues for national science and technology policy. The "South Africa Science, Technology and Innovation (STI) Foresight for 2030" identifies seven STI domains and related priorities (thrusts) to be addressed by national innovation strategies in the decade to come.

The review of current state and prospects of national capacities in these domains and thrusts presented in the report is based on a comprehensive semantic analysis of big data as well as on the analysis of publication activities. The country's research landscape, its strengths and weaknesses, benchmark vis-à-vis global leaders and leading research institutions, scientific specializations and international partners: one can find a lot of detailed information on all this issues in the report. A big number of visual presentations, graphs and tables provide a broad picture of the national science and technology capacities and, on the other hand, show key economic and societal problems to be resolved with the use of these capacities.

New technologies must serve needs of society and society is changed by new technologies. The analysis of STI domains and thrusts shows how important is coevolutionary development of science and technology and society. Comparing the scientific output with the priority STI domains and thrusts, the present report indicates that in some areas of STI, such as ICTs, South Africa's position is research is relatively low, despite the fact that almost all seven STI domains are closely linked to the development of ICTs. In this way, the report shows somewhat the feasibility of the STI goals and targets. Strategies need to be developed to implement a balanced development of research, capacities, skills and infrastructures.

Being one of the first examples of using big data for setting STI priorities, the present report can also help to derive new ideas for researchers, setting agendas and promoting systematic planning among policymakers and better use of new technologies by industries, it might also help all of them and ordinary citizens to be better prepared for the future.

Acknowledgments The chapter was prepared within the framework of the Basic Research Program at the HSE University.

Part III
Science, Technology and Innovation Domains for South Africa

The Circular Economy

Petrus Letaba

Abstract Although circular economy (CE) is closely associated with the management of waste, it is a concept that extends beyond the waste. Different concepts and terminologies are discussed in this chapter to develop a common understanding of CE. The sustainable development goals (SDGs) provide a common vision of ensuring sustainability of the planet Earth, and different countries such as South Africa have developed appropriate framework conditions for CE sociotechnical transition. The science, technology and innovation (STI) stakeholders have an important role of accelerating achievement of CE in resource-intensive sectors such as manufacturing, agriculture and mining. Patent analysis shows that South Africa is lacking behind the world's average in developing the innovations that promote circularity of the economy. However, this can be addressed through the revitalisation of the innovation competitiveness as a whole of the South African private sector. In general, CE is not new in South Africa with some successes, and there is a need to deepen and modernise its practices.

1 Introduction

Recently, it was announced that a global population has surpassed the 8 billion mark.[1] It took just above 50 years to double the world's population from 4 billion people. Although the world population growth rate has decreased from 1.94% in 1974 to 1.50% in 2020,[2] any increase in the world population has implications in terms of food security, natural resources availability, water security, energy security,

[1] https://www.weforum.org/agenda/2022.

[2] https://www.worldometers.info/world-population.

P. Letaba (✉)
Department of Engineering and Technology Management, University of Pretoria, Pretoria, South Africa
e-mail: Petrus.letaba@up.ac.za

O. Saritas et al. (eds.), *21st Century Foresight*, Science, Technology and Innovation Studies, https://doi.org/10.1007/978-3-031-69557-5_8

environmental sustainability, etc. In response to this background, the circular economy (CE) is meant to minimise consumption and waste, to encourage doing more with less, to encourage the material recycling and reuse and to protect the planet Earth.

There are various definitions of CE as demonstrated below:

- A regenerative system in which resource input and waste, emission and energy leakage are minimised by slowing, closing and narrowing material and energy loops (Geissdoerfer et al., 2017);
- An economic system that is based on business models which replace the "end-of-life" concept with reducing, alternatively reusing, recycling and recovering materials in production/distribution and consumption processes (Kirchherr et al., 2017);
- Mode of economic development that aims to protect the environment and prevent pollution, thereby facilitating sustainable economic development (Ma et al., 2014); and.
- A cycle of extraction, transformation of resources and the distribution, use and recovery of goods and materials (Prieto-Sandoval et al., 2018).

From these definitions of CE, the emerging key actors are the producers (business) as well as the consumers. Business has a responsibility to optimise the use of the inputs in a way that there is a high efficiency in their conversation to the products, the inputs material are degenerative and recycled waste is used as part of the inputs. The transformation process of the inputs into the products should be in a way such that a waste is minimised, less inputs are innovatively used to produce more products and the environmental pollution is minimised. The producers also have a responsibility on the packaging material used, in a sense that this should also be recyclable.

The following are some of the concepts that emerged from these responsibilities of the producers:

- Regenerative materials.
- Material passport.
- Regenerative agriculture.
- Recycling.
- Regenerative design.
- Sustainable production.
- Net-zero manufacturing.
- Zero waste manufacturing.
- Net-zero emission.
- Green manufacturing.
- Regenerative products.

The role of consumers is to ensure that they minimise the consumption, minimise waste and recycle the materials, products and packaging. The minimisation of consumption and waste also speaks directly to an issue of inequality. Those who are wealthy tend to consume more and generate more waste. As an example, it has been shown that the wealthiest 10% of the population consumes about 20 times as much energy as the poorest 10% of the world population (Wiggins, 2020). UNICEF (2022) recently warned that if everybody in the world consumed resources at the rate people do in Organisation for Economic Co-operation and Development (OECD) and European Union countries, the equivalent of 3.3 earths would be needed to keep up with consumption levels. In summary, the following concepts are linked to the CE in context of the consumers:

- Responsible consumption.
- Reduce, reuse and recycle (3Rs).
- Circular consumption.
- Green consumption.
- Sustainable consumption.

There are many other stakeholders that are involved with the CE, the obvious one being the government. The role of government will be explored more in detail in this chapter. The waste pickers are an important component to the CE as they collect, sort, recycle, repurpose and/or sell recovered material to the middlemen or the recycling industry (Gutberlet & Carenzo, 2020). The recycling industry is responsible for repurposing of waste into the input material that is ready for production. Lastly, the NSI stakeholders have a role to play in CE. These include design and development of cleaner production technologies, materials research, development of innovative waste management technologies, etc.

2 Circular Economy in a Global Context

Recently, the world has moved beyond the millennium development goals (MDGs) to the sustainable development goals (SDGs). Although one of the eight goals of the MDGs (2000 to 2015) was sustainable development (Goal 7), a focus was mainly on the development assistance of the underdeveloped countries with a contribution from more developed countries through a surcharge of up to 0.7% of the country's Gross National Income coordinated by the Official Development Assistance (ODA) agency (Wysokińska, 2017). The SDGs (2015–2030) builds on the successes of the MDGs for the global sustainable development agenda, leaving no one behind. One of the five Ps of SDGs is Planet, which is a pillar that CE seeks to address.

Although SDG 12 (Sustainable Consumption and Production) is at the core of CE, several other SDGs embrace this concept. These are SDG 2 (End Hunger), SDG 6 (Clean Water), SDG 7 (Affordable and Clean Energy), SDG 13 (Climate Action) and SDG 15 (Life on Land). The SDG 12 indicators emphasise the desired changes for sustainable production and consumption:

- Number of countries developing, adopting or implementing policy instruments aimed at supporting the shift to sustainable consumption and production.
- Material footprint, material footprint per capita and material footprint per GDP.

- Domestic material consumption (DMC) and DMC per capita, per GDP.
- Establishment of food waste index: to measure the amount of food wasted in tonnes.
- Number of parties to international multilateral environmental agreements on hazardous waste, and other chemicals that meet their commitments and obligations in transmitting information as required by each relevant agreement.
- Hazardous waste generated per capita; and (b) proportion of hazardous waste treated, by type of treatment.
- National recycling rate, tons of material recycled.
- Number of companies publishing sustainability reports.
- Degree of sustainable public procurement policies and action plan implementation.
- Amount of fossil-fuel subsidies per unit of GDP (production and consumption).

It is clear that not all of these targets are at a country level, as others can be implemented at the international level. SDG 12 introduces issues such as elimination of hazardous waste, elimination of food waste, sustainability reporting by the companies, sustainable public procurement policies and reduction of fossil-fuel subsidies in order to promote green economy and to reduce the carbon emissions.

A specific target in SDG 2 that relates to the CE is the one of Sustainable Food Production and Resilient Agricultural Practices. This entails the activities such as the increased resilience and adaptation to climate change, extreme weather, drought, flooding and other disasters. The targets of SDG 6 that relate to the CE are Improve Water Quality, Waste Water Treatment and Safe Reuse; as well as Increase Water-Use Efficiency and Ensure Freshwater Supplies. For the SDG 7, the relevant targets are Increase Global Percentage of Renewable Energy, Double the Improvement in Energy Efficiency and Promote Access to Research, Technology and Investments in Clean Energy. A relevant plan of the SDG 13 is Strengthen Resilience and Adaptive Capacity to Climate Related Disasters. Lastly, several SDG 15 targets embrace the CE, namely: Conserve and Restore Terrestrial and Freshwater Ecosystems, End Deforestation and Restore Degraded Forests as well and Desertification and Restore Degraded Land.

3 The Future of Circular Economy in South Africa

In absence of an overarching CE policy or a master plan in South Africa, the newly approved STI Decadal Plan provides a pathway for the medium-term future of CE in South Africa in terms of the STI capabilities developed within the NSI. With a background that the circularity of the South African economy is estimated at 7% and the goal being to reach about 17% by 2030, the role of the STI initiatives is that of accelerating transition to the CE. According to the recent Circularity Gap

Reports,[3] the world is on average 8.6% circular, with New Zealand being a global frontrunner at 24.5%. New Zealand has an ambition to increase the circularity of its economy to 50% by 2030 and 100% by 2050.

As directed by South Africa's STI Decadal Plan, the STI contribution on circularity of the three resources-intensive industries (manufacturing, agriculture and mining) as well as the energy sector is anticipated to be as follows:

- Agriculture: adoption of reformative agricultural practices and usage of precision agriculture and digital tools to improve resource utilisation. Furthermore, it is anticipated that STI can be used within the agricultural sector to reduce waste and to reduce chemical pollution by embracing biorefining innovation opportunities for processing agricultural by-products. On the use of precision agriculture and digital tools, the CSIR report (Wadiwala et al., 2022) provided some case studies on the use of local and remote sensing decision support platforms to drive transformation of the agricultural sector. Through this CSIR study, the following were determined as some application requirements relating to the precision agriculture for the emerging farmers: (i) weather patterns and warnings, (ii) soil and crops parameters, (iii) crop growth information throughout the growing season; and (iv) community of practice for sharing of precision agriculture information. In response, the CSIR is collaborating with the Department of Science and Innovation (DSI) and Agricultural Research Council (ARC) to develop a precision agriculture information system that will support farmers with a near real-time information to enhance decision making of the farm.[4] It can be deduced that these farm decision support systems can have immense impact on the circularity of the agricultural sector by optimising the usage of scarce resources such as water and energy. On the reduction of agricultural waste, a study by Bala et al. (2023) has shown that agro-wastes contain high concentrations of complex hydrocarbonates, proteins, fibres, polyphenolic components, bioactive components, etc. This study concluded that despite a fact that organic compounds waste poses a threat to the environment, it could be used as raw material in a wide range of agricultural, food and pharmaceutical goods.
- Mining and minerals: improvement of the sector's resources utilisation efficiency (energy and water). Some of the useful circular practices suggested for the mining and minerals sector through the STI capabilities are optimisation of the recycling efforts and fostering of sustainable automated mining and mineral processing efforts. Figure 1 shows how the mining and minerals sector can achieve a complete circularity in collaboration with the manufacturing sector through a process stewardship (mining and minerals) and product stewardship (manufacturing). According to Khan et al. (2022), the South African STI community should consider the following initiatives in order to improve the circularity of mining and minerals process include (i) designing out waste and pollution (through increased ore extraction efficiency, water recovery and recycling and

[3] https://www.circularity-gap.world/about.

[4] https://www.csir.co.za/videos/introducing-smart-farming-solutions-assist-emerging-farmers.

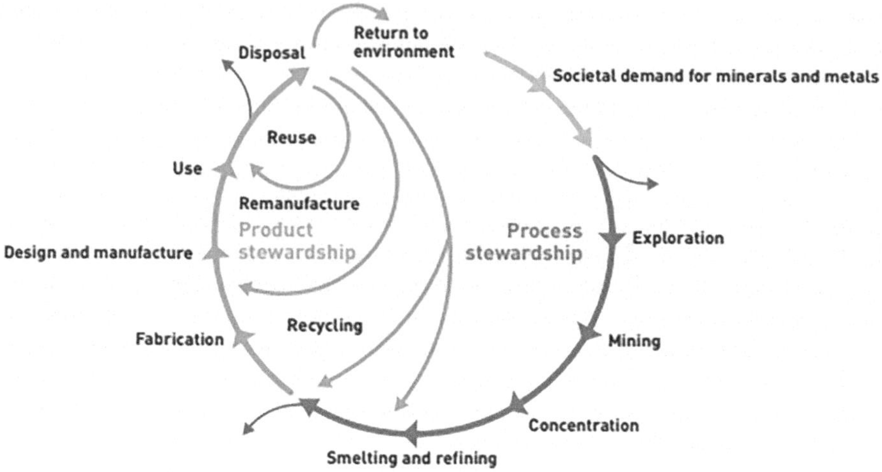

Fig. 1 Main life cycle stages for minerals and metals. Source: International Council on Mining and Minerals (2016)

substitution of raw materials), (ii) keeping products and materials in use (reduction, reusage and recycling of materials through the initiatives such as finding alternative usage of waste rocks, tailings, mine water, slag and mining sludge; repurposing of old tyres; and acceleration of technological developments and innovation to improve recycling of mining and mineral processing waste) and (iii) regeneration of natural systems through integration of renewable energy sources in mining operations (e.g. green hydrogen and solar), green hydrogen production, repurposing of post mining landscapes (e.g. rehabilitation and repurposing of old mining sites into educational training centres, agricultural land or tourist attractions) and eradication of alien plants for water reutilisation (e.g. Working for Water programme's aim of making available about 2.9% to 6% of water that is captured by alien invasive plants).

- Manufacturing: reduction of the resource input usage by integrating the fourth industrial revolution-based manufacturing technologies and innovations. In addition, promotion of the circularity in manufacturing is expected to be achieved by embracing the design thinking and reinforcing the principles of regenerative manufacturing. As shown in Fig. 1, the circularity can be achieved by focusing on processes of fabrication, design and manufacturing as well as products usage, disposal, reuse and recycling. The CE in the South African manufacturing sector is discussed in more details in Sect. 5 of this chapter.
- Energy: development of renewable energy sources to decouple economic growth from non-renewable fossil fuel dependency. The energy landscape of South Africa is evolving rapidly with a rapid adoption of solar energy solutions, increased power cuts through loadshedding and aging, dwindling business profits due to interrupted energy supply, a search for alternative renewable energy

sources such as green hydrogen, etc. The supply and demand of energy in South Africa is well summarised by the Department of Mineral Resources and Energy (2021) as follows: too much reliance on coal with primary energy supply mix composed of 65% coal, 18% crude oil, 11% renewables and waste, 3% gas, 2% nuclear and 1% geothermal; and more energy demand from the industry sector as shown by distribution of energy demand by sector: Industry (51%), transport (26%), Commerce and Public (11%), Residential (7%), Non-specified (3%) and Agriculture (2%). The South African Renewable Energy Masterplan (SAREM) is a recent implementation plan that is being developed by government (draft document at the time of writing) to drive transition towards the renewable energy. SAREM implements a vision of the Integrated Resource Plan (IDP2019) of by 2030 adding 14,400 MW of wind, 400 MW of solar PV, 4000 MW of embedded generation and 2000 MW of energy generation. As shown in Fig. 2, SAREM seeks to increase the local content of the renewable energy manufacturing through a three-horizon landscape, focusing on what can be done in a short-term (current situation), medium-term (expansion of existing capacity) and long-term (development of new capacity). SAREM is a good reminder that a transition to CE can be managed in a responsible way that leads to job creation, improvement in balance of payment and growth in economy (George et al., 2015).

Patents provide a good indication of the readiness of the STI community in transitioning the economy of South Africa to be more circular. As shown in Fig. 3, the percentage of patents promoting circular economy filed in South Africa used to be higher than that of the world's average until 2005. Following the global economic recession, the private sector's innovativeness went low, as seen by drastic decline of newly published patents in South Africa. In fact, the CE promoting patents filed in the country decreased from a high of 58 in 2003 to zero from 2009 to 2017. In average, the circularity of the world's patents is on the increase, reaching 1.3% in 2020. What is clear is that for South Africa to drastically develop the STI capacity for circular economy (especially on complex waste such as the electronics), the declining innovation within the private sector should be addressed holistically.

4 Enabling Environment for Transition into Circular Economy

The country's 2008 National Plan for Sustainable Development provides a national vision for sustainable development and to indicate strategic interventions to re-orientate South Africa's development path in a more sustainable direction. It is worth noting that this Framework precedes the SDGs, and it was in response to the MDGs' Goal 7. The vision of this Framework is as follows:

> South Africa aspires to be a sustainable, economically prosperous and self-reliant nation state that safeguards its democracy by meeting the fundamental human needs of its people, by managing its limited ecological resources responsibly for current and future generations,

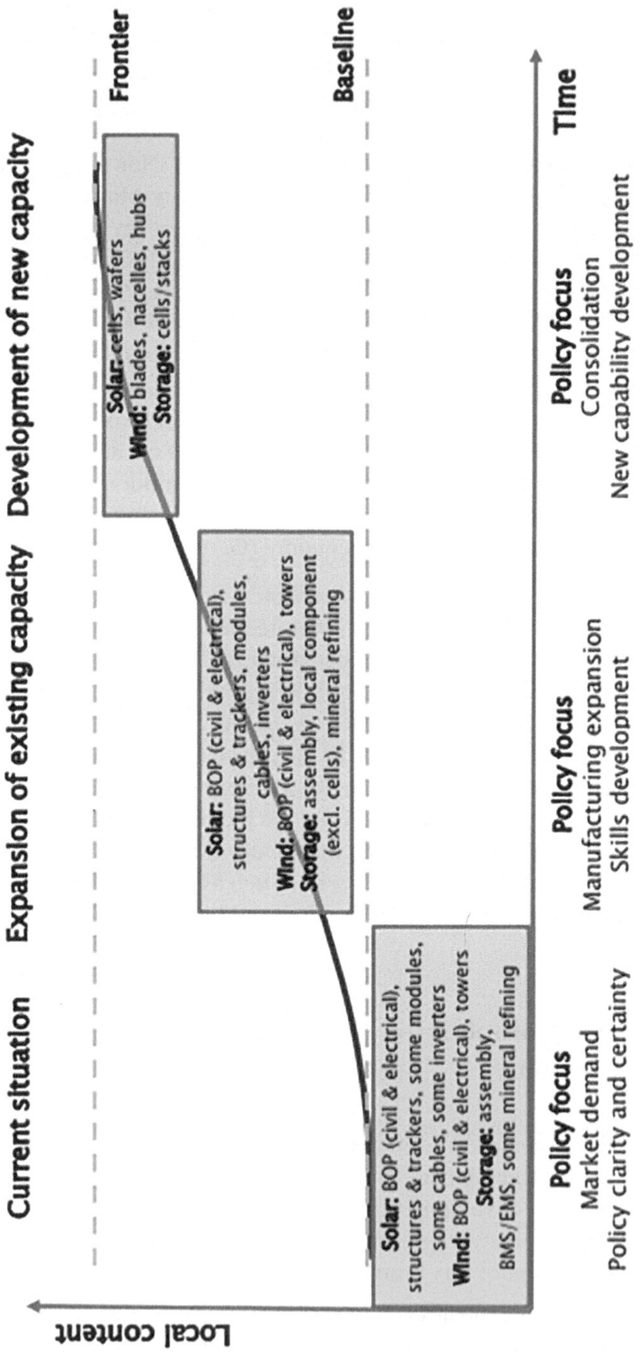

Fig. 2 South African renewable energy masterplan. Source: DMRE, DTIC and DSI (2023)

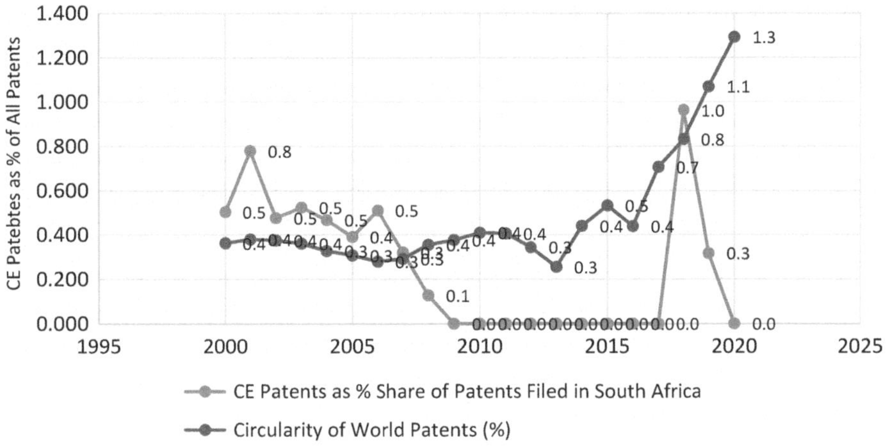

Fig. 3 Circularity of South Africa and world patents. Source: computed from WIPO "Patentscope"

and by advancing efficient and effective integrated planning and governance through national, regional and global collaboration.

A key message out of this vision, that is directly aligned to the principles of CE, is that the natural resources are limited which necessitate their responsible usage for the current and future generations. Indeed, this Framework emphasises a cyclical and systems approach to achieve sustainable development through:

- Efficient and sustainable use of natural resources.
- Socio-economic systems are embedded within, and dependent upon, eco-systems.
- Basic human needs must be met to ensure resources necessary for long-term survival are not destroyed for short-term gain.

The 2011 National Waste Management Strategy (NWMS) identified the responsibilities for different stakeholders to minimise and manage waste. These stakeholders are government (to lead on legislation, regulations, standards, infrastructure, coordination, multilateral agreements, etc.), the private sector (mainly responsible for cleaner technologies development, waste recycling and products' lifecycle management) and civil society (responsible for waste separation, recycling initiatives and to comply with regulations). Since the public research institutes and the universities are not listed as one of the key stakeholders, an assumption should have been that their agenda will be led by the government and the business. To accommodate the changes in the waste landscape, the 2020 National Waste Management Strategy was developed, and it identified the three main pillars for South Africa's future of zero waste in landfills, cleaner communities, well managed and financially stable waste services, and a culture of zero tolerance of pollution, litter and illegal dumping (DEFF, 2020). These pillars are Waste Minimisation; Effective and Sustainable Waste Services; and Compliance, Enforcement and Awareness.

A recent waste landscape discussed by the 2020 NWMS includes the following key issues:

- Organic waste as shown in Fig. 4 is the largest, with a recovery rate of 49.2%. The recycling/recovery rate is high for metals (80.0%), followed by glass (71.2%), paper (58.0%) and construction and demolition waste (52.0%). Tyres' waste is unique in a sense that most of it (76.4%) is kept on storage/stockpile, resulting with only 23.6% of recycled/recovered waste. The high disposal rate of waste comes from the following types: municipal waste (100.0%), commercial and industrial waste (100.0%), slag (100.0%), fly ash and dust (96.9%) and bottom ash (96.9%). Although the most imported waste type is paper, its recycling/ recovery rate is not so high, with a disposal rate of 42.0%.

- South Africa generated about 67 million tonnes of hazardous waste in 2017, of which over 93% of that is landfilled. The biggest contributor of hazardous waste by tonnage is fly ash (66%) and by volume it is a waste from the generation of electricity (75%).
- The main drivers of food waste include population growth and urbanisation. It is observed that about one-third (33%) of food produced for human consumption is lost or wasted. Out of this 33% loss, 50% takes place during harvesting, with processing, packaging, distribution and retail accounting for a further 45%, and consumers accounting for a remaining 5%.
- In terms of the waste services, some improvements since 2011 NWMS are noted to be improvement in waste collection and disposal services (including a programme to licence the landfill sites) and initiation of waste separation at source programmes in metropolitan areas. In general, the waste management challenges are reported to be: (i) littering and illegal dumping, (ii) low levels of separation at source, (iii) lack of infrastructure for recycling, lack of recycling culture, backlogs in waste service delivery, inconsistent waste collection, pertinent challenges due to non-compliance to permit conditions, burning in landfills, lack of support and corporation for service providers working with waste in some municipalities, lack of education and awareness in some districts and waste sector not prioritised in some municipal integrated development plans (IDPs) and budgeting.

CSIR has a repository[5] of STI roadmaps and strategies that are aimed at transitioning South Africa towards a more circular economy. The following are the relevant STI roadmaps:

- Waste research, development and innovation (RDI) roadmap: to provide strategic direction and to coordinate and manage South Africa's portfolio investment for the period 2015–2025.
- Water RDI roadmap: a high-level planning intervention that facilitates and guides refocusing of research, reprioritisation of funds, synergising of existing initia-

[5] https://www.circulareconomy.co.za/.

Waste type		Local generation	Imports	Exports	Total tonnage managed	Storage / stockpile	Recycling / recovery	Treatment	Disposal	LOC
GW01	Municipal waste	4 821 430			4 821 430	0.0%	0.0%	0.0%	100.0%	
GW10	Commercial and industrial waste*	360 884			481 179	0.0%	0.0%	0.0%	100.0%	
GW14	Fly ash and dust	4 346 080			4 346 080	0.0%	3.1%	0.0%	96.9%	
GW15	Bottom ash	6 489 080			6 489 080	0.0%	3.1%	0.0%	96.9%	
GW16	Slag	4 859 025			4 859 025	0.0%	0.0%	0.0%	100.0%	
GW20	Organic waste	19 247 851	4 048	298	19 251 600	0.0%	49.2%	0.2%	50.8%	
GW30	Construction and demolition waste	4 482 992			4 482 992	0.0%	52.0%	0.0%	48.0%	
GW50	Paper	2 211 225	57 855	129 374	2 139 706	0.0%	58.0%	0.0%	42.0%	
GW51	Plastic	1 113 362	6 748	20 856	1 099 254	0.0%	43.7%	0.0%	56.3%	
GW52	Glass	2 752 636	38 378	11	2 791 003	0.0%	71.2%	0.0%	28.8%	
GW53	Metals	4 035 929	24 168	527 037	3 533 059	0.0%	80.0%	0.0%	20.0%	
GW54	Tyres	174 640		12 473	162 167	76.4%	23.6%	0.0%	0.0%	
GW99	Other	729 615			729 615	0.0%	9.0%	0.1%	91.0%	
	TOTAL	**55 624 746**	**131 196**	**690 050**	**55 186 188**	**0.2%**	**34.5%**	**0.1%**	**65.2%**	

*percentages may not add up to 100% due to rounding off

** level of confidence (LOC) of the values in the last column is indicated by the intensity of shading

Fig. 4 Waste generation and management in South Africa. Source: DEFF (2020)

tives and ring-fencing of new resources to facilitate a more optimal water innovation system.
- Sustainable human settlements roadmap: to deploy innovations and technology to achieve green, smart and sustainable human settlements in the context of fourth Industrial Revolution.
- Hydrogen society roadmap: to contribute to the goal of a just and inclusive net-zero carbon economic growth for societal wellbeing by 2050.

5 Circular Economy in the South African Manufacturing Sector

In response to a declining manufacturing sector in South Africa (from 20% contribution to the GDP in 1993 to 13% in 2020), Fazluddin et al. (2021) view CE as central to the resuscitation of the country's manufacturing sector. The potential applications of CE for this sector are outlined as that of designing out waste (redesigning manufacturing processes and products to enhance resource efficiency, coupled with sharing economy business models), keeping materials in use (remanufacture, refurbish, repair and recycle materials and products across value chains) and regenerating natural systems (transition to green energy (solar, wind, hydrogen) and decouple resource utilisation). The use of the fourth industrial revolution technologies such as digitisation and internet of things are seen as vital in speeding up CE adoption within the South African manufacturing sector.

It should be noted that a practice of keeping materials in use is not new in South Africa as in 1993 ArcerlorMittal formed a Collect-a-Can (Pty) Ltd. joint venture with Nampak to focus on the promotion of the recovery and recycling of used beverage cans and other steel packaging material in Southern Africa.[6] Another example is Sasol that used to recycle its spent catalyst from its industrial chemical reaction processes with Umicore, a global materials technology and recycling group of companies. According to Umicore,[7] around 35 tons of rhodium are produced annually, and about one-third of that comes from recycling.

There are various Extended Producer Responsibility (EPR) organisations or Producer Responsibility Organisations (PROs) in South Africa that are established under the directive of Section 18 of the National Environmental Management: Waste Act 59 of 2008 (NEMWA) for establishment of programmes within a value-chain for a specific product class in order to reduce, reuse, recycle, recover, treatment and disposal of waste. Under the EPR regulations published by the Department of Forestry, Fisheries and Environment (DFFE) on the 05 May 2021, it became mandatory for all the producers in South Africa to become a member of an approved

[6] https://www.arcelormittalsa.com/Portals/0/Collect-a-Can.pdf.

[7] https://www.umicore.com/en/about/our-metals/rhodium/#:~:text=Around%2035%20tons%20of%20rhodium,in%20rhodiums%20global%20market%20supply.

EPR Scheme or PRO by 05 November 2021.[8] An establishment of the EPR Schemes is therefore no longer voluntary, but it obligates all the producers to be fully responsible for the impact on environment of their products at the end of their lifecycles. Although currently there is no data to show an extent of success in establishing the EPRs and their success in South Africa, it has been shown that the EPR programs worldwide drastically increase recycling of packaging (including plastic, glass, aluminium and steel) and printed paper to the rates of about 75%).[9]

It is important to emphasise that CE is not new in the manufacturing sector of South Africa. What is needed is deepening and modernisation of its application, including being more proactive through designing of products for circularity.

6 Conclusion

Although a concept of CE might be relatively new in South Africa, it has been shown that certain circularity practices such as recycling have been part of the South African economy. An example of this achievement has been to be the metals of 80%. A study of recycling rates in Iran (Norouzian Baghani et al., 2016) shown achievement of 11% and 12% for disposable metallic containers and other types of metals respectively. The policy landscape, through the leadership of DFFE, supported by sister departments such as the DSI, also seems to be proactive in promoting the environmental sustainability. A clear alignment between the SDGs and 2020 NWMS has been shown in this updated strategy which allows for a better alignment and coordination at the global, continental, regional and local levels.

That said, there are still various challenges in various sectors such as energy that is still much reliant on the fossil fuels. Lack of recycling of the municipal waste, with a huge reliance on landfills seems to be a major challenge in driving CE. Xevgenos et al. (2015) documented the success stories of municipal solid waste recovery/re-use from the 19 case studies. The successes were generally attributable to the implementation of kerbside waste collection schemes, the provision of economic incentives, the level of public awareness and engagement achieved, the population size as well as duration of the implementation of the scheme.

Indeed, there is an ongoing challenge in South Africa of educating consumers and households about the CE. Lack of separation of waste by the consumers at a point of collection can dampen the circularity efforts. Human capital and skills are also needed in order to develop the champions that can lead CE initiatives in South Africa. The more advanced countries in CE have introduced the specialised CE courses at the postgraduate level, an area which is still lacking for South Africa. A typical masters course of CE also covers the sustainability issues. In other instances,

[8] https://www.packagingsa.co.za/epr-regulations/.

[9] https://recyclingpartnership.org/study-of-extended-producer-responsibility-policy-across-7-jurisdictions-worldwide-shows-it-dramatically-increases-recycling-rates/.

CE is covered as part of a sustainability course. Several universities in South Africa have some sustainability postgraduate programmes covering the issues such as energy efficiency, sustainable design, ecological intelligence, sustainability reporting, sustainability practices in the new fourth industrial revolution world, Pan-African sustainable growth and development, etc.

Overall, the CE initiatives should be owned by all the stakeholders within the innovation system. Most cases of CE at the sectoral and firm-level require full contribution of the private sector to transform their processes, design out waste in their products and to optimise the full product lifecycle (including the reuse of waste). The public research institutes and the universities have an important role of increasing the scientific and technological advancements in favour of the CE. The municipalities have a role to play in encouraging and incentivising the residents to separate waste, in reusing waste for the applications such as energy generation and in incorporating transition towards the CE as part of their IDPs. The consumers as stated have a role of separating waste which they produce and to transition towards the responsible consumption.

With all the challenges and opportunities stated, South Africa and the rest of the world do not have an option but to act with speed in implementing the CE. This chapter will end with these words:

> Imagine a scenario at some point and time in the future in which most of the natural resources are depleted. Circular economy will be forced at us as a way of survival rather than an option. We need to act now to secure wellbeing of the future generations.

References

Bala, S., Garg, D., Sridhar, K., Inbaraj, B. K., Singh, R., Ramma, S., Tripathi, M., & Sharma, M. (2023). Transformation of agro-waste into value-added bioproducts and bioactive compounds: Micro/ nano formulations and application in Agri-food-pharma sector. *Bioengineering (Basel), 10*(2), 152. https://doi.org/10.3390/bioengineering10020152

DEFF. (2020). *National waste management strategy*. Department of Environment, Forestry and Fisheries.

Department of Mineral Resources and Energy. (2021). *South African energy sector report 2021*. Energy Economics and Statistics Department.

DMRE, DTI, DSI (2023). South African renewable energy masterplan: An industrialisation plan for the renewable energy value chain to 2030.

Fazluddin, S., Ojijo, V., Nahman, A., & Godfrey, L. (2021). Supporting the development of a globally competitive manufacturing sector through a more circular economy. In L. Godfrey (Ed.), *The circular economy as a development opportunity: Exploring circular economy opportunities across South Africa's economic sectors* (pp. 27–35).

Geissdoerfer, M., Savaget, P., Bocken, N. M., & Hultink, E. J. (2017). The circular economy: A new sustainability paradigm? *Journal of Cleaner Production, 143*, 757–768. https://doi.org/10.1016/j.jclepro.2016.12.048

George, D. A. R., Lin, B. C., & Chen, Y. (2015). A circular economy model of economic growth. *Environmental Modelling and Software, 73*, 60–63. https://doi.org/10.1016/j.envsoft.2015.06.014

Gutberlet, J., & Carenzo, S. (2020). Waste pickers at the heart of the circular economy: A perspective of inclusive recycling from the global south. *Worldwide Waste, 3*(1), 1–14. https://doi.org/10.5334/wwwj.50

International Council on Mining and Metals. (2016). *Mining and metals and the circular economy*. ICMM.

Khan, S., Maphalala, B., Nontso, Z., Magweregwede, F., & Godfrey, L. (2022). South Africa's resource availability as a driver for transitioning into circular economy. CSIR report number: CSIR/MIN/MMR/IR/2022/0001/A, Pretoria, South Africa.

Kirchherr, J., Reike, D., & Hekkert, M. (2017). Conceptualizing the circular economy: An analysis of 114 definitions. *Resources, Conservation and Recycling, 127*, 221–232. https://doi.org/10.1016/j.resconrec.2017.09.005

Ma, S. H., Wen, Z. G., Chen, J. N., & Wen, Z. C. (2014). Mode of circular economy in China's iron and steel industry: A case study in Wu'an city. *Journal of Cleaner Production, 64*, 505–512. https://doi.org/10.1016/j.jclepro.2013.10.008

Norouzian Baghani, A., Farzadkia, M., Azari, A., Zazouli, M. A., Vaziri, Y., Delikhoon, M., & Shafi, A. A. (2016). Economic aspects of dry solid waste recycling in Shiraz, Iran. *Journal of Mazandaran, 25*(133), 330–334.

Prieto-Sandoval, V., Jaca, C., & Ormazabal, M. (2018). Towards a consensus on the circular economy. *Journal of Cleaner Production, 179*, 605–615. https://doi.org/10.1016/j.jclepro.2017.12.224

UNICEF Office of Research. (2022). *Places and spaces: Environments and children's Well-being*. Innocenti Report Card 17. UNICEF Office of Research—Innocenti.

Wadiwala, E., Schoeman, C., Main, R., Madonsela, S., & Cho, M. (2022). Transformative farming through local and remote sensing decision support. CSIR Report Number: DST/CON 0094/2020, Pretoria, South Africa.

Wiggins, B. (2020). The world's wealthiest consume 20 times more energy than the world's poorest. Global citizen. Accessed December 02, 2023, from https://www.globalcitizen.org/en/content/wealthy-people-climate-change-energy-consumption/

Wysokińska, Z. (2017). Millenium development goals/UN and sustainable development goals/UN as instruments for realising sustainable development concept in the global economy. *Comparative Economic Research, 20*(1), 101–118.

Xevgenos, D., Papadaskalopoulou, C., Panaretou, V., Moustakas, K., & Malamis, D. (2015). Success stories for recycling of MSW at municipal level: A review. *Waste and Biomass Volarization, 6*(5), 657–684. https://doi.org/10.1007/s12649-015-9389-9

Education for the Future (Educating for 2030)

Nisha Sewdass

Abstract The South Africa Foresight Exercise for Science, Technology and Innovation (SAForSTI) (2019, NACI Synthesis Report, Department of Science and Innovation. Republic of South Africa) identified the education for the future domain as one of the priorities for the new decadal plan for STI to achieving the vision of the National Development Plan 2030 which aims to improve the quality of life of all South Africans by addressing social and economic challenges in an inclusive and sustainable way. This chapter discusses the findings from the second phase of SAForSTI (2019, NACI Synthesis Report, Department of Science and Innovation. Republic of South Africa) exercise. Specifically, the chapter will focus on the issues and drivers for the education for the future domain and the findings from the second phase of SAForSTI (2019, NACI Synthesis Report, Department of Science and Innovation. Republic of South Africa) exercise focussing on the three STI thrusts that were identified and prioritised, namely, Skills for the Fourth Industrial Revolution (4IR), Inclusive Innovation and Development and Curriculum Development 2030. The impact of Covid-19 on the education for the future domain will also be discussed. It is also encouraging to see that while there are still several challenges in the education system in South Africa and in providing the skills required by the economy, the South African government and industries are already busy deliberating and planning for developments and challenges that are expected with 4IR. South Africa is on the right trajectory, and with political will and commitments in terms of funding and policy implementation, the education for the future looks positive.

N. Sewdass (✉)
Department of Business Management, University of South Africa, Pretoria, South Africa
e-mail: sewdan@unisa.ac.za

O. Saritas et al. (eds.), *21st Century Foresight*, Science, Technology and Innovation Studies, https://doi.org/10.1007/978-3-031-69557-5_9

1 Introduction

1.1 Overview

Nelson Mandela stated that "education is the great engine of personal development, no country can really develop unless its citizens are educated, and education is the most powerful weapon which can be used to change the world" (Nelson Mandala, posted in Patterson, 2013). He went further to state that "young people must take it upon themselves to ensure that they receive the highest education possible so that they can represent us well in future as future leaders". Education for the future has been a focus for South Africa as will be seen from this chapter.

The South Africa Foresight Exercise for Science, Technology and Innovation (SAForSTI) (2019) report identified the education for the future domain as one of the priorities for the new decadal plan for STI to achieving the vision of the National Development Plan 2030 which aims to improve the quality of life of all South Africans by addressing social and economic challenges in an inclusive and sustainable way. Furthermore, it addresses the needs to enhance access to learning and improve learning outcomes using technologies, while also harnessing neuroscience research outputs relevant to learning. The focus of this domain is also on the urgent need to develop educators, learners and graduates who are ready to meet the demands of Fourth Industrial Revolution (4IR) and to address inclusivity and sustainability challenges related to education.

The *Constitution of the Republic of South Africa* (1996:1256) advocated that "Everyone has the right to a basic education, including adult basic education; and the right to further education". In addition, education should be equitable and practicable and redress the results of past racially discriminatory laws and practices. UNESCO (2019:3) suggests that "education is a fundamental human right, a precious public good and an indispensable tool in building peaceful, sustainable and fairer societies".

The exact impacts of 4IR technologies on society and economies are still unknown, but the fact that they will bring profound and rapid changes are certain. Hence it is essential for education and higher education in particular to respond urgently as the power of 4IR technologies could either have positive social impacts or devastating environmental damage (Penprase, 2018:217). There is a widening gap between the sophistication in skills required by the 4IR and the relatively poor education that many learners currently receive. The public education system in South Africa has, repeatedly, been measured as one of the world's least effective, especially in mathematics and science. Access to education and training is not equitable, particularly in rural areas, and is falling further behind each year. Given that 2030 is one school generation away, addressing this domain is of critical importance.

The overall objective of this chapter is to discuss the findings from the second phase of SAForSTI (2019). Specifically, the chapter will focus on the issues and drivers for the education for the future domain; the findings from the second phase of SAForSTI (2019) exercise focussing on the three STI thrusts that were identified and prioritised for the education for the future domain, namely, Skills for the Fourth Industrial Revolution, Inclusive Innovation and Development and Curriculum

Development 2030; and the impact of Covid-19 on the education for the future domain.

2 Issues and Drivers

The World Economic Forum found that the major drivers of transformation affecting global industries are expected to have a significant impact on jobs, ranging from significant job creation to job displacement and from heightened labour productivity to widening skills gaps. What is more interesting in their findings was that in many industries and countries, the most in-demand occupations or specialties did not exist 10 or even 5 years ago, and the pace of change is set to accelerate. In such a rapidly evolving employment landscape, the ability to anticipate and prepare for future skills requirements, job content is essential (WEF, 2016:3). South Africa has to overcome the skills gap swiftly to optimise new technologies for national development and global competitiveness.

The Fourth Industrial Revolution (4IR) offers opportunities for countries to advance their economic prospects through innovative technology that drive growth. Success in the 4IR will depend heavily on having the right skills, at a personal, business and national level. The challenge is that many jobs in the 4IR do not yet exist and the skills needed have not yet been defined. The World Economic Forum (2016:3) estimated that 65% of children entering primary school today will ultimately end up working in completely new job types that don't yet exist. These challenges were also noted in several interviews during the SAForSTI exercise in 2019. Input included the following: (a) future graduates need to be prepared for a world in which AI and AI-assisted technology will play an increasingly dominant role; (b) the best strategy would be to teach skills that make humans human, rather than training students to outcompete new technologies; and (c) neuroscience research could contribute towards a better understanding of how learning outcomes could be enhanced and also for the design and developing effective teaching strategies that can engage students in higher-level thinking and deep learning.

The present South African system of education and skills is more focussed on manual work and lagging in skilled jobs; hence there is limited availability of skilled labour in the manufacturing sector, for example. South Africa will need to rectify this in order to mitigate future job losses as a result of the introduction of 4IR technologies and to have an adequate skills pool to work with and leverage advances that come with 4IR in manufacturing and other sectors such as automotive, retail, agriculture, food and beverages (NPC, 2020:50). There is a dramatic shortage of scientists, engineers, data scientists and specialised software developers and coders. The current education and skills base of the country continues to reflect extreme inequality particularly for those at the intersection of various forms of historical and traditional discrimination and marginalisation. There is also a need to address the lack of human capital, specifically the skills deficits associated with digital skills ranging from basic literacy to the high-level engineering and advanced coding, as well as for data scientists.

The NSTF (2018) concept paper also noted that South Africa continues to suffer from problems in its education system especially with regard to the teaching of science, technology, engineering and mathematics (STEM) subjects. The university graduation rate for life sciences, physical sciences, mathematics and statistics was reported as a meagre 3%, while STEM-related courses, including engineering and information technology, was only about 18%, contributing to a dire need for high level skills in STEM areas in both the economy and academia.

As workforce transformations accelerate, the window of opportunity for proactive management of this change is closing fast, and business, government and workers must proactively plan and implement a new vision for the global labour market.

Four technological drivers to change that can positively affect business growth have been identified as ubiquitous high-speed mobile Internet, artificial intelligence, widespread adoption of big data analytics and cloud technology (WEF, 2018). In response to these drivers, companies indicated that they are likely or very likely to expand their adoption of technologies such as the Internet of Things and app- and web-enabled markets and to make extensive use of cloud computing. Considerable investments were considered by these companies for machine learning and augmented and virtual reality for the period 2018–2022 (WEF, 2018:vii). In addition to machine learning algorithms and artificial intelligence, robots and fully automated aerial drones have been adopted by several businesses.

The 2018 WEF report also found that companies expect automation to lead to some reduction in their full-time workforce, while some businesses surveyed expect automation to extend their workforce to new productivity-enhancing roles or to the creation of new roles in their enterprise. In addition, businesses expressed their intension to expand their use of contractors doing task-specialised work, with many respondents highlighting their intention to engage workers in a more flexible manner, utilising remote staffing beyond physical offices and decentralisation of operations.

As it stands operating in the current industrial revolution requires higher education levels and cognitive skills. If the 4IR is really to deliver on the promise of economic growth, job creation due to innovation, improved safety and security, better education and skills transfer, South Africa has to rapidly and immediately change its education focus and delivery model to be responsive to the benefits and opportunities of the 4IR (Khathu, 2020).

3 Findings from the Second Phase of SAForSTI (2019) Exercise

During the second phase of SAForSTI (2019) exercise, three STI thrusts were identified and prioritised for the education for the future domain:

- Skills for the Fourth Industrial Revolution.
- Inclusive Innovation and Development.
- Curriculum Development 2030.

3.1 Skills for the Fourth Industrial Revolution

Education and training must keep pace with the changes going on globally specifically with technological innovations and the 4IR. In order for South Africa to be a global role player, it needs to ensure there is sufficient motivation to change the way things are done. Education can no longer be offered as it is currently. The new generation is very technologically motivated, and it is anticipated that they will be the leaders of the future, hence they need new ways of doing things. Business environments are transforming very rapidly and in order to keep pace with this, new methods of education are required to develop individuals for the skills needed for the future.

This thrust within the education for the future domain was prioritised for addressing the need to develop the human capacity and skills required to implement the 4IR. To accomplish this, it will be essential to identify and highlight the various skill sets required across sectors for the future. This can be achieved through a skills audit or mapping of the industries that are either starting to use or preparing for using 4IR technologies in the future. With greater automation in industries in South Africa, fewer people would probably be employed in traditional areas (NPC, 2020). Recent reports of job losses at large organisations such as Standard Bank, MultiChoice, etc. have also been attributed to digitisation and are evidence of how technology is affecting the labour market. For example, Standard Bank chief executive indicated "that customer needs and competitive pressures has forced the group to rapidly digitise its business – resulting in job losses and branch closures". In addition, Standard Bank reported a loss of 2423 permanent employees in its latest annual report for the year ended December 2019 (Businesstech, 2020). Hence, there will need to be large-scale development of entrepreneurial skills to enable those that lose their jobs to create their own work or to work with multiple employers on freelance and micro work. Khathu (2020) suggests that entrepreneurship in the 4IR will open up new opportunities that don't yet exist. Therefore, it is important that the education that learners receive equips them with the necessary skills to succeed as entrepreneurs. Some of the skills identified are creativity, problem-solving, critical analysis, independent thinking and analytical skills, and these align well with the emerging skills identified by the WEF (2018:97) for South Africa with the introduction of 4IR:

- Analytical thinking and innovation.
- Creativity, originality and initiative.
- Active learning and learning strategies.
- Technology design and programming.
- Complex problem-solving.
- Leadership and social influence.
- Reasoning, problem-solving and ideation.
- Critical thinking and analysis.
- Resilience, stress tolerance and flexibility.
- Emotional intelligence.

It will appear that a substantial number of businesses in South Africa have already embraced new technological trends and therefore can play a significant role in helping educators tailor and update the curricula to meet future requirements and, most importantly, fast track training to meet the supply shortage for appropriate skills (NPC, 2020:69). Some skills may be available in these industries, while others will have to be developed and appropriate education and training will need to be provided.

High-technology industrialisation in South Africa would require new ways of teaching using a range of technologies such as augmented reality (AR), artificial intelligence (AI), machine learning and virtual reality (VR) to enable children, students, the employed and unemployed individuals to be trained appropriately for a future digital world. A culture of lifelong learning is also essential for adults and those already employed, and this needs to be supported through diverse learning channels such as distance/remote learning using apps and web-based tools.

Mezied (2020) suggests that 4IR not only urges us to think creatively about the manufacturing process, value chain, distribution and customer service processes but it also compels us to focus on the strategic use of the Internet of Things (IoT) and other technologies rather than attendance at educational institutions to prepare the coming workforce for the challenges that lie ahead.

It is evident that the boundaries between the Internet, physical world and people are becoming more blurred. This will mean that education, specifically higher education that is currently "place-based", is diminishing and being connected to mobile devices through applications in the cloud and is no longer limited to knowledge but extended to skills acquisition (Mezied, 2020). With the expansion in networking services nationally and globally, physical boundaries are no longer barriers to education, and individuals no longer have to physically attend a specific institution to gain the required education and skills. Mezied (2020) further observed that education is increasingly becoming "just in time" rather than "just in case", and it is more about what you need to know for a certain time than compiling knowledge that may never be needed. Academics/teacher competency and reskilling will become essential as educators must possess skills and tools necessary to prepare learners for opportunities in the 4IR. A revised curriculum for educators will also be an important consideration.

It is anticipated that this thrust will produce a more highly skilled and self-driven workforce that is the direct outcome of a reformed education system with enhanced teaching methods. It will produce academics and educators who can deliver on new curriculum requirements and deliver globally competitive individuals and industries, as well as improved access to training opportunities by the marginalised. The outcome will be a society that is more resilient and equipped with the appropriate skills to live and work effectively in the 4IR, with higher levels of employment for all through improved education and skills development.

3.2 *Inclusive Innovation and Development*

While education institutions play an important role in shaping future technology by being the testbeds for innovation and educating future generations, it is essential to understand how higher education institutes would be affected by the Fourth Industrial Revolution and how the delivery of education will be transformed. Innovation can be regarded as a key driver of economic growth in a country. However, although technology is advancing rapidly, a large share of the population in South Africa has often been excluded from the innovation process, further contributing to increased economic and social inequalities. In order to ensure economic development for all, it is essential that systems are designed to benefit everyone and improve the welfare of those excluded; hence it is essential for education institutions to focus on Inclusive Innovation and Development.

Considering the above, the purpose of this thrust is to ensure that education institutions align and develop STI initiatives and activities to develop individuals that can take responsibility for their own lifelong learning to meet the demands of the society in which they find themselves in. It addresses several requirements that are not dealt with effectively in the current innovation ecosystem and that need more attention if lifelong learning is to become a reality: (a) more early childhood development (ECD), adult basic education and training (ABET) and artisan development, (b) a strong drive to establish a sustainable circular economy, (c) the creation of more industry-driven capacity development and (d) entrepreneurship education and training embedded in all curricula, across all disciplines.

3.2.1 Early Childhood Development (ECD), Adult Basic Education and Training (ABET) and Artisan Development

Early years are the opportune time to create the environment and scaffolding to facilitate independent, lifelong learning which must ultimately become self-directed by individuals. There is ample research to suggest that children who participate in quality ECD programmes have higher levels of cognitive development and are better prepared to learn when they enter primary school (WEF, 2018, Gravett, 2019, UNICEF, 2019). Afrika Tikkun's youth and career development programmes acknowledge that when young people choose STEM subjects and the ability to be self-directed are not only powerful competencies in the 4IR but essential for survival. They give children a path of hope and they subsequently have better employment prospects. In addition, children need to develop the competencies that will equip them for a world that is often characterised by volatility, uncertainty, complexity and ambiguity; hence the ability to deal with complexity and ambiguity is crucial. Cultivating the four Cs (critical thinking, creative thinking, communication and collaboration), which are widely touted as fundamental twenty-first-century skills, could go some way towards preparing the youth to deal with complexity and ambiguity (Gravett, 2019).

ABET is usually provided at technical and vocational education and training (TVET) colleges, and it is important to ensure alignment between curricula of TVET colleges and industry needs. Improving education is critical in the context of the changing nature of work and equipping young people to take advantage of the opportunities created by the Fourth Industrial Revolution (4IR). TVET colleges also offer artisan development to meet the needs of industry. Hence it is also important for the public and private sector industries to assist in identifying the needs and ascertain how they can be met. This can also assist in greater cooperation in the development and delivery of education programmes, as well as the development of cross-campus programmes or the establishment of multidisciplinary institutes. According to the WEF (2018), the educational challenges we face are complex, and this includes the rise of inequalities, demographic change and climate change. In addition to this, the world is also changing drastically and quickly. Therefore, if we do not adapt and enhance adults' skills, they will be left behind.

3.2.2 Sustainable Circular Economy

The WEF White Paper (2018:7) defines the circular economy as "an economic system that is based on business models which replace the 'end-of-life' concept with reducing, alternatively reusing, recycling and recovering materials in production/ distribution and consumption processes, thus operating at the micro level (products, companies, consumers), meso level (eco-industrial parks) and macro level (city, region, nation and beyond), with the aim to accomplish sustainable development, which implies creating environmental quality, economic prosperity and social equity, to the benefit of current and future generations".

The transition to a circular economy requires rethinking market strategies and models that encourage competitiveness in different sectors and the responsible consumption of natural resources. This would entail changing consumer behaviour, regulating production processes, generating new avenues for employment and reducing demand for new raw materials (WEF White Paper, 2018:11). In addition, the sustainable use of raw materials, closed-loop manufacturing and production processes, adoption of sustainable methods of consumption and the development of more extensive commodity markets for keeping products in use have all become essential parts of being circular. Kirchherr et al. (2017:229) advocated that the circular economy "operates…with the aim of accomplishing sustainable development, thus simultaneously creating environmental quality, economic prosperity and social equity, to the benefit of current and future generations. It is enabled by novel business models and responsible consumers". The benefit for emerging economies moving towards a Circular Economy strategy is that they may be able to leapfrog to more sustainable development approaches by learning lessons and avoiding pitfalls of resource-intensive practices of the linear economy.

Education and curriculum design to inspiring students to rethink the future through the circular economy framework would include the radical redesign of products and services around ecological, economic and social considerations for a sustainable future (Andrews, 2015:312). Andrews (2015:313) also found that research into learning about sustainability in order to practice sustainable product design and manufacture can have an impact on the behaviour of students in their personal lives, and therefore it can be considered education for sustainability and important to include in education curriculum across the various disciplines.

3.2.3 Industry-Driven Capacity Development

South Africa has ongoing challenges with respect to labour market outcomes. Many explanations are offered to explain this persistent challenge including the absence of foundational education, issues of quality of teaching and the lack of relevance of the programmes (qualification and curricula) (NPC Draft Report, 2020). Therefore, it is essential that vocational skills development (VSD), which include the continuum of vocational skills through TVET and through to higher education, is responsive to the demands of industry. The adoption of technologies implies a change in labour market demand and a dramatic realignment of the skills base. However, in South Africa, the labour and skills supply are currently not aligned with the requirements of the industrial and private sectors. Therefore, it is essential to build a pipeline of future talent that can embrace this dynamic and increasingly digitised environment.

Van der Berg et al. (2020) indicate that there will be a "mismatch" between the skills required and the skills available, between the demand for skills and the supply of skills. For skills acquisition to be beneficial, there needs to be a "match" between the skills acquired, the available skills supply and the skills that are needed in the labour market: the demand for skilled labour, employer or labour market demand. This demand is articulated in the number and type of jobs available in the economy.

Report for the Economic Task Team of the National Planning Committee (Van der Berg et al., 2020) further points out that while the focus on acquiring a post-school qualification as a means of enhancing the employability of youths is important, it is important to remember that three-quarters of employed South Africans currently hold no post-school qualification and hence the skills required by industry. Even though the NDP's targets, plan on raising the proportion of youths who obtain a post-school qualification from the current 19% to about 40%, it is an optimistic scenario that by 2030 over half of youths would not go beyond having a school education. This makes it even more important for education institutions to maximise the employability of school-leavers, for instance, through access to more vocational and technical education at the secondary level, on-the-job training and post-school training and development.

3.2.4 Entrepreneurship Education and Training

The current education offered in South Africa provides little scope for adults and young people to acquire occupational skills or skills for self-improvement and self-employment. The shift to community education and training (CET) was viewed as a more effective way to promote lifelong learning and offer skills (in addition to general education programmes) that could contribute to sustainable livelihoods outside of the formal sector. There is a growth in the township/village economy or better known as the informal sector to train and create jobs for young people. Hence it is essential for education institutions to take the nature and profile of informal enterprises into consideration and focus on providing education programmes through its CET colleges that are responsive to the immediate demands of this sector such as entrepreneurship and small, medium and micro enterprise (SMME) development.

Entrepreneurship in the 4IR will open up new opportunities that don't yet exist; it is important that the education that learners receive equips them with the necessary skills to succeed as entrepreneurs. Creativity, problem-solving, critical analysis, independent thinking and analytical skills are some of the skills needed to exploit opportunities presented by the 4IR (Khathu, 2020). It is also suggested that entrepreneurship skills be embedded in all fields of education.

3.2.5 Intervention Already in Place to Address the Thrust

A range of internal and external factors contribute to South Africa's inequitable higher education system, not least of all the legacy of apartheid. There is a recognition that the inequality cannot be fully addressed until conditions both inside and outside of the education system are repaired. The state recognises the need for investment in the whole education "pipeline"; as a result there has been increased attention to early childhood development and primary and secondary schooling (Shay, 2017:3).

The following interventions regarding this thrust have already been noted in South Africa:

- According to the VNR Report (2019:8), support for early childhood education has been significantly increased, and every South African child will be provided with digital workbooks and textbooks on a tablet device over the next 6 years. The rate of early childhood learning level is high (with almost 95% of the country's children participating in organised learning in the year before the official primary school entry age (VNR, 2019:10)). When compared globally, the UNICEF's State of the World's Children report (2019) indicates that in South Africa 48% of children aged 3 and 4 attend some form of preschool, and this figure seems not to stand out as exceptionally high or low.
- Expanded artisan development programme enables the country to produce more artisans such as plumbers, welders and electricians to ensure that young people have skills for self-employment as well as skills to be employed by other companies (VNR, 2019:60). ABET and artisan graduates must be increased by 2030, hence funding has been provided to enable those who have not finished school to pursue ABET/artisanship opportunities.
- Expansion of TVET colleges to ensure that young people have vocational skills that will help them to find work easily. TVET enrolments have doubled without

concomitant capacity development. Some companies have introduced joint offerings with both TVET colleges and universities that effectively expand the responsiveness of the PSET system. These should be encouraged and expanded (Van der Berg et al., 2020).

- Introduce entrepreneurship courses in the country's high schools, and include entrepreneurship as part of curricula in all disciplines and not just in business courses.
- Build recycling and refurbishment plants for ICT waste (possibly at educational centres) in all provinces, underpinned by appropriate training that will generate jobs in the circular economy.
- As a means of providing citizens with the requisite skills for participating in South Africa's digital economy and protecting their employment potential, a comprehensive turnaround strategy was developed by the Department of Higher Education and Training (DHET) for 50 further education and training (FET) colleges. The NDP aims to increase graduation levels to 75% in 2030, with all new courses and curricula in the ICT field being evaluated closely. On-the-job training and short courses have been identified as the approach to take for workplace training (NPC, 2020:73).
- In order to ensure that technological advances benefit all, and not a select few and to fully harness the potential of technological innovation to grow our economy and to uplift the people, South Africa has established a Presidential Commission on the Fourth Industrial Revolution to develop an integrated national response strategy. The commission is composed of representatives of tech start-ups, academia, cybersecurity specialists, researchers, social scientists, trade unionists and other representatives from key economic sectors. This commission has various workstreams to report on such as infrastructure and resources, research, technology and innovation, human capital, industrialization and policy and legislation (Foresight Africa, 2020).

The outcome of these interventions will be a larger pool of economically active participants resulting from the increased number of people trained in entrepreneurship and innovation and an increased number of SMME startups. This is fully aligned with the NDP vision of significantly increasing job creation. It will also seek to promote social innovation to support social well-being and sustainable livelihoods, knowledge-sharing and the promotion of cooperation and partnerships between stakeholders, including between governments, business, academia and civil society, in sectors contributing to the achievement of the SDGs and global partners especially in the African continent to develop African solutions for African challenges.

3.3 Curriculum Development 2030

Traditional education as we know will be affected by the 4IR, and that is about what we teach, not just how we teach it (Reaves, 2019). Existing jobs and definitions of work are also expected to change fundamentally or disappear completely and the new jobs replacing them may be radically different or perhaps not even be defined as work. The educational curricula, career and training programmes, assessments, degrees and certifications that are currently being offered by educational institutions could become largely irrelevant to the challenge of creating value and earning income in a constantly evolving market. This will force individuals, industries and educational institutions to prepare for a new environment with constant fluctuations and where having one career for life or working for one organisation is not guaranteed. According to Reaves (2019), carefully structured curriculum from a previous year may include the wrong or inappropriate and outdated elements. This could mean that unlearning outdated material could become almost as important as learning. It was also noted that the more detailed the learning programme, the more likely it is to be disrupted by changes brought about by 4IR. Operations and procedures are expected to be flexible and remain relevant for shorter time spans such as months rather than years. In this changing environment, Reaves (2019) points out that the emphasis will be on doing/learning/doing, empowering individuals to take charge of their own education and career strategies, with maximum flexibility and minimum lag time between the learner and real-world opportunities.

Many of the programmes and curricula offered at South African universities and TVET colleges are not adequately responsive to the needs of the economy or those expected to emanate because of 4IR (VNR, 2019:60). Hence this thrust addresses the need to transform the education curriculum so that it is responsive to the needs of the industries and the economy. Research and development are needed to identify methodologies for teaching in the future (interactive, facilitated and innovative teaching methods). Investigation is also needed into how a blended learning approach could allow for the marginalised and those from poorly developed areas to have access to education and lifelong learning. Industry should be engaged to establish sector needs (using foresight) that can be incorporated into curricula. South Africa will need to engage global experts and researchers to establish and benchmark against global trends.

Blended instruction and optimization of flipped and online courses have been proposed as more efficient learning environments that can adapt for diversity in preparation of students (Penprase, 2018). 4IR has also been suggested to place emphasis on adaptability and in self-directed learning and thinking. Furthermore, it has been noted (Reaves, 2019) that the shelf life of any skill in the present-day environment has become increasingly short, thereby requiring future workers to continuously update their skills and teach themselves about new technologies and new industries that may not have existed while they were being trained for their initial academic qualifications. A further design requirement for education within 4IR would be to include a strong overlay of ethical thinking, intercultural awareness and critical thinking to enable for thoughtful and informed application of the exponentially developing technologies (Penprase, 2018:220).

The NPC (2020:82) further suggested that the school curriculum must be revamped, not only to advance the take-up of science, technology, engineering and mathematical courses but also to promote critical thinking, flexibility and creativity. It acknowledged that while these challenges are not specific to the Fourth Industrial Revolution, they are exacerbated by its emergence. Penprase (2018:222) also proposed that a new 4IR curriculum will need to reduce the divisions between humanities and STEM to create a more integrated system of education which can explore the newly emerging conceptions of self and identity within the 4IR, including discussions of autonomy, free will and genetic vs. social determinism. The changing nature of social relations and interactions—social media, obligations to identity groups, society, nation and world—needs to be central in the 4IR curriculum as all of these identities and loyalties are shifting rapidly due to increased globalization. More than ever, higher education in the 4IR age must develop the capacity not just for analysing and breaking a technical or scientific problem into its constituent parts but also must emphasise the interconnections between each scientific problem across global scales and interrelations between physical, chemical, biological and economic dimensions of a problem.

An integrated post-school education ecosystem system including various stakeholders and partners such as universities, science councils, private sector, public sector, government, labour unions, research institutions, experts, SMMEs, youth, women and non-governmental organizations will be useful for the designing of multidisciplinary curriculum and one that can be regarded as a more "fit-for-purpose" curriculum that provides opportunities for "breadth", rather than the focus being on one discipline only, and a wider skill set than higher education has traditionally produced. This "breadth" would include key graduate attributes, opportunities for electives outside the discipline and the promotion of interdisciplinary thinking (Shay, 2017:5).

Mezied (2020) noted that higher education institutions are embracing data mining in order to gain better understanding of student performance and deliver "education for you" that is tailored to meet the demand of the job markets while considering the students' needs. The data regarding student performance, behaviour, development and interaction inside classrooms and on the online platforms can then assist higher education institutions to analyse and integrate this information into smart data to assist them in intelligent decisions regarding curriculum design and the delivery of customised education and personalised learning experience for students.

Exponential growth and rapid change are the hallmark of the 4IR. This will mean that curriculum will have to update content frequently to match and keep up with the rapid changes of scientific and technological advances. A more responsive curriculum of this sort places an extremely high premium on faculty development and curriculum renewal, as well as the mandate to develop students who can think and reinvent themselves within the changing world. Hence, to enable faculty to maintain expertise based on the latest discoveries and technologies, more proactive and creative forms of faculty development will also be required (Penprase, 2018:224).

The outputs of this thrust are the creation of accessible electronic learning platforms and the implementation of new teaching methods using technology-enhanced and blended learning environments. This would help to create an STI culture in the learning sphere and within the community at large. While some resources are available at existing educational institutions for curriculum development, more funding will be needed to develop a new curriculum for 2030. More collaboration needs to be initiated with global experts and research institutions.

4 The Impact of Covid-19 on the Education for the Future Domain

A novel coronavirus, SARS-CoV-2, referred to as Covid-19, emerged in December 2019 in the central Chinese city of Wuhan and surrounding Hubei province, where more than half of global cases had been initially recorded (Hageman, 2020). Covid-19 has taken everybody by surprise across the globe, especially the education sector including universities and colleges.

Covid-19 resulted in a national lockdown in South Africa on 27 March 2020. With very little time to prepare, teacher and educators had to move to online teaching and assessing as campuses had to be closed. During this time students and staff had to study and work from home, university/college activities became virtual events, and new ways of communicating with students and staff had to be considered. While all this took place spontaneously, it presented challenges to educational institutions in terms of resources available, technical issues such as availability of access to computers and mobile devices and data and electricity as well as other pedagogical challenges as they faced the immediate task of adapting their teaching material to an online modality, while at the same time maintaining the educational purpose and academic integrity of their modules (Robinson & Rusznyak, 2020:1). This move to online teaching and learning (also referred to as emergency remote teaching) took place against a set of conditions that risked exacerbating socio-economic inequalities between students and teachers in South Africa especially those from disadvantaged and rural communities.

While highly innovative and creative solutions were developed and managed very quickly by the education sector in response to the Covid-19 outbreak, there remains enormous creative challenges that lie ahead as we look to transform learning and teaching into something that is stable, sustainable and high quality for both existing and new students in a socially distanced and Covid-reactive world (Parkin & Brown, 2020:18) as this pandemic is expected to remain with us for a long time in the future and even new pandemics are expected in time to come. Institutions, which have solid infrastructure and internal capacity, have managed the transition from traditional classroom teaching to online delivery relatively easily. However, the inequalities and inefficiencies that existed before Covid-19 became very pronounced and may be exacerbated going forward because of social lockdown measures and the physically distanced campus. To avoid compounding existing and new inequalities, we need to build sustainable solutions which are capable of accommodating new ways of doing things and learn from this, so we don't return to a previous unequal and marginalised societies.

Emphasis is now on the leaders of academic institutions to put the appropriate strategies, policies, bylaws, resources and health and safety measures in place. This may require a mindset shift, as they will need to consider how to deal with students who have been under enormous psychological pressure resulting from having to stay at home for a prolonged period and the ever-present fear of contracting the virus while at the same time stressing about completing their education qualifications to seek employment. The challenge here will be of taking an entire university into a completely new way of delivering teaching and learning, and this has meant revisiting the strategy for the future of education. We have learned many lessons which would have normally taken months and probably years of careful experimentation (Parkin & Brown, 2020). Corvid-19 has acted as a driving force towards digital transformation in the education sector. Education institutions were catapulted into adopting 4IR technologies almost instantly without hesitation in an attempt to save the academic year, while at the same time abiding by the lockdown regulations.

It is obvious that the world after Covid-19 will not be the same as the one before, and universities are no exceptions. There will be drastic changes in demographics, local and international alliances, economic direction and ways of working and learning. It is expected that online teaching will become better understood and widespread across the globe and blended learning will become more popular (Parkin & Brown, 2020). The improved collaboration with global alliances could result in changes in the global higher education landscape. Students will seek better educational opportunities rather than searching for jobs, which could become scarce. The subject of entrepreneurship will become more important as stronger international collaboration will be engaged in to benchmark lessons learned. It is anticipated that there will be a big increase in demand for medical and related courses, and new majors will start to emerge in the areas of pandemic management, e-commerce, data mining techniques for health management, foresight and scenario planning. In addition, more research centres on future studies are expected to appear as a result of interest created during Covid-19 (Parkin & Brown, 2020). We have learned from Covid-19 that research teams made up of experts from various disciplines are essential to research and find solutions for global problems.

In a changing world, universities will need to adjust to face new challenges. The impact of this pandemic will be long-lasting, and it will be felt across many sectors. The potential impact on jobs, skills and ability to adopt new technological innovations are already being felt by industries globally. As such, foresight and scenario planning will play a major role in driving future universities and their strategic plans to create a supportive, inclusive and sustainable culture, which will enable us to prepare better for a changing higher education environment.

5 Conclusions and Key Policy Considerations

This chapter focussed on discussing the findings from the second phase of SAForSTI (2019) exercise focussing on the education for the future domain. The education for the future domain proposes an integrated suite of interventions, including skills development for the future, both technical (such as for AI) and soft (such as critical thinking and creative thinking), along with lifelong learning (ECD, ABET and artisan development) through industry-driven capacity development, curriculum transformation (embedding entrepreneurship education and training in all curricula across all disciplines) and also conducting R&D to identify new methodologies for teaching in the future.

South Africa has a serious deficit of research in the STEM education areas. Current efforts are largely in the arts, social sciences and psychology, with some capacity in the neurosciences. More funding for STEM educational research is needed, including on the most relevant and appropriate solutions to build human capacity that is resilient and ready for the 4IR.

While South Africa already has various education-related policies, regulatory frameworks, reports and regulations in place and specifically for the improvement of the education systems and to ensure that it delivers on its constitutional mandate of providing equitable, practicable education to all its citizens and to redress the results of past racially discriminatory laws and practices, thereby creating a foundation for a sustainable and fair society, it is essential that these policies be revisited and updated to be aligned with the requirements of the 4IR and now the impact of Covid-19. For example, entrepreneurship could be embedded as part of the curriculum for all relevant fields, and entrepreneurship courses could be introduced at all high schools in the country.

It is also encouraging to see that while there are still several challenges in the education system in South Africa and in providing the skills required by the economy, the South African government and industries are already busy deliberating and planning for developments and challenges that are expected with 4IR. A high-level Presidential Commission on the Fourth Industrial Revolution has already been established to develop an integrated national response strategy. This will include the facilitation of a skills audit to determine what 4IR skills are already in place and in which industries and what is still needed, what capacity is available to undertake the required training and development, and the monitoring and evaluation of the progress that has already been made. With the impact of Covid-19, it will also be important to investigate alternative multidisciplinary, multimodal, agile education programmes that will produce workers with appropriate skills enabling them to thrive in the workplace of the future and the ability to continue to retrain throughout their lives. In addition, policy-makers, regulators and educators in South Africa will need to play a fundamental role in improving education and training systems, as well as updating labour policy to help those who are displaced from jobs and education programmes as a result of Covid-19 to repurpose their skills or retrain to acquire new skills that match the realities of the Fourth Industrial Revolution (WEF, 2018:

xi). In doing this, South Africa will ensure that the Fourth Industrial Revolution will be an opportunity for all.

It is also recommended that more research and discussions emphasising the negative consequences of 4IR apart from the loss of jobs be undertaken. Such discussions could include the psychological and other health consequences, environmental consequences and energy demands especially in South Africa where severe challenges regarding the supply of electricity are present and especially to rural and disadvantaged communities.

In terms of achieving the vision of the National Development Plan 2030 and in meeting the Sustainable Development Goals, in particular the SDG 4 to "ensure inclusive and equitable quality education and promote lifelong learning opportunities for all", it is evident that South Africa is on the right trajectory, and with political will and commitments in terms of funding and policy implementation, the education for the future looks positive.

Acknowledgements The author wishes to acknowledge the discussions and thoughts of members of the working group that participated in the *STI Foresight 2018 Exercise f*or the STI Domain: Education for the Future. However, none of the members of the working group were available for contributing to this chapter.

References[1]

Andrews, D. (2015). The circular economy, design thinking and education for sustainability. *Local Economy, 30*(3), 305–315.

Businesstech. (2020). Standard Bank boss talks job cuts and branch closures in pursuit of digital transformation. https://businesstech.co.za/news/banking/391631/standard-bank-boss-talks-job-cuts-and-branch-closures-in-pursuit-of-digital-transformation/

Foresight Africa. (2020). *Top priorities for the continent 2020–2030*. Brookings Institution's Africa Growth Initiative (AGI).

Gravett, S. (2019). Industry 4.0 is being taken seriously. https://mg.co.za/article/2019-01-18-00-industry-40-is-being-taken-seriously/

Hageman, J. R. (2020). The coronavirus disease 2019 (COVID-19). *Pediatric Annals, 49*(3), e99–e100. https://doi.org/10.3928/19382359-20200219-01

Khathu, R. (2020). 4IR and the SA education system. https://www.bizcommunity.com/Article/196/371/193213.html

Kirchherr, J., Reike, D., & Hekkert, M. (2017). *Conceptualizing the circular economy: An analysis of 114 definitions*. Innovation Studies Group, Copernicus Institute of Sustainable Development, Utrecht University.

Mezied, A. A. (2020). What role will education play in the Fourth Industrial Revolution? https://www.weforum.org/agenda/2016/01/what-role-will-education-play-in-the-fourth-industrial-revolution/

National Planning Commission (NPC). (2020) Draft digital futures: South Africa's digital readiness for the fourth Industrial Revolution. https://www.tralac.org/documents/resources/by-country/south-africa/3902-draft-digital-futures-south-africas-digital-readiness-for-the-fourth-industrial-revolution-npc-july-2020/file.html

[1] (Below are examples to indicate formatting of references).

NPC Draft Report. (2020). Analysis of PSET Trends, focuses on post-school higher and vocational education. https://www.nationalplanningcommission.org.za/assets/Documents/Post%20School%20Education%20and%20Training%20(PSET)%20Trends%20towards%202030.pdf

NSTF. (2018). STEM Education with a focus on Maths, the ultimate building block of the sciences, engineering and technology. Concept Paper. http://www.nstf.org.za/wp-content/uploads/2018/07/Concept-document.pdf

Parkin, D., & Brown, G. (2020). The advance HE creating socially distanced campuses and education project: Final Capstone Report. https://www.advance-he.ac.uk/knowledge-hub/creating-socially-distanced-campuses-and-education-project-final-capstone-report

Patterson, A. (2013). The top 10 Nelson Mandela quotes on education. https://writerswrite.co.za/rest-in-peace-nelson-mandela/

Penprase, B. E. (2018). The Fourth Industrial Revolution and higher education. In N. W. Gleason (Ed.), *Higher education in the era of the Fourth Industrial Revolution* (pp. 207–229). https://doi.org/10.1007/978-981-13-0194-0_9

Reaves, J. (2019). 21st-century skills and the Fourth Industrial Revolution: A critical future role for online education. *International Journal on Innovations in Online Education, 3*(1) https://onlineinnovationsjournal.com/download/4f6fa2666b8cd098.pdf

Republic of South Africa. (1996). *Constitution of the Republic of South Africa, Act 108 of 1996.* Government Printer.

Robinson, M., & Rusznyak, L. (2020). Learning to teach without school-based experience: Conundrums and possibilities in a South African context. *Journal of Education for Teaching, 46*(4), 517–527. https://doi.org/10.1080/02607476.2020.1800408

Shay, S. (2017). Educational investment towards the ideal future: South Africa's strategic choices. *South African Journal of Science, 113*(1/2), 1–6. https://doi.org/10.17159/sajs.2017/20160227

South Africa Foresight Exercise for Science, Technology and Innovation 2030 (SAForSTI). (2019). NACI Synthesis Report, Department of Science and Innovation. Republic of South Africa.

UNESCO. (2019). *4th Global report on Adult learning and education. Leave no one behind: Participation, equity and inclusion*. UNESCO Institute for Lifelong Learning.

UNICEF. (2019). The state of the world's children 2019: Children, food and nutrition.

Van der Berg, S., Gustafsson, M., & Malindi, K. (2020). Education and skills for the economy and links to labour markets in South Africa. Report for the Economic task team of the National Planning Committee. https://resep.sun.ac.za/wp-content/uploads/2020/06/Education-and-skills-for-the-economy-and-links-to-labour-markets-in-South-Africa.pdf.

VNR. (2019). Voluntary National Review Report 2019. South Africa's implementation of the 2030 Agenda for Sustainable Development. https://sustainabledevelopment.un.org/content/documents/23402RSA_Voluntary_National_Review_Report__9_July_2019.pdf

WEF. (2016). *The future of jobs: Employment, skills and workforce strategy for the Fourth Industrial Revolution*. World Economic Forum. http://www3.weforum.org/docs/WEF_Future_of_Jobs.pdf

WEF. (2018). *The future of jobs report 2018*. Centre for the New Economy and Society. http://www3.weforum.org/docs/WEF_Future_of_Jobs_2018.pdf

WEF White Paper. (2018). *Circular economy in cities evolving the model for a sustainable urban future*. World Economic Forum. http://www3.weforum.org/docs/White_paper_Circular_Economy_in_Cities_report_2018.pdf

Sustainable Energy

Mkhulu Mathe

Abstract The United Nations Environment Programme has Sustainable Development Goal 7, which is communicated as energy access for all by 2030. A summary of a chapter on sustainable energy from the SA Foresight Exercise for STI 2030 opens the scope for the chapter. Several types of sustainable energy are discussed with a special focus on solar and wind due to their different relevance to Africa. The COVID-19 pandemic and climate change impact on sustainable energy rollout towards equal access for all is presented. The updated Integrated Resource Plan (IRP) 2019 is presented to show the 2030 energy mix of South Africa. Elements critical for sustainable technologies are presented. The energy innovation system showed people, catalysts and the innovation process as key elements with four primary challenges of regulatory risk, financing, enabling infrastructure and market access, plus the social and cultural. All this leads to the future of society.

1 Preface: Summary of the Foresight Chapter on Sustainable Energy

The 2019 synthesis report (Synthesis Report, 2019) on South Africa Foresight Exercise for Science, Technology and Innovation (STI) 2030 has a chapter on sustainable energy as summarised. The chapter gave an overview indicative that sustainable energy technologies have the potential to leapfrog old technologies. Sustainable energy technologies will create decentralised energy resources resulting in the reduction of the dependency of the national grid against the ongoing load shedding problems experienced in South Africa (SA). The sustainable energy focus is on clean, affordable and renewable energy solutions, energy efficiency and distributed generation. The issues of climate change impact, COVID-19 pandemic and

M. Mathe (✉)
Deanery, Institute for Catalysis and Energy Solutions (ICES), University of South Africa, Pretoria, South Africa
e-mail: mathemk@unisa.ac.za

O. Saritas et al. (eds.), *21st Century Foresight*, Science, Technology and Innovation Studies, https://doi.org/10.1007/978-3-031-69557-5_10

water scarcity challenges for Africa are issues and drivers noted for the introduction of renewables, solar and wind. Other problems affecting the energy access disparities are illegal connections coupled with a culture of non-payment of electricity bills.

The status of the South African research capacity and knowledge generation including the quality of publications was assessed, such will not be covered in the chapter. Of note, the significance is that the overall scientific output stays low. The South African R&D spent as a percentage of GDP is still below 1%. South Africa as a BRICS member is positioned in the middle for her contribution according to both the GDP and the population size. Across the energy-related Science, Engineering and Technology (SET) disciplines, South Africa contributes well to the areas of materials science, chemistry, biochemistry and chemical engineering. Generally, South Africa has an opportunity to increase investment in the expansion of the research capacity.

Finally, four STI thrusts for the energy domain were named which will be expanded during the second phase. The thrusts are presented without elaboration:

- Clean, affordable and sustainable energy for all.
- Renewable energy sources and technologies.
- Energy efficiency solutions for industry and household use.
- Distributed energy generation and storage.

The conclusions highlighted the ongoing South African energy crisis, with its impact on the broader society; this can be addressed for the growing population by bold investments for sustainable economic growth.

2 Introduction

Over 640 million Africans have no access to energy, corresponding to an electricity access rate for African countries of just over 40%, the lowest in the world (https://www.afdb.org/en/the-high-5/light-up-and-power-africa-%E2%80%93-a-new-deal-on-energy-for-africa, n.d.). A sustainable energy future is central to the energy sector with energy and carbon efficiency central to this goal.

The six pillars of sustainable energy solutions are identified as cures for the fossil fuel problem: (1) better efficiency, (2) better cost-effectiveness, (3) better resources use, (4) better design and analysis, (5) better energy security and (6) better environment. Alternative fuels are needed for sustainable energy development, regarded as a cleaner means of chemical energy storage concerning fossil fuels.

In the article, "A New Industrial Revolution for a Sustainable Energy Future", Majumdar (Majumdar, 2013) provides a techno-economic snapshot of the current energy landscape and names R&D challenges with opportunities related to materials science engineering for the creation of the sustainable "new industrial revolution". Humanity continues to receive help from 250 years of the industrial revolution with a question of how electricity access will change for a global population of 10 billion in the next 100 years. The answer could be in the energy systems of the future, largely for stationary power and transportation. The future envisaged stationary power system will differ in infrastructure for generation, transmission and distribution as it will have increased use of natural gas for heating and manufacturing. For transportation, other less popular modes like electric vehicles and hydrogen propulsion will come into play. Shale gas will be a welcome choice where available with low extraction costs.

With renewables, wind and solar costs per kilowatt-hour continue to compete with that of coal. The intermittency problem will be countered by the inclusion of storage to continuously reduce costs. Permanent magnets used in turbines depend on rare-earth elements from China, neodymium and dysprosium posing a risk to the security of supply. This material issue opened opportunities for rare-earth-free magnets under the ARPA-E programme. Iron boride magnets discovered in the 1970s and early 1980s are finding applications as alternatives. Materials science thus contributed to the replacement of rare earths.

Solar continues to be reported as the fastest downward trending technology in electricity cost without precedent in the energy business. Literature suggests that there are increasing opportunities to lower the cost especially as shown in the residential solar photovoltaic market. The efficiency improvement of 20% and higher in the material work on copper indium gallium selenide (CIGS), cadmium telluride (CdTe) and multi-crystalline silicon will be a notable change for solar.

To successfully dispatch electricity in a predictable manner, storage will play a vital role as it will allow for the integration of wind and solar with the grid. The material issues on power electronics are still a challenge. Pumped storage stays the popular storage option with a 90% round trip efficiency. Future grids will in time move from centralised architectures to distributed ones.

The role and importance of water in energy production are well understood. Water trends in sustainable energy are reported by Ku and Shapiro (Ku & Shapiro, 2012) in the energy-water nexus publication. The review illustrates options for managing water requirements related to power generation. Their work focus on thermal systems will be highlighted. Future water scarcity challenges have been experienced globally with changed and changing weather patterns attributed to climate change. Globally, electricity demands are projected to grow with population growth.

The publication highlights opportunities for the materials research community to address the R&D needs at the interface of energy and water treatment, i.e. water use for electricity generation. The article concludes with a proposal for an evaluation framework for different technology options for water reduction in power production from the conserved water cost to the levelised cost of electricity for improved water use.

The work of Dincer and Zamfirescu (Fig. 1) introduces the relation between energy conversation and sustainability as presented in Fig. 1. In the figure, energy resources and their use are intimately related to sustainable development. In societies that reach or try to attain sustainable development, much effort must be devoted not only to discovering sustainable energy sources but also to increasing the energy efficiencies of processes using such resources.

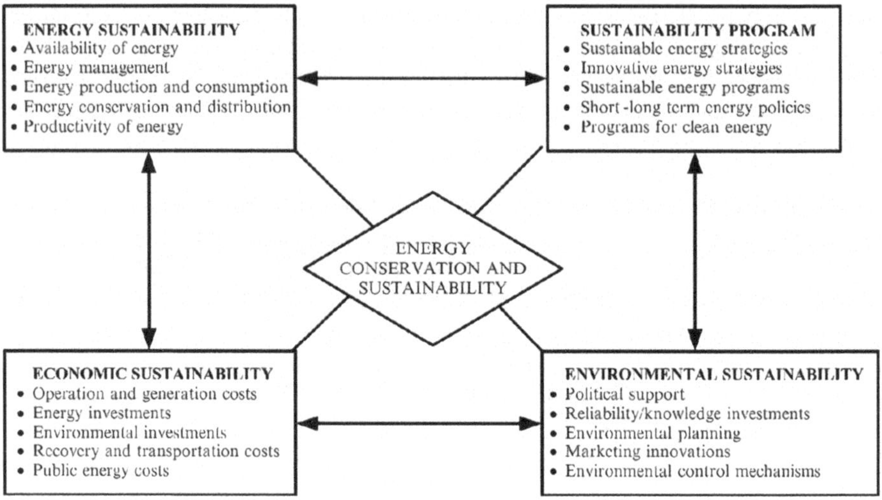

Fig. 1 Linkages between energy conservation and sustainable development

The above figure shows that energy sustainability is possible with guaranteed productivity of energy and its availability. The sustainability programme supports clean energy programmes. For economic sustainability, affordable costs are necessary to ensure access for all. Political support is essential for environmental sustainability. The linkages above show interdependencies.

The United Nations Sustainable Development Goal 7 is affordable and clean energy. This goal is divided into the following environmental targets (https://sdgs.un.org/goals/goal7, n.d.):

Target 7.1: By 2030, ensure universal access to affordable, reliable and modern energy services.

Target 7.2: By 2030, increase substantially the share of renewable energy in the global energy mix.

Target 7.3: By 2030, double the global rate of improvement in energy efficiency.

Target 7. a: By 2030, enhance international cooperation to ease access to clean energy research and technology, including renewable energy, energy efficiency and advanced and cleaner fossil fuel technology, and promote investment in energy infrastructure and clean energy technology.

Target 7. b: By 2030, expand infrastructure and upgrade technology for supplying modern and sustainable energy services for all in developing countries, least developed countries, small island developing states and landlocked developing countries, following their respective programmes of support.

The elimination of energy poverty is linked to the above targets which will contribute to sustainable energy development, in a low carbon future. Some factors to be considered for the transition to long-term sustainable energy for humanity are as follows:

- Low carbon future R&D.
- Behavioural change—people can be more careful in their energy use.
- Modelling in understanding the options for a sustainable future.
- Role of regions in moving the world towards a sustainable energy future.
- Reducing energy consumption through efficiency and conservation supplies the most important and economically strategic projects for translation to a sustainable energy system.

According to Africa Energy and Economy Review and Outlook. (2021–2022), Sustainable energy is defined by many descriptors as: "the use of any type of energy that can meet demands without putting the resources in danger of running out". Sustainable resources are known to cause minimal environmental depletion. Known benefits from sustainable resources are in the form of healthy, safe, long-lasting and self-replenishing energy sources. Sustainable sources have ecological benefits as they pose a minimal environmental risk and are reusable with quick replenishment.

3 Sustainable Energy Can Sustain Humanity for the Long Term

Households can use sustainable renewable/green energy sources like wind, solar, geothermal and hydropower. Wind turbines and solar panels can be used all over the world to provide wind and solar power.

Some of the reasons why we need sustainable energy are the following:

1. **Sustainable energy can reduce (and eventually eliminate) our reliance on fossil fuels**. Current unsustainable energy sources can be replaced globally by harnessing some form of sustainable energy, which are safer healthier options.
2. **Renewable energy will never deplete or run out**. Resources like wind energy, solar power, geothermal energy and hydropower are infinite resources.
3. **Sustainable energy does not harm the environment and can help improve public health**. Renewable energy resources benefit climate change relief.
4. **Renewable energy will eventually be the more affordable option**. Wind and solar power solutions continue to be cheaper by the year. The trend points to eventual competitiveness with other non-renewable resources.

The energy consumer has a future option of switching to renewable and green energy utility companies, to enjoy clean sustainable energy.

3.1 Types of Sustainable Energy

Sustainable energy sources can be used in homes and industries with their environmental benefit.

3.1.1 Solar Energy

Solar energy as the best form of sustainable energy is seen as the future of energy. The cost of solar panels continues to drop as installation increases in home and commercial properties. Solar energy benefits both man and plants. The solar radiation of Africa positions her favourably to have an accelerated uptake/installation.

3.1.2 Wind Energy

The wind is a naturally available sustainable energy source and finds application in many ways and places. The wind energy generators/utility companies can be sold to power homes and commercial properties.

3.1.3 Geothermal Energy

Geothermal energy is harnessed in areas of high seismic activity prone to volcanoes. Geothermal power stations use the heat coming from inside the earth to generate electricity. Geothermal energy is geographically specific.

3.1.4 Ocean Energy

Water covers 70% of the earth presenting ocean energy as a resource able to generate superior power compared to other sources of energy. Ocean energy can be harnessed in three ways, i.e. wave, tidal or ocean thermal energy conversion (OTEC).

3.1.5 Biomass Energy

Biomass energy is completely renewable without the production of harmful gases. Biomass energy is generated by the burning of wood, timber, landfills and municipal and agricultural waste.

3.1.6 Hydroelectric Power

Water from the rivers and waterfalls is captured to turn turbines to generate power, known as hydroelectric power. The hydroelectric power is transmitted through power lines within the country of origin and to export markets. Water-stressed countries benefit from the importation of hydroelectric power.

3.1.7 Hydrogen Energy

Hydrogen is emerging as future fuel as incorporated into the national energy strategies of several countries. South Africa has a Hydrogen South Africa (HySA) strategy to exploit the country's competitiveness in the platinum group metals. The HySA programme is divided into the three phases of skills development, the establishment of the infrastructure and creation of intellectual property (IP); phase two is the pilot and demonstration of technologies, while the third phase is the creation of industry. Hydrogen is delivered as a clean fuel from compounds that contain it. Hydrogen is used in a fuel cell to give water, electricity and heat. Hydrogen and fuel cells find application in transportation, commercial, industrial, residential and portable sectors.

Hydrogen and fuel cells can provide energy for use in diverse applications, including distributed or combined heat and power; backup power; systems for storing and enabling renewable energy; portable power; auxiliary power for trucks, aircraft, rail and ships; speciality vehicles such as forklifts; and passenger and freight vehicles including cars, trucks and buses.

Sustainable cities need dependable, affordable and renewable sustainable energy to function. Cities can change their energy profile by influencing legislation and regulations relating to urban form, energy efficiency, standards, public transport, energy distribution and tariffs.

4 COVID-19 and Climate Change

In 2020, South Africa experienced one of the region's steepest declines in power demand both due to the COVID-19 pandemic and to scheduled load shedding. South Africa accounts for about 50% of sub-Saharan Africa's (SSA's) total power demand and experienced declines exacerbated by the country's worsening power supply crisis. Power demand rebounded in 2021 and will continue to grow in 2022, but at a slower pace. However, South African power demand recovered to pre-pandemic levels in 2021, where demand increased 8% year on year thanks to the swift recovery in the industrial sector. Coal and oil are two of the major sources that produce a large amount of carbon dioxide in the air. This led to an increase in global warming.

Power demand across SSA in 2021 grew by over 5%, following a 4% drop in 2020 due to the disruptions and economic slowdown caused by the COVID-19 pandemic. The commercial and industrial segments led the demand recovery which, respectively, grew at around 4% and 8%. Residential demand also rebounded in 2021 despite the 2020 weak growth, unlike the other demand segments. The year 2022 is projected for an overall power demand growth, albeit at a slower pace than 2021 with the assumption of little economic impact by the waves of COVID-19 variants. The per capita power demand across SSA remains among the lowest in worldwide levels with an average of less than 500 kWh.

Many sub-Saharan countries rely on hydropower as an important source of their power generation. Climate-related risks such as drought, with important consequences for economies, make hydropower susceptible with Zambia being a case in point. Multilateral financing institutions' investment in renewable solar projects provides a popular energy diversification option. South Africa and Botswana continue to hold the largest shares of coal-fired generation in their national power mix unlike the less coal-intense SSA's overall energy mix. Energy transition objectives are increasingly implemented across SSA countries.

Eskom, the South African utility, has problems with decarbonization, and the Department of Environment recently denied its request for a pollution exemption for 16 GW of coal-fired capacity, exposing Eskom to the risk that its power plants will be forced to shut down even as the country suffers from chronic load shedding. Recently at COP-26, South Africa signed an initial pledge of an $8.5 billion finance package with France, Germany, the United Kingdom, the United States and the EU to help the country to decarbonise and take advantage of new energy transition economic opportunities such as green hydrogen and electric vehicles.

The lifting and reduction of COVID-19 restrictions enabled the implementation of renewable projects gradually in 2021 across SSA. COVID-19 affected the global supply chain resulting in chronic bottlenecks leading to higher equipment costs plus supply shortages for solar PV and onshore wind projects.

5 Africa in Focus: A Time of Transition to an Uncertain Destination

The emergence of solar power as the common clean energy of choice is empowering Africa as endowed by its generous solar radiation footprint. Small-scale installations grow on farms and individual houses. Large solar power plants like Khi Solar One in South Africa are examples of initiatives led by energy ministries.

Wind as the other renewable source of choice for many coastal African regions has seen the Eastern Cape in South Africa becoming home to wind turbine farms with all year electricity generation capacity. SSA can exploit hydro-powered electricity plants from Kariba and Cahora Bassa along the Zambezi River with respective outputs of 1626 and 2075 megawatts at full capacity.

One advantage that Africa has over the rest of the world is its lower number of fossil fuel energy production facilities. The year 2021 was a challenging year for sub-Saharan Africa's energy and economy sectors, complicated by the COVID-19 pandemic. COVID-19 variants continue to disrupt the recovery from the historic 2020 recession. These complications put SSA on a multitude of uncertainties without a clear line of sight ahead. The 2021 rebound in power demand is expected to continue with slow growth in 2022. The uptake of renewables continues to gain momentum away from oil and coal towards the diversification and decarbonization of energy sources.

Between 2016 and 2019, 43% of climate finance went to Asia, while only 26% went to Africa according to the Organisation of Economic Co-operation and Development (OECD).

South Africa's sustainable energy plan according to the 2019 IRP (Integrated Resource Plan (IRP), 2019) to 2030 for the grid is shown in Fig. 2.

5.1 *Sustainable Energy Innovation*

The Sustainable Energy Fund for Africa (SEFA) (https://www.afdb.org/en/topics-and-sectors/initiatives-partnerships/sustainable-energy-fund-for-africa, n.d.) was established in 2011 through a partnership with the government of Denmark and has since received contributions from the governments of the United States, United Kingdom, Italy, Norway, Spain, Sweden, Nordic Development Fund and Germany. SEFA is a multi-donor special fund managed by the African Development Bank. It

	Coal	Co,l (Decommissioning	Nuclear	Hydro	Stora1e	PV	Wind	CSP	Gas & Diesel	Other Distributed Generation, CoGen, Biomass, Landfill)
Current Base	37149		1860	2100	2 912	1474	1980	300	3830	499
2019	2 155						244	300		Allocation to the extent of the short-term capacity and energy gap.
2020	1 433					114	300			
2021	1433					300	818			
2022	711				S13	400 1000	1600			
2023	750					1000	1600			500
2024			1860				1600		1000	500
2025						1000	1600			500
2026							1600			500
2027	750						1600		2000	500
:ZO:Z8						1000	1600			500
2029					1575	1000	1600			500
2030				2500		1000	1600			500
TOTAL INSTALLED CAPACITY by 2030(MW)		33364	1860	4600	5000	8288	17742	600	6380	
% Total Installed capacity (% of MW)	43		2 36	5.84	6.35	10 52	22.53	0 76	8 1	
% Annual Energy Contribution (% of MWh)	58.8		4.5	8.4	1.2*	6.3	17.8	0 6	1.3	

Installed Capacity
Committed/ Already Contracted Capacity
Decommissioned
New Additional Capacity
Extension of Koeberg Plant Design Life
Includes Distributed Generation Capacity for own use

2030 Coal Installed Capacity is less capacity decommissioned between the years 2020 and 2030
Koeberg power station rated/ installed capacity will revert to 1926 MW (original design capacity) following design life extension work.
Other/ Distributed generation includes all generation facilities in circumstances in which the facility is operated solely to supply electricity to an end-use customer,
within the same property as the facility
Short term capacity gap is estimated at 2000 MW

Fig. 2 South Africa's sustainable energy plan till 2030

provides catalytic finance to unlock private sector investments in renewable energy and energy efficiency.

6 Water Scarcity in Energy

Electricity is often taken for granted, just like the water supply. The goal 6 of the United Nation's Sustainable Development Goals recognises that access to clean, affordable and safe drinking water is a fundamental human right. Sadly, in Africa, one in three Africans face water scarcity, and approximately 400 million people in sub-Saharan Africa lack it.

The 2019 report by the World Resources Institute (WRI) (Mason et al., 2019) revealed that addressing climate change and poor management of water resources and services is paramount to tackling Africa's water stress. The equatorial and tropical regions around the Democratic Republic of the Congo also enjoy a significant surface area with noticeably less water risk than their continental neighbours (Fig. 3).

7 Energy-Critical Elements for Sustainable Development

Science decision-makers continue to formulate risk mitigation supply policies for renewable energy technologies for the complexity of crucial factors of unanticipated market response to policy on "energy-critical elements" (ECEs) (Hurd et al., 2012). The term ECEs was first coined by a joint committee of the American Physical Society (APS) and Materials Research Society (MRS) in 2009. This was informed by the emerging relationships on abundance, markets and geopolitics of rare earths that could constrain supply.

The United States named 14 critical elements: cerium, cobalt, dysprosium, europium, gallium, indium, lanthanum, lithium, neodymium, praseodymium, samarium, tellurium, terbium and yttrium. These were selected based on supply risk factors due to a small global market and lack of diversified supply plus market complexities of risks of co-production and geopolitics (Fig. 4).

The publication of the report attracted the attention of China which accounted for more than 95% of the worldwide production of rare-earth oxides at the time. The export price of neodymium, used in high-strength magnets for windmills and electric vehicles, nearly tripled in 2011 from July 2010.

ECEs, as defined in the report, is not restricted to rare earths; they include broadly chemical elements necessary for emerging transformative energy technologies whose supply risk could limit research, development and deployment of such. ECE lists devolve over time as from the 1940s when nuclear fission was an emerging energy technology. The 2050s will see the new list of ECE appear as some elements fall off the list. The ECE list is important to ensure stable supplies of elements needed to achieve global sustainability in energy. The list speaks to the international

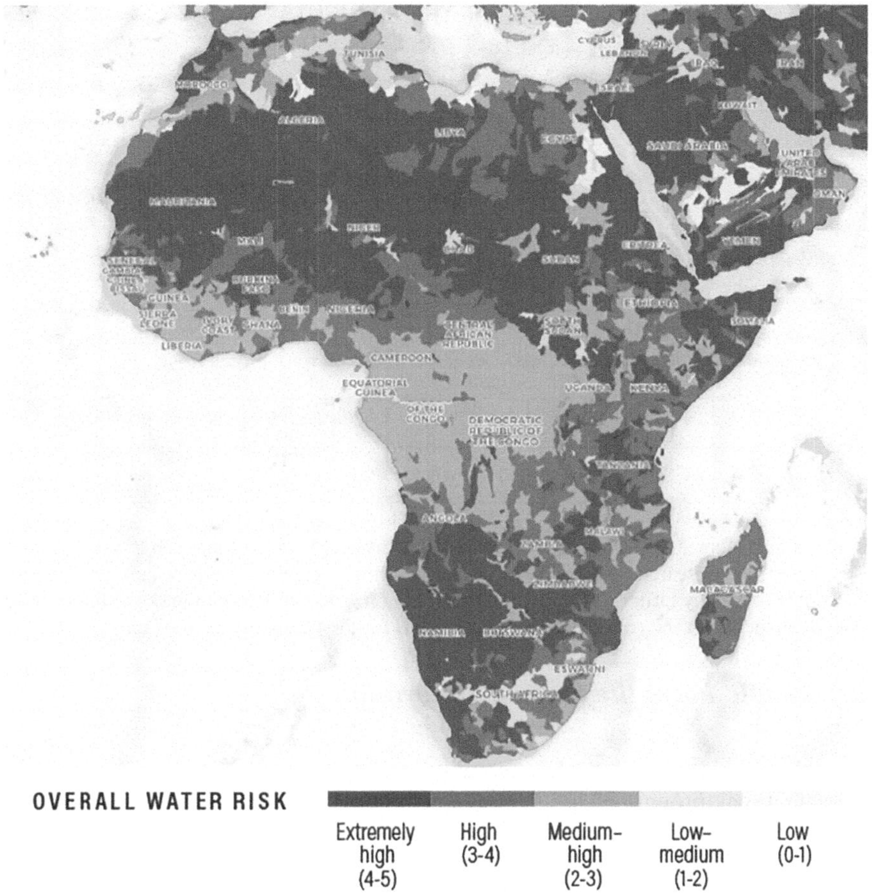

Fig. 3 Africa faces some of the highest water risks in the world. Source: "Climate Change Is Hurting Africa's Water Sector, but Investing in Water Can Pay Off", World Resources Institute, 2019 (Holtz & Golubski, 2021)

research community's R&D effort on ECEs with possible substitutes and recycling programmes.

8 The Energy Innovation System: The Future of Society—Next Chapter

According to the World Economic Forum (2018), people, catalysts and the innovation process are key elements of the sustainable energy innovation system. The understanding of key processes showed components of a healthy energy innovation system and roadblocks to accelerated innovation in sustainable energy.

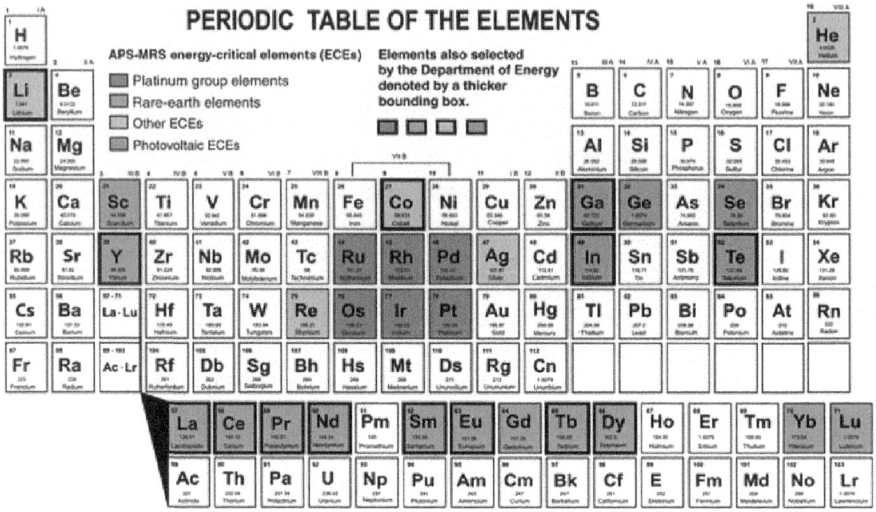

Fig. 4 Critical elements are chosen by the American Physical Society. American Physical Society (APS)-Materials Research Society (MRS) energy-critical element study panel and the US Department of Energy Office of Energy Policy. Selection criteria differed in the two studies, leading to 29 elements for the APS-MRS and 14 elements for the US Department of Energy

8.1 People Form the Core of Innovation

Inventors and researchers explore, combine and test new and existing technologies to create breakthrough ideas. Politicians function as influencers, affecting the enabling frameworks by enacting regulatory policies or dedicating public money to innovation programmes, while scientists and academics fill the sustainable energy pipeline through basic and applied research, development and demonstration.

Within the innovation ecosystem, entrepreneurs often bear notable risks in bringing innovation to market through startups, which are often spin-offs of projects by universities, research institutes or national labs. They work together with private and institutional investors to raise capital to fund progress.

8.2 Innovation Process

Figure 5 presents the innovation process that encompasses basic research, applied research, development, demonstration, deployment and upscaling including the Technical Readiness Level (TRL) model as articulated in the report (World Economic Forum, 2018).

The key elements of the energy innovation system are people, the innovation process and the catalysts building the system's enabling framework.

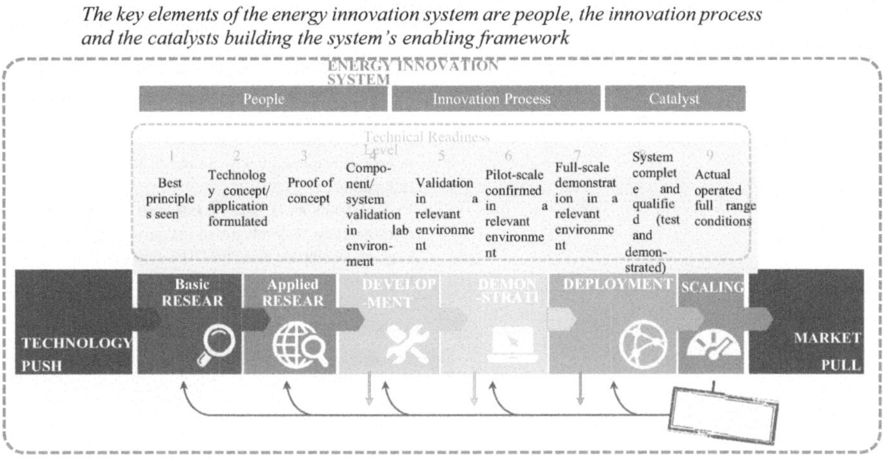

Fig. 5 Energy innovation system, innovation process and TRL 12. Source: World Economic Forum

Figure 6 summarises the key barriers to bringing modern technologies and solutions to market: regulatory risks, access to financing, lack of enabling infrastructure and market access and social and cultural challenges.

Relatedly, the main catalysts that can help overcome these barriers in an enabling framework for innovation are the following:

1. Social and industrial trends influencing consumer demands.
2. A positive climate for overall innovation13.
3. Regulatory policies.
4. Public funding programmes.
5. Multistakeholder collaboration.

Systemic barriers to sustainable energy innovation systems fall into four primary challenges.

9 Innovation Alliances

Innovation alliances can play a significant role in accelerating energy innovation, and their potential is far from fully exploited as presented below.

9.1 Membership

Successful alliances need to involve "the right players" as informed by the alliance's purpose and targets with different focus areas. For the RD&D focus, the alliance will receive help from a strong heterogeneous membership that might include corporates, startups, research institutes and universities.

Fig. 6 Barriers to the energy innovation system. Source: World Economic Forum

Regulatory risks

- Lack of credibility in policy frameworks creates uncertainty about future policy stringency, timing, nature, and durability
- Inability for institutional investors to depreciate investments more quickly in sustainable energy innovation
- Lack of flexibility in policy framework hampers testing of new technologies in real environment
- Lack of perceptible price for GHG emissions in many jurisdictions
- Lack of standard metrics to evaluate sustainability at a circular economy level

Enabling infrastructure and market access

- System inertia and energy security concerns limit the integration of innovative technologies, such as zero emission generation and demand response in existing networks and markets
- Lack of standardization for better integration and faster scale-up of innovative hardware and digital solutions
- Non-uniform pace of innovation in different components of the system solutions, e.g., electric vehicles and charging infrastructures
- Cyber-security risks and privacy concerns of 4IR solutions are less understood

Financing

- Policy framework still does not make business plans profitable for many technologies
- Financial valley of death: Lack of patient risk capital available in the sustainable energy innovation system
- Technical risks: Development, deployment, and system integration costs exceed available budget
- Lack of public money available for demonstration projects
- Lack of international funds to invest in international flagship projects

Social and cultural

- Lack of sufficient public-private partnerships
- Pre-competitive long-term collaborations in RD&D are still rare
- Consumer uptake of sustainable energy products is still not satisfactory although clean and economic solutions exist
- Key decision-makers appear to lack knowledge about the crucial importance of speeding up innovation

9.2 Involvement of Key Decision-Makers

Key decision-makers can function as champions of the alliance, such as the ministers in Mission Innovation or the CEOs involved in the Oil and Gas Climate Initiative.

9.3 Impact Orientation

The importance of communication with a clear and transparent purpose contributes to the success of the alliance. The purpose communicated will have defined targets with goals and progress milestones.

A more symbiotic relationship for the public and private sector collaboration around TRL 6–8 is essential for a "sweet spot" focus on technology innovation.

9.4 Learning from the Semiconducting Industry

The Semiconductor Research Cooperation (SRC) and SEMATECH (Semiconductor Manufacturing Technology) successful collaborative R&D approach is an ideal model to be used for a sustainable innovation system.

The US semiconductor industry was revitalised by SEMATECH founded in 1988. The collaborative groups were made up of universities and national labs.

The US DoE adopted the SEMATECH success for their SunShot programme, and the EU Fuel Cells and Hydrogen Joint Undertaking (FCH JU) followed that initiative.

9.5 Examples of Good Practices in Innovation Alliances

i. Mission Innovation (MI).

MI with 22 countries and the EU among its members, including the world's leading economies, is a public alliance focused on the acceleration of global sustainable energy innovation by doubling early-stage RD&D investments for this sector until 2021. MI stands for 70% of the global population, including the five most populous countries, with over 80% of clean energy RD&D budgets.

i. Oil and Gas Climate Initiative (OGCI).

The OGCI is a voluntary CEO-led alliance, founded in 2014 often leading oil and gas companies that collaborate to reduce greenhouse gas emissions by jointly investing in research projects and combining research activities.

ii. China Energy Storage Alliance (CNESA).

The CNESA is China's first professional organization to promote energy storage in China, while remaining technology-neutral. It was founded in 2009. Membership includes various important energy storage companies, power sector companies and national as well as international experts focused on the development of industrial policies.

iii. iv. Energy Web Foundation (EWF).

The EWF is a global non-profit organization founded in 2017 and focuses on the demonstration and market release of early blockchain applications.

9.6 Mission Programmes

"Mission programmes" as defined (Foray et al.) denote a systemic, bold and long-term approach by governments to develop or deploy a certain technology area by aligning policies, public RD&D programmes and public-private collaboration with a concrete aim ("mission"). Both the Japanese hydrogen (1908s) and Danish wind programmes (the 1990s) are excellent examples of mission programmes.

High-level criteria for mission programmes design amongst others include the following:

- Establishment of intensive public-private partnership from the outset.
- Definition of clear targets (mission) and a comprehensive road map for the partnership.
- Employment of several different technical solutions for the mission and fund RD&D in public research as well as in the private sector.
- Ongoing address of found roadblocks through tailored RD&D programmes.

Global initiatives like the hydrogen economy, GHG free transport, ocean energy, etc. are examples of technology areas that fit the mission programme's strategy. Normally, mission programmes are big national initiatives and could become global imperatives as learned by the race to the vaccine from COVID-19.

10 Opportunities for Step Changes to Accelerate Sustainable Energy Innovation

Conditions for a successful step change needed in energy innovation require effective regulatory policies and public innovation programmes plus innovation alliances. Examples of good practices for exploration are listed.

10.1 Create Institutions for Energy Innovation

An institutional approach takes a systemic perspective of the energy system and encourages strong collaboration among many stakeholders. The US, ARPA-E agency and the EU's KIC-InnoEnergy programmes are such good examples of institutional approaches to energy innovation.

With these agencies, seed-funded by the public sector, they have managed to spin off several companies, forge industry partnerships and raise substantial private capital as well.

It is a worthy consideration to replicate such in different countries depending on national circumstances that may require customization of rules and procedures accordingly.

Highly qualified government officials as investment stewards are key to the success of these institutions for energy innovation in their "high-risk, high-reward" projects.

10.1.1 Establish an Independent International Fund to Finance Energy Technology Projects, Blending Public and Private Sources of Capital

Establishing a global entity could also help show high-potential investment opportunities across different technology areas globally and promote the effective allocation of RD&D funds to solve common problems too large, costly or risky for any one nation or company.

Managing how to pool RD&D funding through an international and independently from interested countries, including from the private sector and philanthropists, will have a significant impact when well designed and supported by a critical coalition of interested stakeholders.

Capital intensive innovation areas in their precompetitive developmental stages will immensely meet the RD&D needs with pooled funds and resources.

Strategic diversion of a part of domestic clean energy RD&D funding to an international fund pool can provide countries with better market access to innovative technologies and solutions. This will result in collaboration on innovation agendas in specific technology areas and offer a platform for private sector investment in sustainable energy innovation.

10.1.2 Develop Instruments for Public-Private co-Investment

Co-investment with venture capitalists can target recipients with a higher commercial success chance, and the VC employs their experience and knowledge to help choose RD&D grant recipients, particularly in the pilot stages of innovation. Ensure that only top-quality companies receive funding through a transparent procedure.

Support semi-automatic grant co-financing for investments by venture funds specialising in science- and RD&D-based companies. Pre-qualify for the VC funds to help them access the co-financing automatically without lengthy separate applications for each of their new investee companies. The investment burden on the public sector in the proposed "co-investment" could depend on the maturity stage of the investee company.

If intelligently designed, the instrument would stimulate more private money into breakthrough energy projects and significantly improve the success rate or overall impact of public RD&D grants.

10.1.3 co-Define Energy Technology Road Maps tThrough Public-Private Collaboration

Collaboratively develop technology road maps to better position and enable innovation in areas where a product/service involves different components and supply chains, e.g. electric mobility, which requires simultaneous innovation and standardization across electric vehicles, charging infrastructure, battery modules, inverters, etc. for faster scale-up of low or zero-emission transport.

10.1.4 Mainstream Public Procurement of Pre-Commercial Energy Innovation

Strategically designed public procurement programmes can drive innovation with the potential to scale up the deployment of proven technologies and services as well as stimulate RD&D for innovative solutions.

The proven technologies and solutions, used as a "demand-side" sustainable public procurement policies, have helped in the diffusion of energy-efficient solutions for lighting, buildings and heating and low-emission transport.

10.2 Super-Transparency of Government RD&D Spending

Global acceleration of innovation in sustainable energy is a shared concern. National and multilateral initiatives across countries continue to express an urgency to promote innovation in transformative sustainable energy solutions. More transparency on sustainable energy RD&D expenditure is a necessary condition to check grand initiatives. Transparency will bring awareness to companies and innovative start-ups of the exploitable multiple government funds and programmes.

Acknowledgements The author acknowledges the University of South Africa, College of Science, Engineering and Technology (CSET) for the opportunity to contribute, NACI to participate in the discussions and Russia for the guidance in the synthesis.

References

Africa Energy and Economy Review and Outlook. (2021–2022). A time of transition to an uncertain destination. https://ihsmarkit.com/research-analysis/africa-energy-and-economy-review-and-outlook-2022.html

Holtz, L., & Golubski, C. (2021). Addressing Africa's extreme water insecurity. https://www.brookings.edu/blog/africa-in-focus/2021/07/23/addressing-africas-extreme-water-insecurity/

https://sdgs.un.org/goals/goal7

https://www.afdb.org/en/the-high-5/light-up-and-power-africa-%E2%80%93-a-new-deal-on-energy-for-africa#:~:text=Over%20640%20million%20Africans%20have,the%20lowest%20in%20the%20world

https://www.afdb.org/en/topics-and-sectors/initiatives-partnerships/sustainable-energy-fund-for-africa

Hurd, A. J., Kelley, R. L., Eggert, R. G., & Lee, M. H. (2012). Energy-critical elements for sustainable development @2012 Materials Research Society. https://www.cambridge.org/core/terms. https://doi.org/10.1557/mrs.2012.54.

Integrated Resource Plan (IRP). (2019). Department of Energy, South Africa.

Ku, A. Y., & Shapiro, A. P. (2012). The energy-water nexus: Water use trends in sustainable energy and opportunities for materials research and development @2012 Materials Research Society. https://www.cambridge.org/core/terms. https://doi.org/10.1557/mrs.2012.39.

Majumdar, A. (2013). A new industrial revolution for a sustainable energy future @2013 Materials Research Society. https://www.cambridge.org/core/terms. https://doi.org/10.1557/mrs.2013.261.

Mason, N., Nalamalapu, D., & Corfee-Morlot, J. (2019). *Climate change is hurting Africa's water sector, but investing in water can pay off.* World Resources Institute.

Synthesis Report. (2019). *South Africa Foresight exercise for science, technology and innovation, 2030.* Department of Science and Innovation.

World Economic Forum. (2018). Accelerating sustainable energy innovation.

The Future of Society

Ozcan Saritas

Abstract This chapter delves into the Future of Society domain within South Africa's National System of Innovation (NSI), highlighting the urgent imperative to tackle inequality and unequal access to technology. It scrutinizes the challenges within the supportive framework against the backdrop of recent technological advancements, shedding light on the stark digital gap and how it compounds existing disparities. The significance of the Future of Society domain is underscored not only as a technological progression but as a fundamental stride toward attaining social and economic equity. Through the lens of big data analytics, the chapter evaluates the current status of the Future of Society domain and delineates key thrusts identified through the SAForSTI 2030 initiative. These encompass policies and metrics for science, technology, and innovation (STI), the evolving South African populace, and STI for an inclusive, community-driven development. Recommendations for actionable steps are outlined, with an emphasis on adopting a socially attuned approach. Three policy considerations are articulated, namely: (i) reevaluating existing STI-related policies, (ii) fostering active engagement from all sectors in policy formulation, and (iii) facilitating collaborations among NSI stakeholders, including communities.

1 Introduction

The Future of Society domain is a critical facet that delves into the meta-dimension of inequality and disparate access to technology within South Africa. The challenges within the enabling environment must be comprehensively addressed to pave the way for a more inclusive and equitable society. Recently, the digital divide has

O. Saritas (✉)
Rochester Institute of Technology Dubai, Dubai, UAE

National Research University Higher School of Economics, Moscow, Russian Federation
e-mail: oxscad1@rit.edu

O. Saritas et al. (eds.), *21st Century Foresight*, Science, Technology and Innovation Studies, https://doi.org/10.1007/978-3-031-69557-5_11

become even more pronounced, exacerbating existing disparities and creating new challenges that need urgent attention.

Recent technological developments have seen an unprecedented acceleration, with innovations such as 5G networks, artificial intelligence, and the Internet of Things transforming the global landscape. However, the benefits of these advancements are not uniformly distributed, and the digital divide persists, particularly in developing countries like South Africa. This divide is not only about access to devices and the Internet but also encompasses issues like digital literacy, education, and the ability to leverage technology for economic empowerment.

In South Africa, where some historical inequalities remain, addressing the digital divide is paramount for ensuring all citizens have equal opportunities to participate in the digital economy. The COVID-19 pandemic has further highlighted the importance of digital inclusion, as remote work, online education, and digital health services became essential for continuity during lockdowns. To overcome the challenges of unequal access to technology, it is imperative to identify and dismantle the barriers that hinder widespread adoption. These barriers may include infrastructural limitations, high costs of technology, and insufficient digital skills among specific demographics. Recent initiatives aimed at providing affordable Internet access, promoting digital literacy programs, and encouraging the development of local digital solutions are steps in the right direction.

The relevance and importance of the Future of Society domain in South Africa cannot be overstated. Bridging the digital divide is not just ensuring equal access to technology; it is a fundamental step toward achieving social and economic justice. A digitally inclusive society is better equipped to harness the benefits of technological advancements, fostering innovation, economic growth, and improved quality of life for all citizens. Moreover, the Future of Society domain serves as a crosscutting framework that influences and enhances the implementation of initiatives in other domains. With a social lens, it ensures that progress in various sectors, such as healthcare, education, and governance, is socially responsive. This holistic approach is essential for the sustainable development of South Africa, promoting inclusivity and ensuring that the benefits of technological progress reach every corner of the nation.

The Future of Society domain is essential for South Africa's journey toward a more equitable and technologically advanced future. Tackling the challenges of unequal access to technology is not only a matter of technological advancement but a crucial step in building a society that values inclusivity and social responsiveness. As South Africa navigates the complexities of the digital age, addressing the issues within the Future of Society domain will be pivotal for achieving comprehensive and sustainable development.

This chapter begins with a discussion on the issues and drivers regarding the Future of Society in South Africa. The current state of the domain is analyzed based on the findings of the big data analytics. Key thrusts identified for the domain through the SAForSTI 2030 exercise, including policies and indicators for STI in a changing South African community and STI for inclusive and people-led development, are discussed in detail, with recommendations for actions. Finally, conclusions are drawn with specific policy considerations for the domain.

2 Issues and Drivers

The barriers to equal access to technology in South Africa are multifaceted, spanning patriarchal, legal, institutional, economic, sociocultural, and class dimensions. These barriers disproportionately affect the poor, vulnerable, and marginalized communities, exacerbating existing social inequalities. As South Africa grapples with the Fourth Industrial Revolution, questions emerge regarding how these marginalized groups can effectively appropriate technology to their benefit. The discourse has expanded to explore the technological opportunities that could be harnessed to bridge the digital divide and, crucially, how equal access can translate into improved employment opportunities for those on the fringes of society.

An essential concern revolves around the fragmented efforts of policy-makers, government departments, the private sector, and civil society players. The lack of collaboration across domains impedes effective responses to complex societal challenges and inhibits the maximization of technology for the social good. The prevailing market-driven nature of innovation further complicates matters, as it often operates outside traditional public policy-making processes and tends to overlook the social aspects of technology.

The inadequacy of previous national science, technology, and innovation (STI) policies has been underscored, with a minimal emphasis on local, community-based research and development (R&D) and social innovation. To address this, there is a growing recognition of the need for policy instruments that facilitate the inclusive appropriation of technology. This may entail restructuring or redesigning existing institutions or even establishing new ones strategically positioned within the economy to drive equitable technological access.

The complexity is further heightened by the call for a pro-African/pan-African National System of Innovation (NSI). This necessitates research, development, and innovation (RDI) programs that encourage collaboration across African nations. Such pan-African initiatives can foster knowledge exchange, resource sharing, and collective problem-solving. By establishing a more collaborative framework, South Africa can leverage the strengths of neighboring countries and create a synergistic approach to technological development, ensuring that the benefits of innovation are shared across the continent.

Overall, the future of South Africa's society confronts many challenges related to unequal access to technology. From patriarchal and legal barriers to institutional and economic obstacles, these challenges disproportionately affect the most vulnerable segments of society. Tackling these issues requires a concerted effort from policy-makers, the private sector, and civil society, emphasizing collaboration across domains. Additionally, there is a pressing need for policy instruments that prioritize inclusive technology appropriation and a reevaluation of national STI policies to incorporate better local, community-based R&D. The vision for the future must also embrace a pro-African/pan-African NSI, fostering collaboration and innovation across the continent to ensure a more equitable and sustainable technological landscape for South Africa.

3 The Current Status of the Domain

Figure 1 depicts the involvement of South African researchers in global knowledge generation within the Future of Society domain spanning from 2008 to 2018. Overall, there has been a noticeable upward trajectory in research emphasis concerning the Future of Society domain in South Africa, particularly evident since 2016. South Africa's contribution to the global scientific output in this domain has also risen, albeit with a notable decline between 2008 and 2009, as well as a slight dip from 2014 to 2016. Globally, the country holds the 11th position, with 1164 publications (NACI, 2019).

Figure 2 illustrates publication activity in comparison to other nations from 2008 to 2018. The number of publications generated has been adjusted based on both GDP and population size. Within the BRICS countries, South Africa stands out for producing the most publications pertaining to the Future of Society domain, significantly surpassing others both in relation to GDP and population size (NACI, 2019).

Figure 3 illustrates the publication quality within the primary areas of specialization pertinent to the Future of Society domain. The size of each bubble corresponds to the volume of articles published from 2011 to 2015 within the specified area of specialization. Given the interdisciplinary and intersecting nature of this domain, publications from diverse disciplines hold relevance. Broadly, publications in the natural and physical sciences exhibit higher quality, with a larger proportion published in first quartile journals, in comparison to those in the social sciences and

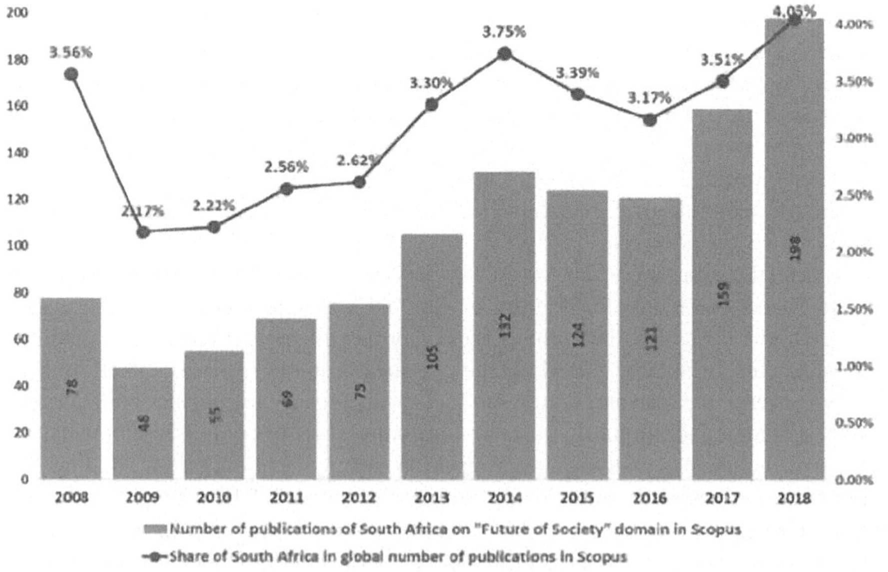

Fig. 1 Publication activity related to the Future of Society domain

Country	Publ ('000)	GDP (USDbn)	Publ/GDP
South Africa	1164.0	348.9	3336.5
Finland	629.0	252.3	2493.0
Netherlands	1712.0	830.6	2061.2
Sweden	1020.0	535.6	1904.4
Denmark	594.0	329.9	1800.7
UK	4487.0	2637.9	1701.0
Norway	638.0	399.5	1597.0
Belgium	746.0	494.8	1507.8
Australia	1802.0	1323.4	1361.6
Malaysia	426.0	314.7	1353.6
Austria	532.0	416.8	1276.3
Nigeria	429	375.7	1141.7
Switzerland	726.0	679.0	1069.3
Canada	1637.0	1647.1	993.9
Spain	1248.0	1314.3	949.5
Italy	1303.0	1943.8	670.3
Russia	974.0	1578.4	617.1
Germany	2217.0	3693.2	600.3
France	1354.0	2582.5	524.3
Brazil	904.0	2053.6	440.2
India	1148.0	2650.7	433.1
US	8121.0	19485.4	416.8
South Korea	471.0	1530.8	307.7
China	3102.0	12237.7	253.5
Japan	1113.0	4872.4	228.4

GDP (2017)

Country	Publ ('000)	Pop (m)	Publ/Pop
Norway	638.0	5.3	12037.7
Finland	629.0	5.5	11419.3
Denmark	594.0	5.8	10241.4
Sweden	1020.0	10.1	10099.0
Netherlands	1712.0	17.1	10011.7
Switzerland	726.0	8.5	8541.2
Australia	1802.0	24.6	7325.2
UK	4487.0	66.0	6798.5
Belgium	746.0	11.4	6543.9
Austria	532.0	8.8	6047.1
Canada	1637.0	36.7	4460.5
Germany	2217.0	82.7	2680.8
Spain	1248.0	46.6	2678.1
US	8121.0	325.1	2498.0
Italy	1303.0	60.5	2153.7
South Africa	1164.0	56.7	2052.9
France	1354.0	67.1	2017.9
Malaysia	426.0	31.6	1347.1
South Korea	471.0	51.5	914.6
Japan	1113.0	126.8	877.8
Russia	974.0	144.5	674.0
Brazil	904.0	209.3	431.9
Nigeria	429	190.9	224.7
China	3102.0	1386.4	223.7
India	1148.0	1339.2	85.7

Population (2017)

Fig. 2 Number of publications relevant to the Future of Society domain (normalized by GDP and population)

economics. Conversely, the level of specialization within the social sciences and economics surpasses that within the natural and physical sciences (NACI, 2019).

4 Recommendations

Two STI thrusts for the Future of Society domain were identified and elaborated on as recommendations based on the foresight exercise.

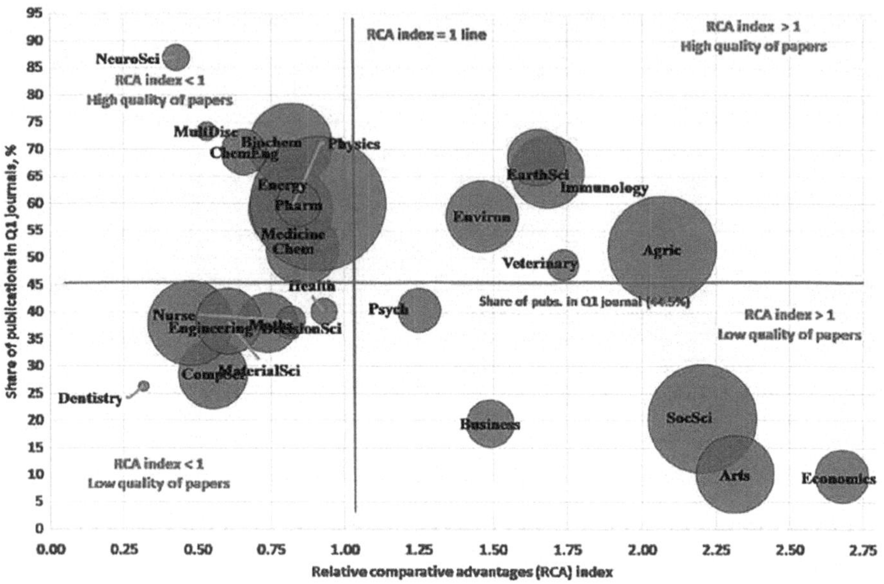

Fig. 3 Quality of publications relevant to the Future of Society domain

4.1 Policies and Indicators for STI in a Changing South African Society

The thrust on policies and indicators for science, technology, and innovation (STI) in a changing South African society recognizes the shortcomings of current interventions in adequately addressing societal challenges. It advocates for a proactive "science for all" approach that ensures the benefits of STI reach everyone, focusing on addressing national-, local-, and community-level challenges. The goal is to develop STI policies that are not only relevant but also responsive to the unique context of South Africa's societal landscape (NACI, 2019).

Developing appropriate STI indicators is crucial, and the thrust emphasizes the need for a robust research base to underpin these indicators. This includes incorporating societal indicators and reflecting the broader impact of STI interventions on communities and individuals. Establishing a well-resourced monitoring and evaluation system is essential to track STI policy outcomes effectively. This system should place equal emphasis on both traditional STI indicators and societal indicators, ensuring a comprehensive understanding of the societal impact of technological advancements (NACI, 2019).

The crosscutting nature of this thrust is highlighted, acknowledging its interconnection with other identified STI domains and their associated thrusts. By aligning with the Sustainable Development Goals (SDGs), particularly those focused on poverty reduction, inequality reduction, and employment promotion, this thrust positions itself as a catalyst for positive societal change through science and technology. The call for research based on international good practice and experience underscores the need for a global perspective in shaping South Africa's STI indicators. Learning from successful international models can inform the development of a more relevant set of indicators that accurately capture the societal dimensions of STI (NACI, 2019).

A key proposal is the reassessment and modification of existing STI policies to include societal issues as crosscutting concerns within relevant domains explicitly. This realignment ensures that the societal impact is considered from the outset, fostering a more integrated and holistic approach to STI policy formulation and implementation. Moreover, the thrust advocates for policy instruments that support community self-sufficiency to recognize the importance of empowering local communities to harness technology for their development and self-reliance. By incorporating such instruments, the STI framework aims to contribute to technological advancement and the broader goal of community empowerment and resilience (NACI, 2019).

This thrust proposes a comprehensive initiative that seeks to reshape the country's approach to science, technology, and innovation. Addressing societal challenges at various levels and advocating for inclusive policies and indicators aim to ensure that the benefits of STI are accessible to all. The emphasis on crosscutting collaboration, alignment with SDGs, and the inclusion of societal indicators positions this thrust as a pivotal force in driving positive societal change through the strategic use of technology (NACI, 2019).

4.2 STI for Inclusive, People-Led Development

This thrust acknowledges the need for South Africa to harness its scientific and technological capacities more effectively to address persistent economic, social, and political challenges. The imperative for change is emphasized by the call for the National System of Innovation (NSI) to become people-centered and ecologically sensitive. This necessitates holistic and crosscutting initiatives to rebuild trust within the NSI, fostering collaboration between the state and citizens, the public and private sectors, and the knowledge worker and research community (NACI, 2019).

One critical observation is the prevailing bias in current STI initiatives, which predominantly focus on technological innovations for narrow economic competitiveness rather than for social benefit and the public good. The thrust emphasizes the need to rectify this imbalance and pay due attention to social, political, and ecological constraints on utilizing technological innovations. The lack of emphasis on major socioeconomic problem-solving and the need for more focus on targeted research and development (R&D) activities to provide practical solutions are also identified as challenges (NACI, 2019).

To address these issues, the thrust proposes formulating a national framework for prioritizing R&D. This framework should guide all NSI players based on domestic needs, community and grassroots innovation, and consideration of global ecological conditions. By aligning R&D priorities with the actual needs of South Africa, this framework aims to bridge the gap between technological advancements and societal challenges (NACI, 2019).

Furthermore, the thrust advocates for the expansion, extension, and strengthening of programs for technology demonstration and technology-based pilots that promote both social and technological innovation. The proposal of a technology appropriation fund and a preferential procurement program aims to empower communities to become technology providers. Successful pilots in prioritized areas should be identified, and efforts should be made to replicate these models nationally, ensuring widespread benefits and scaling up successful initiatives (NACI, 2019).

Recognizing the imminent labor transition in the Fourth Industrial Revolution (4IR), the thrust highlights the need for STI mobility studies to understand the movement of individuals with STI skills. This understanding is crucial for building, maintaining, and expanding South Africa's scientific and technological capabilities. Additionally, substantial investment is needed to support applied social science and humanities-based research. This investment would enable the examination of the harnessing of STI within the broader context of social innovation, ensuring a comprehensive understanding of the societal impact of technological advancements (NACI, 2019).

Overall, this thrust proposes a comprehensive strategy to reshape South Africa's approach to STI. By prioritizing R&D based on domestic needs, promoting social and technological innovation, and addressing the impending labor transition, the thrust aims to create an inclusive and people-centered NSI that effectively harnesses the benefits of science and technology for the broader societal good (NACI, 2019).

5 Conclusions

In anticipating the future of society in South Africa within the National System of Innovation (NSI), it becomes evident that prioritizing social innovation is important for the optimal harnessing of innovations and new technologies. The significance of this focus necessitates the creation of an enabling environment that emphasizes the human and social sciences. This deliberate attention is pivotal for ensuring that the benefits of technological advancements are directed toward the well-being and progress of society.

Tracking the outcomes of research outputs related to the Future of Society domain becomes crucial in assessing the impact of collaborative research efforts. The aim is to ensure that research does not remain confined to academic realms but translates into implementable and successful scaled-up programs for the social good. This underscores the importance of close monitoring mechanisms to gauge the effectiveness of interventions and adjust accordingly.

Developing policies and associated instruments to promote "science for all" represents a transformative agenda with far-reaching implications across multiple sectors. These policies are about science and technology, societal values, equity, and inclusivity. Assessing and modifying existing STI-related policies to explicitly include societal issues cutting across all sectors is an essential step. Policy instruments supporting community self-sufficiency exemplify the commitment to a more inclusive and equitable society.

Effective monitoring and evaluation systems incorporating social innovation into their scope are essential for gauging the impact of STI-related policies. Strengthening existing systems to include social innovation ensures a comprehensive understanding of the broader societal impact of technological advancements.

Active participation of all sectors of society in developing and updating STI-related policies is crucial for creating a more inclusive and representative framework. Specifically, the inclusion of industries that have been historically excluded ensures diverse perspectives and insights are considered, contributing to more robust and equitable policies.

Another crucial policy consideration is facilitating partnerships between all NSI players, including communities. Establishing forums to promote social innovation or identify action plans to address issues of social inequality can serve as mechanisms for collaboration and shared decision-making. These partnerships are essential for creating a more participatory and inclusive NSI that actively involves various stakeholders in shaping the future of society.

In conclusion, the future of society within the South African NSI hinges on a deliberate and strategic focus on social innovation. The outlined policy considerations represent a road map for creating an environment that not only harnesses technological advancements for societal benefit but also ensures that the benefits are distributed equitably across all segments of society. These considerations allow South Africa to chart a path toward a more inclusive, equitable, and socially responsive National System of Innovation.

Acknowledgments The chapter was prepared within the framework of the Basic Research Program at the HSE University.

Reference

NACI. (2019). South Africa foresight exercise for science, technology and innovation 2030, Synthesis Report, National Advisory Council on Innovation, Department of Science and Innovation, Republic of South Africa. Retrieved June 3, 2024, from https://www.naci.org.za/wp-content/uploads/2020/07/South-African-Foresight-Exercise-For-Science-Technology-and-Innovation-2019.pdf

Health Innovation

Michael Kahn

Abstract This chapter focuses on the critical intersection of health, well-being, and innovation in the Anthropocene era, emphasizing the centrality of these factors to our planet's sustainable future. The COVID-19 pandemic is a stark reminder of the interconnectedness of health, socio-politico-economic-environmental realities, and the urgent need for foresight activities. The author challenges the notion of a precise post-COVID-19 era, suggesting a state of 'with COVID-19' (WC) that coexists with other major health challenges, revealing societal fault-lines and amplifying shortcomings in health management. A framework for understanding health innovation is introduced, followed by a PESTEC analysis identifying deficiencies noted in the appraisal. The anticipatory analysis explores possible futures and their impact on present and near-future health innovation prospects. The chapter concludes by suggesting actionable steps to address the challenges outlined, advocating for a holistic approach that aligns healthcare and economic development for a more resilient future.

1 Introduction

Health and well-being, understood in the broadest sense, encompass the survival of animals, humans, and the environment, in short, the biosphere. Health and well-being stand central to our future in this, the Anthropocene Era.

Any foresight activity must anticipate the various futures that may enable a sustainable planet Earth. Our record to date is dismal, dangerous, and disastrous, as the COVID-19 pandemic so powerfully illustrates. Commentators speak glibly of 'before COVID-19' (BC) and post-COVID-19. It may be that there is no post-COVID-19, rather a state of 'with COVID-19' (WC). Perhaps the two eras are best abbreviated as BC and WC? In addition, one should not claim that COVID-19 is a

M. Kahn (✉)
CREST, Stellenbosch University, Stellenbosch, South Africa

CREST/SciSTIP, Stellenbosch University, Stellenbosch, South Africa
e-mail: mjkahn@sun.ac.za

O. Saritas et al. (eds.), *21st Century Foresight*, Science, Technology and Innovation Studies, https://doi.org/10.1007/978-3-031-69557-5_12

'black swan event'. The portents were there. That was certainty. The only uncertainties were when, who, and how bad. COVID-19 has struck in parallel with other major killers—malaria, tuberculosis, HIV/AIDS, neglected tropical diseases, and the rising toll of non-communicable diseases, trauma, and substance abuse.

It is commonly observed that COVID-19 has revealed the fault-lines that characterize the socio-politico-economic-environmental reality. True enough. These have manifested for all to see, but avoid. COVID-19 did not require a special announcement of disaster, but it has concentrated thinking in regard to politics, economics, health, and society. What was breaking is broken, with recession, cuts in spending, job losses, and reduced tax collection. We are living through a once in a century perfect storm, with clear weather yet to appear.

Given the disruption to society, economy, and politics, COVID-19 obviously serves to highlight shortcomings in the way we think about and manage health, and health innovation. For South Africa this is especially important, given the high burden of disease that she must deal with; the largest HIV-positive caseload or burden; the highest numbers receiving antiretroviral therapy; the extreme incidence of obesity, diabetes, and hypertension; and foetal alcohol syndrome. From the AIDS-induced decline in life expectancy of the early 2000s, a drop from 61 years down to 52 years, there has been a notable increase to 62.5 years as antiretroviral therapy has become widely available. The health system has coped but is under extreme strain. It is important to recall the intense resistance of the Mbeki administration to confront the HIV pandemic (Kahn, 2016).

Through its Vision for 2030 (RSA, 2011), the National Planning Commission presented a desirable and possible path for the country. Vision for 2030 focused primarily on the political, the social, and economic, with health receiving attention in but 23 of the over 400-page document. The thrust of the health section was for the improvement of governance and management systems, to lay the basis for the gradual introduction of a National Health Insurance scheme.

Late 2019, the country adopted a new policy for science, technology, and innovation (DST, 2019). Whilst mentioning health as essential to social and economic development, that policy statement, following the lead of Vision for 2030, remained at the level of generality, save for advocating technological solutions such as e-health, data analytics, and data mining. It offered little guidance on the need for demand-driven, use-inspired applied research that would address the fundamentals of the health crisis.

In contrast, the African Union Science, Technology and Innovation Strategy for Africa (African Union, 2014) displayed a sharper focus through its priority 2: prevent and control diseases and ensure well-being, with the position that 'African countries will not develop economically and socially without substantial improvements in healthcare delivery' (idem: 22).

The reality might be that improved healthcare and economic development walk together. If one lags, so does the other.

It is against this background that this chapter offers an anticipatory analysis of possible futures and how these may impact on the present and near-future prospects for health innovation.

The chapter proceeds through six sections. Following the Introduction, Sect. 2 offers an appraisal of the recent South Africa Foresight Exercise for Science, Technology and Innovation 2030 (DSI, 2020). Next is a framework for understanding health innovation in the broad sense. This is followed by a PESTEC analysis that identifies some of the shortcomings noted in the appraisal. Section 5 presents the anticipatory analysis. A final section suggests some actions.

2 Appraisal: Health Innovation in the South Africa Foresight Exercise for Science, Technology and Innovation 2030

The chapter on health innovation in Foresight Exercise for STI 2030 may be summarized as an exercise that is strongly anchored in present ICT thinking. The chapter reads as offering a range of ICT solutions to health innovation, all of which are based on known areas of application or development. A possible reason for this technical emphasis is that the chapter makes little connection with the current realities of the health sector, its weaknesses and strengths. Furthermore, the chapter exhibits a tendency toward techno-nationalism in decrying foreign imports, whilst simultaneously calling for South African exports to Africa. The reader might thus sense a fundamental confusion in play. It might well be that by avoiding any statements that reflect on the political and social, Foresight Exercise for STI 2030 'floats above the battle' and is content to suggest developments that would be useful to have in the near future, e.g. a health information system or quality infrastructure. Furthermore, the associated bibliometric analysis is skewed, failing as it does to highlight the existence of world-class research on infectious diseases and, in public, environmental and occupational health. Both of these subject categories have research activity indices and category normalized citation rates greater than the world average.

In a sense, the chapter on health innovation exhibits continuity of thought with the earlier R&D strategy (DST, 2002), and the Ten-Year Innovation Plan (DST, 2008), both of which avoided engagement with the HIV/AIDS pandemic, whilst pursuing inward-looking agendas, such as a declared resistance to clinical trials even as the Department of Health authorized large-scale global clinical trials, especially into the aetiology of HIV/AIDS. Despite this implicit reticence, the health sciences academic community, in response to the HIV crisis, mobilised international resources to develop frontier research in infectious diseases, immunology, and public, environmental, and occupational health. These in turn built on generations of research into tuberculosis and industrial diseases. The expertise developed in spite of, not because of, direct government support. It is a salutary lesson that the partial containment of the COVID-19 epidemic in the country was enabled by this prior investment. Government and the research community came together in a Ministerial Advisory Committee on COVID-19 that set out to navigate the complexities of the virus, its transmission, and mitigation. However bad the outcomes have been, they would have been far worse but for this engagement.

In all fairness, health innovation was scoped prior to the COVID-19 pandemic. But it was significantly off target and lacking in ambition.

3 A Framework

The unifying framework for this health innovation foresight lies in the concept of health security, as illustrated in the following schematic (Diagram 1). The attainment and maintenance of health security requires innovation in at least six interrelated themes, namely:

- Socio-political.
- Biological.
- Radiological.
- Chemical.
- Food.
- Environmental.

These must be placed in the context of the dominant and emergent techno-economic paradigms (Diagram 1). Each of these themes rests on sub-themes that demand innovative thinking and anticipation. It is to state the obvious that these

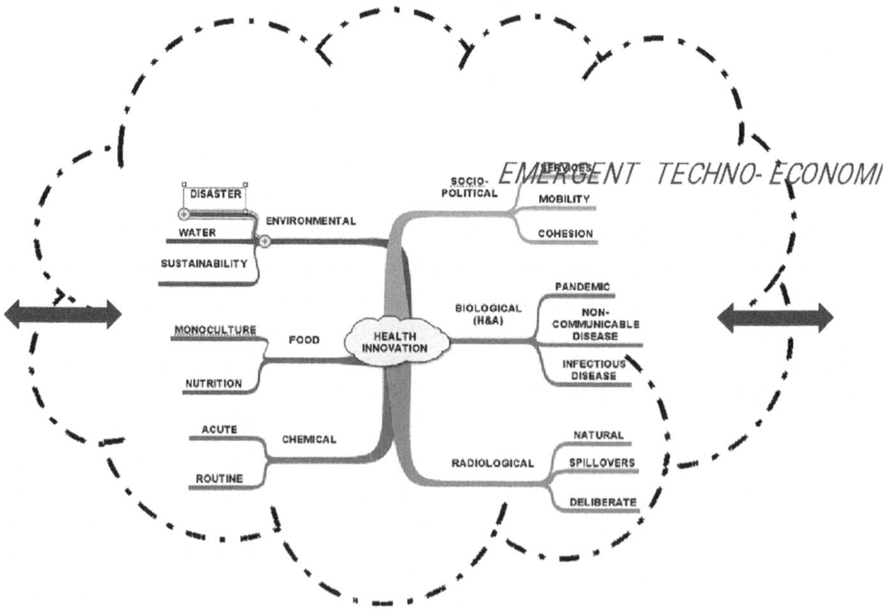

Diagram 1 Health innovation and health security (Author)

themes are not only interlinked, but some 'obvious' themes appear to be absent, for example, energy, employment, and the oceans. But these concerns are present, within the environmental, chemical, and social.

Societies largely find themselves grappling with the ICT techno-economic paradigm (TEP), a long-run revolution that took off a century ago, first through electromechanical devices such as generators, motors, relays, and actuators, then followed by the vacuum tube, and finally solid-state devices, starting with transistors and then the microprocessor. The ICT trajectory is still unfolding, possibly becoming a wholly new techno-economic paradigm. This is variously denoted as Industry 4.0 (European Union, 2020) or the Fourth Industrial Revolution (Schwab, 2016).

As prior industrial revolutions shaped and were shaped by societal forces, so too with the post-1970 phase of the ICT revolution that enabled dematerialization of value chains, robotic manufacturing, remote working, social media, social freedoms, and the surveillance state (Zuboff, 2019). An earlier revolution saw the rise of exploitative capitalism, totalitarian regimes, and organized labour. The present revolution has enabled new geopolitics, extreme within-country and between-country wealth inequality, and environmental destruction, even as the middle class has expanded to enjoy a quality of life unimaginable a century before.

On the economic front, the post-war Keynesian approach gave way to the Chicago School monetarism of the 1980s. That paradigm then yielded to flexible inflation targeting through variable interest rates that held sway to the financial meltdown ca. 2008. The adjustments and disruptions of the ongoing financial crisis and the current pandemic demand fundamental change. This is already evident in constraints on mobility and association. Video streaming enables remote meetings; the office as focus of twentieth-century working life is in question; a flight to the suburbs beckons; the bicycle has become a favoured mode of transport; the local is being rediscovered. A new awareness of personal health, vulnerability, and fitness is evident. The tourism, sport, and leisure sectors are wrecked. Whether this signifies even medium terms gains toward environmental sustainability remains an open question.

4 PESTEC

A PESTEC analysis offers a convenient means to conduct a situation analysis of the political, economic, social, technological, environmental, and cultural (legal and values). The analysis serves to highlight the specifics of the 'now' and shapes the potential to engage with future thinking, and anticipation in particular. As with the above health security themes, the six PESTEC dimensions necessarily overlap, for example, in the form of political economy, the socio-legal, the techno-economic, the enviro-cultural, and so on. The importance of the PESTEC analysis lies in how this depiction of current reality points to the need for, and the potential for, health innovation. The health security themes are blended into the analysis.

4.1 Political

Constitutional democracy was instituted in 1994, bringing three centuries of minority rule to a close. That polity had rested upon social and economic exclusion that to a varying extent prevailed across the land from the southern Cape to the Tropic of Capricorn. This dramatic political change created a new South Africa, but not 'South Africans'. Such nation building is a process of generations, not a single generation. The majority now enjoys political power, but limited economic power, and the previous low-intensity civil war has morphed into 'lawfare', whereby the courts serve as a battleground for the contestation of interests between a corporatist state and the private sector. On the one hand is strong legislation designed to advance redress; on the other, a weak state co-exists with what is in effect one-party rule, with low accountability, regular elections notwithstanding.

For more than a century, the state has played a powerful role in the economy. Developmental state 1 prevailed over 1920 to 1980. Developmental state 2, much discussed, remains a work in progress, with intense and often poorly conceived regulation, oft honoured in the breach. The gestation of developmental state 2 might yet be catalysed in the WC epoch.

4.2 Economic

The economy comprises of a large state sector and a diversified private sector. Nominally based on the minerals-energy-financial complex, the economy has shifted to a dominance of services, the contribution of manufacturing to GDP halving over the last two decades. The country exhibits the most diversified industrial sector on the continent. In many years the country is a net exporter of foods, being a world leader in citrus, berries, and wine. Whilst President Mbeki spoke of a country of two economies, the haves and the have nots, a large offshore third economy now exists.

Until some 8 years ago, she was the largest provider of foreign direct investment into Africa, since then the Peoples Republic of China has assumed this role. Since 2012 she has been a member of the BRICS grouping, where she plays the role of the gateway to Africa. The establishment of the African Continental Free Trade Area is intended to create prospects for intra-African economic development.

Regarding aggregate performance, the early 2000s of the commodity super-cycle generated strong economic growth and the expansion of the public service that together doubled the size of the African middle class. Consumption-led growth waned with the onset of the financial crisis and rising corruption and inefficiencies. The decline is recorded in the fall of a range of indices as computed by the World Economic Forum, World Bank Doing Business, Transparency International, and the UN Development Programme.

Suggestions for WC-induced import substitution hark of pre-1980 industrial policy and may well provide the driver for a modicum of reindustrialization. Dependence on foreign suppliers for crucial medical supplies is a vulnerability best avoided. Latent industrial capacity may be repurposed.

4.3 Social

The spatial separations of the Apartheid era persist, and though there has been upward mobility for those with post-secondary education, inequalities still follow the previous group lines. Simply put, the social fabric is frayed, and social cohesion absent without leave. Measured by the Gini coefficient, Human Development Index, or Palma index, the country is a world outlier exhibiting the greatest income and asset inequalities on record, to some extent mitigated by social grants that reach 44 percent of households. Structural inequality persists and is exacerbated the capital-intensive nature of much of industry. The prospects for export-led growth based on low wages are simply impossible (Rodrik, 2006). Whilst education and basic health are free, government services are generally of low quality. A well-paid and large public service does not deliver to the very people that need this most. Patronage and dependency networks maintain this status quo. Un- and underemployment remain drivers of poverty, morbidity, and mortality.

That the world would face a zoonotic pandemic was long anticipated, the experience of SARS, MERS, and H1N1 flu pandemic of 2009 being predictors. The World Health Organization (WHO), through its One Health Agenda, was exemplary in raising awareness of the threat. The WHO 2005 International Health Regulations (IHR) gives non-binding substance to these principles. From 2016 onward the WHO reviewed member states' compliance with the IHR through the Joint External Evaluation (JEE) process. The United States, South Africa, Mozambique, and Ethiopia were among the first to subject themselves to such scrutiny. There was massive buy-in from Africa, but little from the OECD and European Union member states alike.

The JEE scoreboard assesses 48 indicators on a 1 to 5 scale, 5 being highest. South Africa records a mixed bag, scoring mostly 3 s (WHO, 2018). It may be unsurprising that the United States mostly recorded 5 s. For the United States however, there was a telling exception, namely, the management of zoonotic diseases for which she scored a 3. By comparison, South Africa scored 4, with Australia, Finland, Thailand, Switzerland, and the UAE attaining 5. This score reflects the fact that the management of animal diseases has a long history in South Africa, with excellent research and mitigation directed to rinderpest, anthrax, Ebola, foot and mouth disease, and Congo fever. The country has the necessary public research institutions and institutional memory and the protocols to deal with these.

The US JEE concluded that the low score for zoonotic disease control lay in its state-federal structure that restricted interoperability of information systems, specimen sharing, coordination, and even the One Health approach. This finding was a salutary 'portent of a death foretold'. Whilst international comparisons are replete with problems of testing regimes, capacity to test, and accurate mortality data, the United States as of 30 September 2020 had the highest number of COVID-19 infections on record, and among the fifth highest mortality per capita rates. At some time in the future, a more reliable picture will emerge from studies of excess mortality statistics.

The JEE of South Africa found weaknesses in infection control, and this did indeed manifest in the early days of managing the COVID-19 epidemic.

4.4 Technological

South Africa hosts a modest-sized innovation system that is the most advanced in Africa. Her universities are the continental leaders; her public research organizations are generally well-managed and have spun out many sector-specific research organizations in the health, social sciences, environmental, and industrial sectors. Even so, the legacy system has not sufficiently expanded to meet the needs of its much larger population (Kahn, 2019). Up to the onset of the pandemic, her universities served as the postgraduate hub for sub-Saharan Africa (SSA), with close to third of doctoral graduates and 60 percent of postdoctoral fellows being from SSA. She competes with Egypt for the highest volume of scientific articles and leads for patents, trademarks, and the registration of plant cultivars. The research system is open to the world but weakly connected to local industry. The same holds for the technical and vocational education and training sector that was subject to dramatic consolidation and loss of institutional memory. The system is functioning more as a lay bye for youth who cannot enter the universities through the normal route than for career upgrading and artisan training.

Mention was made of health sciences research competence that resides in the 'Republic of Science'. Health sciences R&D accounts for 25 percent of GERD, with mining and defence R&D having halved from their prior levels. Health sciences research competence resides in what may be considered the 'Republic of Science'. Yet this health sciences research has not resulted in a single new active pharmaceutical ingredient (API). The country boasts large pharmaceutical companies that restrict their product lines to the manufacture of generic drugs either under licence or post-patent expiry. They are not radical, 'new to world market' innovators.

Regarding technological development, previous expertise in weapon systems, mining, and metallurgy has weakened, as these sectors have lost their prominence. Energy research has been throttled by the dominance of the monopoly state utility that has sought to restrict renewables, in favour of 'king coal', supplemented by nuclear ambitions. South Africa had developed its own nuclear weapons and uranium enrichment capabilities and, after decommissioning, sought to preserve this know-how by pursuing its own fourth-generation nuclear power technology. That project consumed scarce financial resources but was terminated in 2010 when foreign partners withdrew. The consequence was that when renewables had to be deployed to deal with electricity supply shortfalls, the core technologies were imported. There is no local photovoltaic or wind turbine R&D of any significance. The major local innovator, Johanna Solar Technology, failed to capture the market. In contrast, agricultural R&D has remained steady in relation to the contribution of agriculture to GDP, and the country remains among the top 12 for the registration of plant cultivars that play their role in ensuring food security.

Areas of state-funded research include platinum catalysis and titanium refining. Despite decades of effort, neither has seen major breakthroughs, jeopardized by limited funding and the absence of critical mass. In parallel is much higher-level state support that promotes astronomy and astrophysics as a prestige blue sky activity. The advocacy for this is that such endeavour will attract generations of young scientists whose long run contribution to the stock of knowledge will be worthwhile.

Alongside is the services sector, where considerable innovation occurs but is absent from the conventional metrics. Examples abound in retail, financial services, health insurance, and the public sector.

A world class, vibrant scientific, and technical services sector trades locally, across the continent and globally.

The concentration on basic research (27 percent of GERD) offers one explanation as to why so little scientific activity results in new products. It seems that poor selection and termination mechanisms dominate, so that lobbies and interest groups vie for money and project survival. If funded projects have weak connections to real societal or industrial problems, then it is unlikely that impact will emerge from the work, no matter how diligent the researchers may be. A second mechanism that discourages public-private linkages resides in the regulations of the Intellectual Property from Publicly Funded R&D Act of 2008 that was modelled on the US Bayh-Dole legislation, with modifications peculiar to our polity. Observation is that local business has avoided the imposed restrictions by withdrawing from local collaborative near-market research with the universities.

4.5 Environmental

The country spans a number of climatic regions and ecosystems, with extensive wildlife and unique flora. These have been nurtured through excellent conservation measures, but these are increasingly threatened by demands for arable land, mining, and the predations of poachers on land and sea. The 3000 km coastline is especially vulnerable to factory fishing in the hands of many foreign fleets.

There is also serious threat to human health as a result of nitrous oxides and sulphur dioxide emissions, with levels among the highest recorded. To this may be added indoor atmospheric pollution and industrial disease, especially due to mineral dusts. World-class environmental protection legislation is in place, but conformance is poor.

Whilst Millennium Development Goal 1, to reduce absolute poverty, was realized ahead of schedule, country progress toward attaining the subsequent Sustainable

Development Goals has been weak, at rank 113 in 2019. Morbidity, inequality, and environmental shortcomings explain this. There is strong resistance from the coal lobby to the Paris Climate Accords, and the scoreboard exhibits year on year sideways drift.

4.6 Cultural

The eminent legal scholar Cass Sunstein (2001) has described the Constitution of South Africa as 'the most admirable constitution in the history of the world'. Another oft-quoted view is that the country is brilliant on designing policies, but weak on their implementation (Schwella, 2001; Van der Berg et al., 2011; Cronin & Sadan, 2015). Her courts remain essentially independent, even if the wheels of justice turn very slowly indeed. Yet, despite the enshrinement of second-order constitutional rights, it is these very courts that have had to be mobilised by civil society to compel government to deliver according to constitutional imperatives. This is especially the case in respect of healthcare, with HIV the exemplary case. Other such cases include access to land, housing, water, and clean air.

At the cultural level, Apartheid era race categories remain in use and are deployed to drive various programmes of redress. Apartheid era efforts to promote identity politics gave rise to separate ethnic and tribal authorities, and state-sponsored media in the various tongues of the country. An unintended consequence is that 11 languages today enjoy official standing, and theatre, music, and literature in these tongues abound. Yet many of the minority groups lack basic communications skills in any African language; in counterpoint, many of the African majority are functionally illiterate in the two languages of commerce (and government), namely, English or Afrikaans. The potential for nation building through language is unbounded. One may think of the way that Afrikaans, Swahili, Hebrew, and Italian were used as nation building devices in other times and places.

5 Anticipation

Life, healthy life, is the most precious thing, the views of some religious beliefs notwithstanding.

At the inaugural World Innovation Forum, held in Kuala Lumpur in 2012, participants engaged in future thinking to envisage the world of 2050. The author joined the group tasked with the field of health. Two radical ideas captured imagination. The first was that 'by 2050 health insurers would adjust premiums to incentivize healthy lifestyles'. Done. Been there. This is the founding principle of a South African insurer that had operated its 'vitality' incentive scheme since the early 2000s. Among other rewards, vitality gives discounts for the purchase of whole foods at the point of sale. The second radical idea was that 'in 2050 life expectancy will be 100 years plus one day'. The proposer explained that in some countries life expectancy was already in the high 80 s, so that by 2050, many more would attain centenary status. However, an ageing population would confer a huge financial and social burden on the young, so that upon attaining the milestone, all centenarians would be encouraged to end their lives, in peace. This would call for deep introspection on the part of lay persons and organized religion alike.

Anticipation is the emerging discipline that seeks to develop the insights needed to imagine different futures and to use such insights to shape current actions (Miller, 2018). This approach draws on the experience of other techniques, such as scenario planning and back-casting that gained traction in studies on environmental sustainability. The difference between back-casting and anticipation lies in the importance that is attached to possibilities. In that sense, one might describe anticipation as multicasting to the present.

Anticipation applied to health innovation spans society as a whole. From a health innovation perspective, the analyst must consider animal, human and environmental well-being, life expectancy, healthcare, access to medicines, the politics of pandemics, equity, etc. Whilst there are ecosystem and habitat determinants, humans cannot survive unprotected outside a narrow temperature range. We all need water, food, air, shelter, and Maslow's pyramid plus Wi-Fi. There are known unknowns and unknown unknowns. All PESTEC dimensions are in play, with the technological ever present, but the technological is not and should not be the overarching consideration.

The unique nature of the SARS-COV-2 virus has destroyed many assumptions as to the nature of disease, the power of science and politics, and the received wisdom of economics. In the case of South Africa, societal fractures, the contestation of interest groups, continued dependence as a price taker, divided government, weak education, and a high burden of disease existed before the COVID-19 storm. Given this storm, the country has weathered COVID-19 as well as might be expected.

Anticipation is grounded in the WC reality modulated with hedonic adaptation, plus the realities of a fractured society that will heal but slowly. This is the lesson of our history, and history in general.

Some critical success factors that will shape future health innovation include the following:

- Responsible research and innovation.
- Survival of a unitary South Africa, embracing the rule of law, separation of powers, and accountable government.
- Deployment of a capable and incorruptible public service.
- Health security principles are central.
- Health staff and infrastructure are designated as essential services and are fully protected.
- Talent is sourced globally.
- Decisions on research funding are informed by need, feasibility, and progression decisions.

- Regulation supports rapid development and trialling of drugs and medical devices.

How then for the age of uncertainty? How will market capitalism evolve under the WC crisis? What will the role of the state look like? Is a binary choice the only option on the table? Nationalize with central control versus the mixed economy? Who will decide on the cost of money and the nature of debt? These considerations necessarily underpin solutions other than a return to a barter economy of times long past.

A future out to 2040 envisages a widely accepted social contract in which the (extended and revised) SDGs are close to attainment. Health innovation yields the following:

- Quality, affordable healthcare.
- The incidence of malaria, tuberculosis, HIV, and SARS-COV-2 or variants is reduced to 10 percent of 2020 levels.
- The capacities to address new pandemics and meet the revised IHR are in place, with JEE (or successor) scores averaging 4.5.
- Domestic pharmaceutical companies each annually deploy one new API.
- Extreme drug resistance is mitigated through genomic medicine.
- Remote sensing, including a network of low-orbit microsatellites, allows for monitoring and enforcement of animal, human, insect, marine, and environmental health status and disaster mitigation.
- Micro-robotic health diagnostics and interventions are, as appropriate, deployed to personal and community level.
- Substance abuse, through careful reappraisal of risks, has been reduced to 10 percent of 2020 levels.
- Traditional medicine is regulated and closely integrated into the public health system.
- Life expectancy reaches 70 years, and suicide is legalized.

Alongside are innovations that yield the following:

- Nationwide robust, fast, and free broadband access.
- Arable organic agriculture, including new substrate technologies, expands production by 30 percent to new crops, new farmers, and new markets, providing livelihoods to 10 percent of the workforce.
- Extensive monoculture, pesticide, and chemical fertilizer usage is reduced to half the 2020 levels.
- Food security is universal and, coupled with 'eating well', addresses the issues of lifestyle diseases and stunting that are reduced to half the 2020 level.
- Sixty percent of energy supply is renewable.
- Two thirds of transport is powered by renewable sources.
- All industries practice zero emission, moving toward the goals of a circular economy.
- Non-essential consumption falls to 50 percent of 2020 levels.
- The working week is reduced to 20 h.

- Work from home is the norm.
- Unemployment is below 10 percent of those seeking work.
- Tariffs for intra-African trade are eliminated.
- Technology balance of payments ratio falls from 10:1 to 3:1 import/exports.
- Pre-schooling and the active aged are merged into a single system.
- Education and training entail 50 percent contact time, through problem-based learning where appropriate.
- Post-school education and training are enjoyed by 40 percent of the age cohort.
- All enjoy freedom of mobility.
- Income inequality is reduced to a Gini coefficient of 0.40 with debt written off through decadal jubilee statute.
- Privacy is enshrined in law and upheld.
- All government decision-making, bar extreme security matters, is open to the public, and available online.
- Political expression is bottom-up, with elected officials subject to popular recall.
- Social cohesion rests on universal oral, written, and numerical literacy, across group and ethnicity.

6 Concluding Remarks

In effect, the two decades out to 2040 span a mere generation. By contrast, healthcare and interventions are as old as human self-awareness. This suggests that innovation might proceed more slowly than change agents might wish.

Key to change is responsible research and innovation that were carefully placed as the lead critical success factor.

Fiat. Through a glass darkly.

References

African Union. (2014). *African Union science, technology and innovation strategy for Africa*. African Union.

Cronin, G P and Sadan, M. (2015). Use of evidence in policy making in South Africa: An exploratory study of attitudes of senior government officials. African Evaluation Journal. 3 (1) a145. DOI: https://doi.org/10.4102/aej.v3i1.145

DSI. (2020). *STI foresight 2030*. Department of Science and Technology.

DST. (2002). *National R&D Strategy*. Department of Science and Technology.

DST. (2008). *Ten year plan for innovation*. Department of Science and Technology.

DST. (2019). *White paper on science, technology and innovation*. Department of Science and Technology.

European Union. (2020). Retrieved September 20, 2020, from https://www.i-scoop.eu/industry-4-0/

Kahn, M. (2016). Mitigating South Africa's HIV epidemic – The interplay of social entrepreneurship and the innovation system. *Minerva, 54*(2), 129–150. https://doi.org/10.1007/s11024-016-9293-x

Kahn, M. (2019). The contract between science and society: A south African case study. *Science and Public Policy, 46*(1), 116–125. https://doi.org/10.1093/scipol/scy042

Miller, R. (Ed.). (2018). *Transforming the future. Anticipation in the 21st century.* UNESCO and natN.

Rodrik, D. (2006). Understanding South Africa's economic puzzles. Paper for the Center for International Development Project on South Africa. John F. Kennedy School of Government, Harvard University: Cambridge, MA.

RSA (2011). National Development Plan Vision 2030. Pretoria: The Presidency

Schwab K (2016). The Fourth Industrial Revolution. Geneva: World Economic Forum

Schwella, E. (2001). Public sector policy in the New South Africa: A critical review. *Public Performance & Management Review, 24*(4), 367–388 (22 pages). Published By: Taylor & Francis, Ltd. https://doi.org/10.2307/3381225, https://www.jstor.org/stable/3381225

Sunstein C R (2001). Designing Democracy: What Constitutions Do. Oxford: Oxford University Press

Van der Berg S, Burger C, Burger R, de Vos M, du Rand G, Gustafsson M, Moses E, Shepherd D, Spaull N, Taylor S, van Broekhuizen H and von Fintel D. (2011). Low Quality Education as a Poverty Trap. Stellenbosch: University of Stellenbosch.

WHO. (2018). *Joint external evaluation of IHR Core capacities of the Republic of South Africa.* World Health Organization. Licence: CC BY-NC-SA 3.0 IGO.

Zuboff, S. (2019). *The age of surveillance capitalism – The fight for a human future at the new frontier of power.* Profile Books.

High-Tech Industrialisation

Fiona Tregenna and Alexis Habiyaremye

Abstract The chapter discusses the critical role of high-tech industrialisation in the growth and development of developing countries, emphasising its importance for sustained economic growth. Recognising the transformative impact of the Fourth Industrial Revolution (4IR), the text explores the potential of robotics, artificial intelligence, the Internet of Things, and additive manufacturing in shaping high-tech industrialisation. The discussion focuses on two dimensions: shifts in the sub-sectoral composition of manufacturing towards technology-intensive sectors and the increasing technology intensity within any manufacturing subsector. Highlighting the relevance of industrialisation, the chapter underscores the complexities of policy choices in countries like South Africa, balancing the imperative for high-tech growth with the need for mass employment creation. The analysis incorporates a foresight orientation, considering emerging trends, challenges, and opportunities in pursuing high-tech industrialisation. The chapter emphasises the coexistence of less technology-intensive manufacturing for employment creation. It outlines the complexities associated with growing high-tech industries, especially in the presence of skill and capability gaps.

1 Introduction

Industrialisation has long been recognised as being central to developing countries' growth and development. Both theoretical evidence and empirical evidence of country experiences internationally and over time point to the importance of industrialisation for sustained economic growth.

There has been growing recognition that not only does industrialisation matter but the *nature* of that industrialisation is of critical importance. While there are

F. Tregenna (✉) · A. Habiyaremye
DSI/NRF South African Research Chair in Industrial Development, University of Johannesburg, Johannesburg, South Africa
e-mail: ftregenna@uj.ac.za; alexish@uj.ac.za

O. Saritas et al. (eds.), *21st Century Foresight*, Science, Technology and Innovation Studies, https://doi.org/10.1007/978-3-031-69557-5_13

various dimensions to this, a central aspect is the technology intensity of industrialisation. As discussed below, more technology-intensive industrialisation has particularly strong growth-pulling potential. High-tech industrialisation is closely associated with R&D, innovation, learning, and productive capabilities. It is especially important for productivity, competitiveness, exports, and for capturing shares of globally growing market segments. Thus, many developing countries have increasingly emphasised not only industrialisation but high-tech industrialisation in particular.

This is also brought to the fore with the rapid technological changes referred to under the rubric of the Fourth Industrial Revolution (4IR, also referred to as Industry 4.0). The 4IR is expected to transform products, processes, and services in a revolutionary way through the application of smart and connected systems. More specifically, technologies such as robotics, artificial intelligence (AI), the Internet of Things (IoT), and additive manufacturing are relevant to high-tech industrialisation. These technologies can both transform old industries and give rise to the emergence of new industries. This opens up both opportunities and challenges for developing countries such as South Africa, in their attempts to pursue high-tech industrialisation. At this stage, the actual adoption of 4IR technologies is generally limited in most low- and middle-income countries, with adoption very uneven by industry and types of firms, but the pace of change is likely to accelerate in the coming years.

While there may be different interpretations of high-tech industrialisation, in this chapter we identify two broad dimensions of this concept: first, referring to shifts in the subsectoral composition of manufacturing towards more high-tech (highly technology-intensive) sectors, and second, referring to shifts in the technology intensity of production in any subsectors of manufacturing. We elaborate on each of these dimensions below.

The first dimension refers to compositional changes in manufacturing towards more technology-intensive sectors. There have been various classifications of manufacturing subsectors that relate to technology intensity. An early taxonomy was that of Pavitt (1984), who categorised manufacturing firms as being supplier dominated, scale-intensive, specialised suppliers, and science-based. This classification is not based on the subsectors of manufacturing. Lall (2000) developed a classification of resource-based, low-technology, medium-technology, and high-technology manufacturing.

Current classifications in common applied usage are those of the OECD and UNIDO. These are based on a classification of subsectors of manufacturing into low-, medium-, and high-tech categories (including subcategories of these), based on the average or overall technology intensity of the subsector. Thus, a shift towards medium- and high-tech (MHT) sectors, or specifically high-tech industrialisation, implies a shift in the composition of manufacturing towards more technology-intensive subsectors of manufacturing. The OECD (1995) classification categorises the following subsectors as comprising high-tech manufacturing: aircraft and spacecraft; pharmaceuticals; electronics components; office, accounting, and computing machinery; radio, television, and communications equipment; and medical, precision, and optical instruments.

The second dimension of high-tech industrialisation relates to increasing technology intensity across any subsector of manufacturing (as opposed to a change in the subsectoral composition of manufacturing). More technologically advanced techniques can be used in any manufacturing activity. For instance, "food products, beverages, and tobacco" are categorised as a low-tech industry in the OECD (1995) classification, yet there are firms and segments of the food manufacturing industry that are highly technologically advanced. 4IR technologies can be utilised in any subsector of manufacturing (although there is of course heterogeneity in the scope of adoption of these technologies).

Thus, high-tech industrialisation can involve both changes in the composition of manufacturing (towards more high-tech industries) and technology intensification across manufacturing industries. Certain policy interventions are relevant to both of these dimensions, while other policies are specific to one or the other dimension.

On balance, South Africa is lagging in high-tech industrialisation (as discussed in Sect. 3). A "foresight" orientation to this entails a consideration of emerging trends (nationally and internationally) and of future challenges and opportunities, as well as how these might shape the actions that are needed now. Policy choices around high-tech industrialisation are especially complex in an economy such as South Africa, with its abundance of low- and semi-skilled labour. While it is important to recognise the importance of high-tech industrialisation for sustained rapid economic growth in the future, it is also important to reflect on such a path in relation to the urgent need for mass employment creation. High-tech manufacturing industries tend to be relatively intensive in capital and in technology, and more intensive in high-skilled labour than in low- or semi-skilled labour.

While this chapter focuses specifically on high-tech industrialisation, this is not intended to suggest that high-tech manufacturing is all important. There is also a crucial role for less technology-intensive manufacturing, especially in terms of employment creation; policy support thus needs to be directed at growing relatively low-tech and medium-tech manufacturing as well. Furthermore, it needs to be recognised that growth in high-tech manufacturing is complex and challenging to achieve, especially where there are weaknesses in productive capabilities. We consider these issues, and the associated policy implications, in the conclusion of this chapter.

The next section discusses why high-tech industrialisation matters. This includes a consideration of the importance of industrialisation and the dimensions of innovation and technological upgrading, learning and productive capabilities, export competitiveness, and green industrialisation. Section 3 provides an overview of the South African high-tech industry within the general context of a deindustrialising manufacturing sector. Section 4 uses semantic analysis to discuss the current trends in high-tech-related scientific publications in order to explore how those trends can underpin a strategic high-tech industrialisation policy. Section 5 discusses some crucial policy dimensions of high-tech industrialisation in South Africa and concludes.

2 High-Tech Industrialisation: Why It Matters?

2.1 *The Importance of Industrialisation*

Within development economics, the structuralist tradition in particular emphasises the importance of manufacturing and industrialisation for developing countries' growth and "catching up" (Blankenburg et al. 2008; Chenery, 1955; ECLAC, 1964; Furtado, 1964; Hirschman, 1958; Kuznets, 1965; and Prebisch, 1950, 1963). Industrialisation is seen as crucial for developing countries being able to "catch up" with advanced economies.

A related corpus of literature is that in the Kaldorian tradition (see Kaldor (1966); Verdoorn, (1993 [1949])). The Kaldorian approach emphasises the importance of dynamic increasing returns in manufacturing, and the importance of manufacturing for economy-wide productivity growth and overall economic growth. As with the structuralist approach, this perspective also points to the importance of industrialisation.

A dynamic current literature continues to engage with the role and potential of manufacturing as an engine of growth in developing countries (e.g. see Alcorta et al., 2013; Andreoni & Tregenna, 2021; Haraguchi et al., 2019; Szirmai, 2012; Szirmai et al., 2013; and Tregenna, 2008, 2009, 2015, 2016). Broadly, this research points to the ongoing importance of manufacturing and of industrialisation for growth in developing economies. Empirically, however, there is a high degree of heterogeneity in this "growth-pulling" role, with some evidence showing that this has weakened over time.

The various bodies of literature referred to above provide relevant and insightful theoretical and empirical perspectives and evidence on the importance of industrialisation, especially for developing countries. Next, we turn to the aspect of technology intensity and technological upgrading, as background to the specific theme of high-tech industrialisation.

2.2 *High-Tech Industrialisation*

2.2.1 Innovation and Technological Upgrading

There are intrinsic links between shifts to higher-productivity activities, structural transformation, technological upgrading, and industrialisation. Shifting to more high-tech manufacturing activities is central to a successful industrialisation path, and more broadly to a rapid growth path. Competitiveness in high-tech manufacturing is also important for retaining and growing market shares in products that are growing their shares of global demand. High-tech manufactured products tend to be relatively complex, sophisticated, and R&D- and innovation-intensive (Hobday, 2013; Montobbio & Rampa, 2005).

What Andreoni and Tregenna (2020) refer to as "keeping pace" with rapid technological change, including through high-tech industrialisation, is critical for the ability of countries to avoid being stuck in what they call a "middle-income technology trap", or a "middle-income trap" in general. Furthermore, high-tech manufacturing does not follow the same typical deindustrialisation path as low-tech manufacturing does (Tregenna & Andreoni, 2020), making high-tech industrialisation an attractive path for the future.

Through its capacity to disrupt markets by rendering old technologies obsolete and continuously creating new ones, innovation is at the heart of the industrial mutation that defines long-term growth trajectories and is especially relevant to high-tech industrialisation (Dosi et al., 1988; Fagerberg & Srholec, 2009; Hobday, 2003; Lee, 2013). An extensive literature recognises the importance of technological upgrading and technology intensity for industrialisation and for growth and catching up (Andreoni & Tregenna, 2021; Avenyo et al., 2021; Bell & Pavitt, 1993; Fagerberg, 2000; Figueiredo, 2001; Hobday, 2003, 2013; Lall, 1992, 1996, 2000; Lee, 2013). This literature has implications for the importance of high-tech industrialisation and what is needed to enable this, in particular concerning the links with aspects such as innovation, learning, capabilities, and export competitiveness.

2.2.2 Productive Capabilities

A growing literature recognises the importance of productive capabilities, including for structural transformation, industrial development, and technological upgrading—and hence for a high-tech industrialisation path. For developing countries to benefit from the advantage of innovation and new technological knowledge, it is now increasingly being accepted that they have to develop appropriate "capabilities" that enable them to overcome their technological disadvantage (Fagerberg & Srholec, 2009). According to Khan (2004), the nexus between capability-based innovation and development runs through "positive feedback loops" that lead to a virtuous circle: innovations lead to higher productivity, which leads to growth and prosperity, which in turn form the basis for more innovations.

Capabilities have been presented through different concepts, such as "technological capability" (Kim, 1980), "technological mastery" (Dahlman & Westphal, 1982), "absorptive capacity" (Cohen & Levinthal, 1990), or "social capabilities" (Abramovitz, 1986), and as being fundamental to the process of development and technological catch-up (Fagerberg & Srholec, 2009).

High-tech industrialisation requires more advanced productive capabilities than does a less technology-intensive industrialisation path (or relative to those parts of other sectors such as agriculture or services that are not technology-intensive). Thus, the building of strong productive capabilities—including through investment, skills upgrading, technology upgrading, and changes in the organisation of production—is critical for the success of a high-tech industrialisation path.

In the long run, a country's competitiveness depends on its capacity to harness human capital and technological knowledge in its productive structures, and the development of high-tech industry plays a key role in this process because of the relatively higher income elasticity that characterises the demand of products with high technological content (Stokey, 1988; Hunt & Tybout, 1999; Andreoni & Tregenna, 2021). Technology adoption renders all producers more efficient through knowledge spillovers, but the effect of these spillovers on efficiency is strongest among relatively sophisticated goods. The expected increase in the uptake of technologies associated with the Fourth Industrial Revolution (4IR) represents not only challenges but also a wide range of opportunities for employment creation, for technological catch-up, and for steering the economy towards a green and sustainable future (Tregenna et al., 2021).

2.2.3 Export Competitiveness

Export competitiveness is an important determinant of a country's economic performance. One aspect of this is the capacity of an economy to earn foreign exchange. Export competitiveness also signals the quality and sophistication of the corresponding country's production system and its ability to meet international requirements and to capture and retain shares in international markets. Among the determinants of export competitiveness, the productivity and innovation performance of a country's enterprises play a prominent role. They affect the efficiency with which products and services can be brought to the market, but also their market attractiveness with respect to those from other countries. Trade openness and the attractiveness of the incentive systems put in place to promote quality and support firms in exporting are also important for strengthening export competitiveness. The export-promotion and incentive schemes available to support exporters and help them mitigate the risks they face in foreign markets are important for strengthening the country's export growth.

The diversification and technological content of a country's exports provide important indications of its resilience to external shocks. A more diversified and technology-intensive export composition is better positioned to resist unexpected changes in prices, since different export industries are usually affected differently by price or demand shocks.

An extensive empirical literature points to the importance of innovation for export performance at the firm level (Cieślik & Michałek, 2017; Cieślik et al., 2018; De Fuentes et al., 2020; Guarascio & Pianta, 2017; Lo Turco & Maggioni, 2015). Avenyo et al. (2021) empirically demonstrate two-way causal relationships between innovation and export performance (both export propensity and intensity) among African firms.

There is a close relationship between high-tech industrialisation and export competitiveness. High-tech manufactures are typically more complex, more sophisticated, and more likely to face growing demand internationally. Technological upgrading is necessary for sustained success in international markets. In turn, exporting can contribute to building the capabilities for successful high-tech industrialisation. For an economy to be competitive on the global markets, high levels of science and technological capabilities are indispensable. It is in recognition of this imperative that at the 2002 Barcelona Summit, the European Union set the Barcelona target with the aim of increasing the rate of R&D investment in European countries and bringing this rate close to 3% of GDP (EASAC, 2004). To emphasise the role of industry in driving this ambitious strategy, it was recommended that two-thirds of the funding for these R&D investments be financed by the private sector (EASAC, 2004). High-tech industry is driven by the intensive use of knowledge to spur innovation and technological change in order to keep pace with continuously evolving user demands (Mani, 2004; Falkowski, 2018). Its competitiveness is therefore dependent on its capacity to sustain innovation by investing in R&D activities.[1]

2.2.4 High-Tech Manufacturing and Green Industrialisation

An aspect of high-tech industrialisation that has begun to attract growing attention is its link with environmental sustainability, and its potential to be part of a of a possible "green industrialisation" path. Generally, more technology-intensive manufacturing tends to be less environmentally damaging. High-tech manufacturing has an important role to play in the shift to more environmentally sustainable production and consumption (Seyoum, 2004). Avenyo and Tregenna (2022) provide econometric empirical evidence from across developing and emerging economies that technology intensity in manufacturing matters for the intensity of CO_2 emissions and specifically that medium- and high-tech manufacturing is less emission-intensive than low-tech manufacturing. Thus, in addition to the importance of high-tech industrialisation for growth, it can also be important to a green transition and to an environmentally sustainable growth path.

In this section, we have reviewed the extant literature as background to the importance of high-tech industrialisation. We have discussed the importance of industrialisation for growth and "catching up", and the specificities of why high-tech industrialisation matters. In particular, we have considered the dimensions of innovation, technological upgrading, and technology intensity, productive capabilities, export competitiveness, and the links between high-tech industrialisation and green industrialisation as part of an environmentally sustainable high growth path. In the next section, we provide an overview of the "state of play" of high-tech industrialisation in South Africa.

[1] Even though many innovations take place without necessarily involving R&D investments, research activities remain the most important source of technological innovation (e.g. see Gottardi, 1996; Som, 2012; Mustapha & Kondlo, 2019).

3 Characterisation of the High-Tech Landscape in South Africa

3.1 A Brief Overview of the South African High-Tech Sector

The characterisation of South Africa's high-tech landscape is a challenging exercise for more than one reason. The first difficulty comes from the fluidity and ambiguity in the classification of what constitutes high-tech industries: there is no single authoritative classification method, and existing taxonomies can be dependent on the country's context (Doms & McGuckin, 1992; D'Aurio, 2021; US Department of Commerce, 2017; Mustapha & Kondlo, 2019). While the US Bureau of Labor Statistics (BLS), for example, uses a methodology based on the share of science, technology, engineering, and mathematics (STEM) occupational workers, the Organisation for Economic Co-operation and Development (OECD) identifies high-tech industries based on R&D intensities measured by the ratio of R&D expenditures to industry sales (US Department of Commerce, 2017; D'Aurio, 2021). Some scholars use hybrid methodologies that combine these indicators with the use of modern capital (Doms & Mc Guckin, 1992). As a result of differences in methodological approaches, countries can end up with different classifications of high-tech industries. For instance, some industries that are classified as medium-tech in the OECD context could be viewed as high-tech in a developing country context.

A second challenge lies in identifying the actual operations and processes within industries classified as belonging to the high-tech sector. These may entail unsophisticated levels of technological inputs, in cases such as providing low-technology parts or maintenance/repair services, rather than the actual design and manufacturing of the corresponding hi-tech products. If the production process is broken down into multiple stages, each of which has a different degree of technological sophistication, it may be difficult to accurately identify the technological complexity of an individual production stage when trade statistics are aggregated for the final product. Some developing countries included in the product value chain may therefore end up performing the low-technology components of a high-technology product in the production chain, while the high-tech production stages remain in a different country. South Africa, for example, is integrated into the personal computer value chain, with final assembly taking place there before the computers are exported. Such assembly operations do not involve high-skills intensity, nor do they add much value to the production, but the firms involved in them will still be classified as being in a high-tech industry (Mustapha & Kondlo, 2019).

A third challenge relates to the difficulty in accessing disaggregated data on the primary indicators of R&D intensity and the share of STEM work occupations in employment among firms classified as operating in the high-tech sector. This challenge is especially difficult to overcome in cases where many firms are startups with a limited amount of publicly available information on these indicators (Driver & Hodge, 2000).

Applying the US Bureau of Census approach[2] on export data from the South African Customs and Excise Trade Statistics reveals that South Africa had active export capacity in all ten high-tech fields, even though the weight of their export value relative to total exports remained limited (TIPS, n.d.). In contrast, using a clustering method on the measures of technology intensity based on the OECD's approach, Mustapha and Kondlo (2019) found that only one industrial category (radio, television, and communication apparatus) could be classified as high-technology, while six more industries could be categorised as medium-high technology: motor vehicles, electrical machinery, chemicals, non-electrical machinery, non-metallic mineral products, and petroleum products and nuclear fuel.

Apart from the defence and weapons industry mainly represented by the state-owned enterprise Denel, the high-tech industries in South Africa are to a large extent dominated by multinational companies, whose primary interest is in accessing the local market rather than in exporting what is produced locally (TIPS, n.d.). In terms of geographical distribution, most employment in the high-tech sector is in the Gauteng province, with the Cape Town metropolitan region accounting only for a smaller fraction.

3.2 *The Performance of South African High-Tech Industry*

One of the most significant indicators for assessing a country's long-term competitiveness is a measure of export performance, particularly its capacity to manufacture and export in high-tech products on global markets (Porter et al., 1996; Godin, 2004, Seyoum, 2005; Falkowski, 2018). In this regard, South Africa has been performing modestly, with its imports of high-tech products consistently surpassing its export over the period 1995–2019. Seen within the broader context of the South African manufacturing industry, the weak performance of the high-tech industry is not surprising. Since 1990, the share of manufacturing value added in the South African gross domestic product (GDP) has been steadily declining (see Fig. 1). This reflects the premature industrialisation as described by Tregenna (2009) and Andreoni and Tregenna (2021). The rapid trade liberalisation pursued from the mid-1990s to the early 2000s had strong negative effects on manufacturing, notably in the clothing and textile industries (Erten et al., 2019; Mabugu & Chitiga-Mabugu, 2007).

The expansion of South Africa's manufacturing sector in general, as the engine of the country's industrialisation, has been constrained by its inability to spur the type of structural transformation that could be strong enough to put the country's

[2] In order to overcome the problem caused by the inclusion of high-technology processes in the high-technology product categories, the US Bureau of the Census developed a list of ten technological fields which are commonly defined as producing high-technology products: biotechnology, life sciences, opto-electronics, computers and telecommunications, electronics, computer-integrated manufacturing, materials design, aerospace, weapons, and nuclear technology.

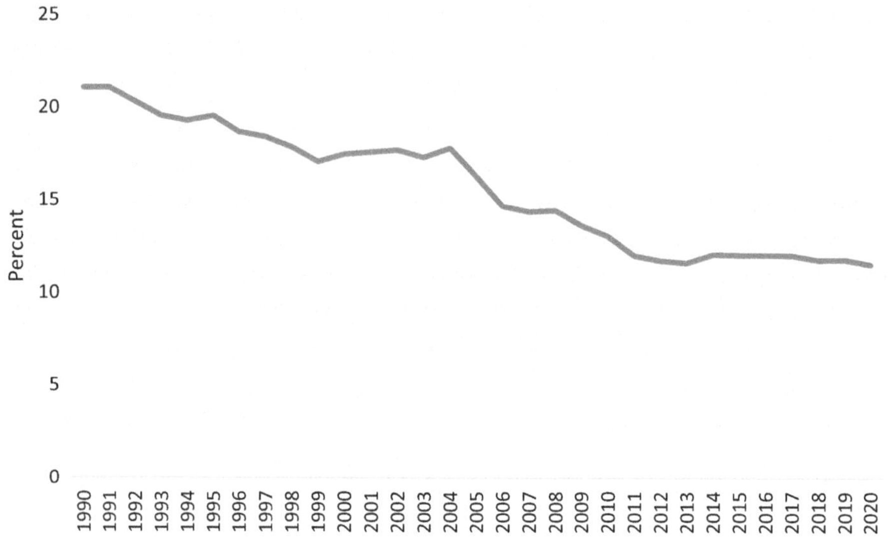

Fig. 1 Manufacturing value added as a percentage of GDP, 1990–2020. Data source: World Bank World Development Indicators, 2021

economy on a higher growth path (Andreoni & Tregenna, 2021). Further to this overall poor performance of South African manufacturing, there has not been growth in high-tech industry. As shown in Fig. 2, there has not been a shift in the composition of manufacturing value added in favour of high-tech manufacturing—actually a slight decline. This bodes ill for technological upgrading and an orientation towards high-tech industrialisation.

In order to form a reliable representation of the current state and possible trends of high-tech industry in South Africa, it is useful to take a close look at the recent dynamics in employment and output performance among industries classified as high-tech by Mustapha and Kondlo (2019)—namely, radio, television, and telecommunication equipment. Data from Quantec[3] show a picture of an industry that has been in decline, with a slight recovery in the wake of the 2007–2008 global financial crisis followed by an inflection in the growth momentum since 2014. Both in terms of output value and contribution of the industry to the total manufacturing value-added, available data indicate an industry in stagnation, with a declining trend in recent years (see Figs. 3 and 4). Some industries classified in the high-tech category according to the OECD approach, namely, pharmaceutical products and aircraft and spacecraft products, have been performing well, with respective export growth rates of 16.4% and 23.1% for the first quarter of 2021, but their export volumes remain relatively modest (TIPS, 2021).

[3] Note that the Quantec data are not official statistics and include various imputations and estimations. In general, it is advisable to avoid basing conclusions on the details of any short-run changes on this data.

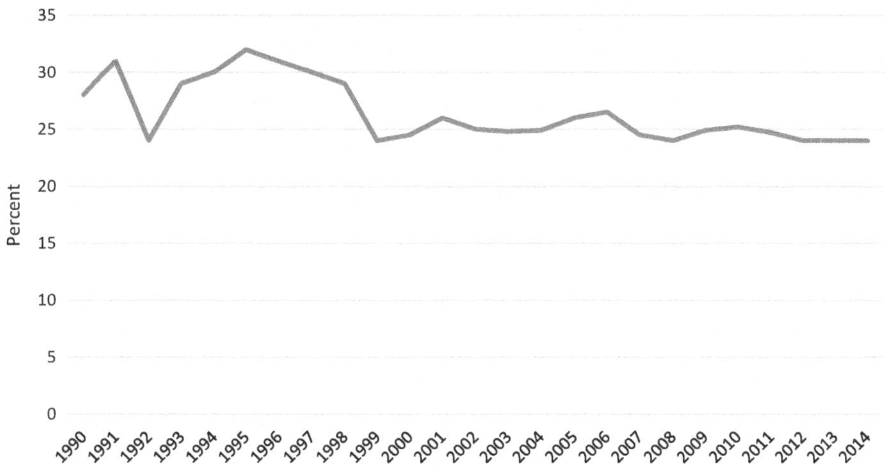

Fig. 2 South Africa's medium- and high-tech manufacturing value added as a percentage of total manufacturing value added, 1990–2014. Data source: UNIDO Competitive Industrial Performance Index, 2021

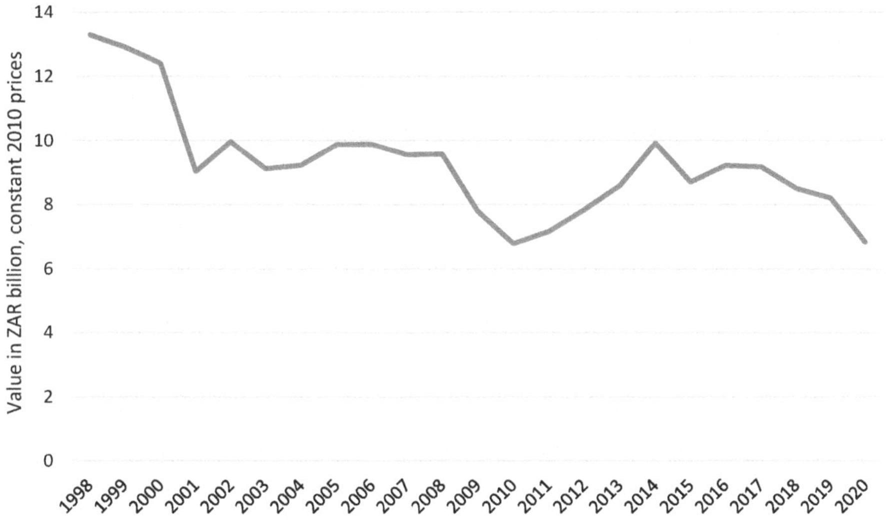

Fig. 3 Gross value added in domestic production of radio, television, and telecommunication equipment, 1998–2020. Data source: Quantec Easydata, 2021

Corresponding data related to industries classified as medium-high technology, including machinery and capital equipment as well as the chemicals and petroleum products, show a similar picture of stagnation since 2014, both in terms of output and employment (TIPS, 2017, 2021). Both global market conditions and domestic demand, as well as supply factors, have contributed to this stagnation. Any success of forward-looking industrialisation strategies in the tech sector is therefore going

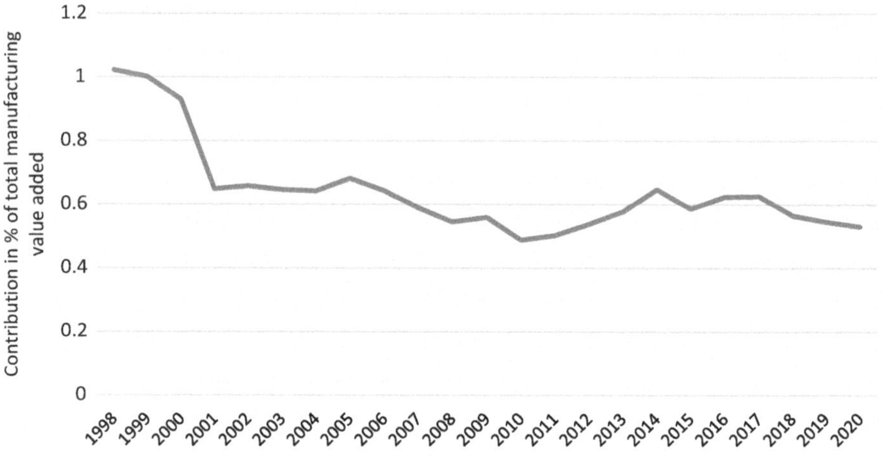

Fig. 4 Contribution of radio, television, and telecommunication equipment to total manufacturing value added. Data source: Quantec Easydata, 2021

to depend on a thorough understanding of the complex factors that have limited the growth opportunities of existing high-tech manufacturing.

The competitiveness of South Africa's exports has long been driven by its endowments in mineral resources, which enable a strong production of ores and metals. Manufactured exports, by contrast, have seen their contribution to export competitiveness slump in the period following the global financial crisis, showing a stagnant and slightly declining trend in recent years (Fig. 5).

The persistent trade deficit in high-tech products is partially explained by the nature of the processes that characterise some high-tech industries. The manufacturing of aircrafts, spacecrafts, and their components, for example, is classified as belonging to the high-tech industry in the OECD taxonomy. South Africa's export revenues in this sector are consistently modest, with companies that operate in this industry mainly performing repairs and maintenance activities instead of the actual production of exportable spacecraft products or component parts (Mustapha & Kondlo, 2019).

As a component of the manufacturing sector, the high-tech industry in South Africa also occupies a modest position in driving the country's competitiveness in global markets. Figure 6 shows that the total export value of South African high-tech products has been on a declining trend since 2015, with its recovery potential hampered by the unfavourable demand conditions in the global markets, exacerbated since early 2020 by the Covid-19 pandemic.

Export growth and the competitiveness of the high-tech sector require sustained investment in innovation to keep products attractive to foreign consumers and advantageous in value for money on international markets. While the real dollar value of South African high-tech exports has been on a declining trend since 2015, their small share in total manufactured exports has also been shrinking—declining

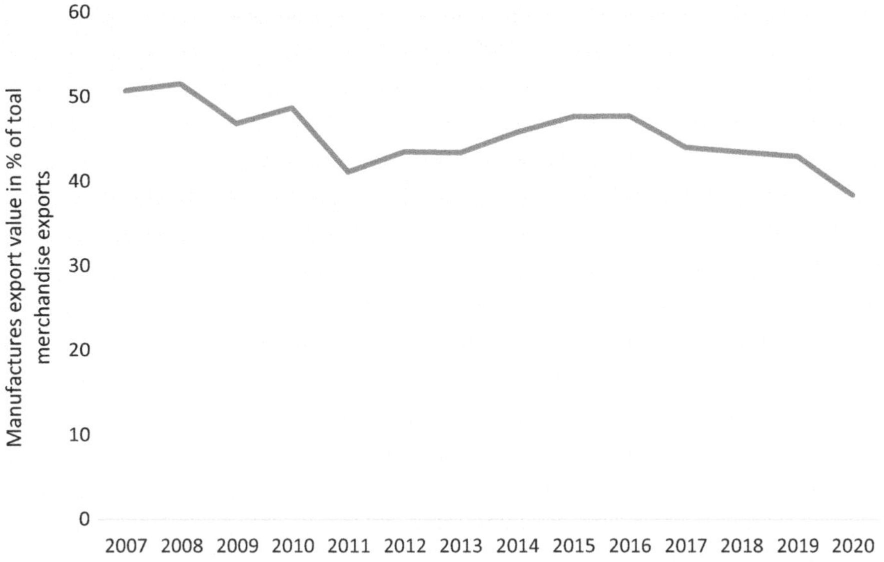

Fig. 5 Manufactured exports as a percent of total merchandise exports, 2007–2020. Data source: Comtrade Statistics, 2021

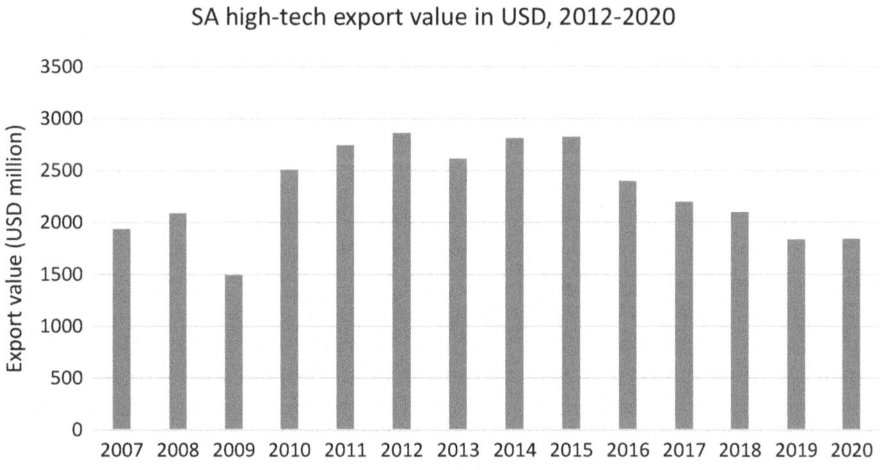

Fig. 6 South Africa's high-tech export value in current USD million, 2012–2020. Data source: Comtrade Statistics, 2021

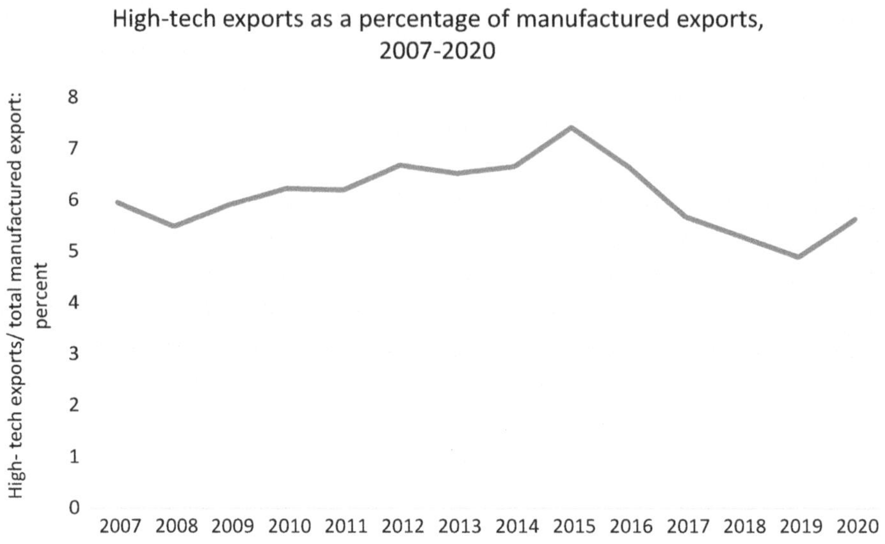

Fig. 7 South Africa's high-tech exports as a share of total manufactured exports, 2007–2020. Data source: Comtrade statistics, 2021

from a high of 7.4% to a low of 4.8% in 2019, before recovering slightly in 2020 (Fig. 7).

3.3 R&D Intensity in South Africa's High-Tech Sector

As stressed by Peña-Vinces and Audretsch (2021), the strong development of human skills at the higher education levels and competence in science and engineering fields in a country are important drivers of a firm's capacity to access the human skills and material resources they need to gain export competitiveness in high-tech industries. The heavy skills intensity of high-tech industries requires firms engaged in such activities to rely on advanced levels of human capital among their managerial and technical personnel to achieve the level of productivity that enables them to compete with their foreign counterparts on global export markets. Intensive investments in R&D, technological capabilities, and innovation diffusion therefore have an important role to play in reigniting high-tech export growth, diversifying the composition of its industries, and supporting their competitiveness in global markets.

The growth of high-tech industry has important spillover effects that enhance the performance of the rest of the economy. This is because a shift in demand towards high-tech products induces additional investment in education and training, which accelerates the rate of knowledge accumulation (Stockey, 1991; Hunt & Tybout, 1999; Mustapha & Kondlo, 2019). For South Africa, the high-tech sector is

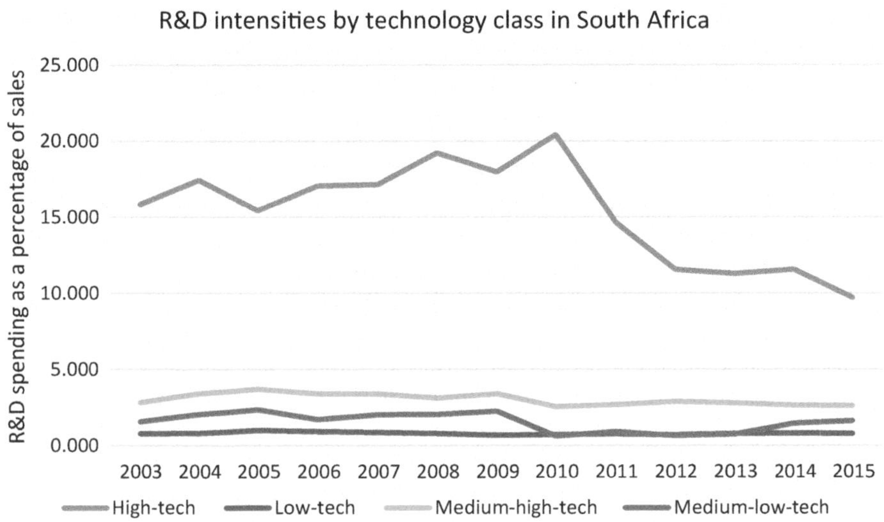

Fig. 8 Evolution of the R&D intensity in South African high-, medium-, and low-technology industries, 2003–2015 (data source: Mustapha and Kondlo, 2019)

characterised by varying degrees of R&D intensity (Mustapha & Kondlo, 2019). The graphical representation of their R&D intensity in Fig. 8 shows a clear difference between the high-tech industries and the rest of the manufacturing industries.

The application of the hierarchical clustering method on the industries that are active in the South African manufacturing sector, as undertaken by Mustapha and Kondlo (2019), shows a substantial degree of heterogeneity in R&D intensities. This approach yields an alternative classification well suited to characterising technology intensity in the context of a developing country such as South Africa.

While export performance and R&D give an indirect measure of the available technological capabilities in South African high-tech industries, a more direct measure of capabilities is difficult to derive. This is inter alia because of the challenge that we have discussed above concerning the classification of activities and processes in terms of categories of technological intensity.

4 Semantic Analysis: High-Tech Industrialisation in South Africa

4.1 Mapping Emerging Trends in New Knowledge Generation

The semantic analysis presented and discussed below is based on the semantic network maps generated by the "Intelligent Foresight Analytics (iFORA)" system developed by the Institute for Statistical Studies and Economics of Knowledge

(ISSEK) Higher School of Economics in Moscow. The iFORA approach uses advanced semantic analysis algorithms, machine learning, and artificial intelligence to combine diverse information sources including scientific articles, patents, news, grants, and analytical reports among the others with the aim of exploring emerging trends in new knowledge generation. Those semantic maps enable the visualisation and the analysis of the semantic connections between numerous concepts featured in the information sources, which are structured according to their respective domain areas and thrusts identified in the South Africa 2030 Science, Innovation, and Technology Foresight reports (DST, 2019; DSI, 2021).

This section therefore draws substantially on the work already carried out in both reports, which were prepared by researchers from the ISSEK's Higher School of Economics. For the high-tech industrialisation domain, the interim National STI Foresight report identified three main thrusts that offer a potential basis for smart strategies aimed to foster the adoption of innovative technologies that will support the necessary structural transformation of the economy: (1) enabling small businesses to adopt new technologies, (2) new thinking for new industries, and (3) new thinking for old industries. The analysis follows the structure of the semantic maps and its different clusters as they are organised into the three thrusts. The process followed to generate the maps as well as the criteria used to for clustering the leading themes of these thrusts are explained in more detail by Saritas et al. (2024) in the introductory chapter of this volume. The final report of the National STI Foresight 2030 study also discusses 7 key high-tech adoption thrusts in relation to 12 main socio-economic indicators of desirable socio-economic outcomes.

The interpretation of these maps produced by the iFORA system is premised on the conditional expectation that the identified network connections between the clustered concepts form leading indicators of actual drivers of new knowledge generation in the high-tech industrialisation domain in South Africa. It is indeed reasonable to expect a close connection between emerging trends in scientific publications and the related knowledge creation that will determine the future of high-tech industrialisation under the strategies identified by the corresponding thrusts. The discussion of the mapping results in the following subsection is also made possible by the harnessing of the ISSEK's Big Data Analysis commissioned by the DSI to identify key bibliometric trends from thousands of documents. The capabilities shown by this foresight exercise suggests that, in the future, scientists will be able to use the algorithms of big data analysis and machine learning to connect the meanings and the implications of concepts to their applicability in the real word based on their identified association in networks of hierarchies, semantic proximity, co-location, and frequency of occurrence in documents.

4.2 *The Future of Industrialisation: Bibliometric Trends*

South African scientists play an active role in research in high-tech-related research. Their output in this domain, as measured by the number of publications in Scopus-accredited outlets, has been rapidly growing over the last decade (see Fig. 9). The relative contribution of South African researchers to the global total remains fairly modest; however, their share grew from a tiny 0.26% of the global output in 2008 to just over 0.5% in 2018.

In terms of ranking, based on the cumulative number of Scopus-indexed publications over the period 2008–2018, South Africa also ranks modestly—in 41st place—just behind Indonesia and Ukraine (Fig. 10).

Table 1 shows the top ten South African institutions contributing to high-tech-related publications, ranked according to the number of distinct publications authored by their affiliated researchers. The University of Pretoria tops the list, followed by the University of Cape Town in second place, and Stellenbosch University in third place.

Research collaboration plays an important role in enhancing the likelihood of success in innovation outcomes. Table 2, which uses data taken from the foresight exercise, shows that South African researchers have been engaged in collaborative research with the global research community, primarily with researchers from the Global North but also with many from the Global South.

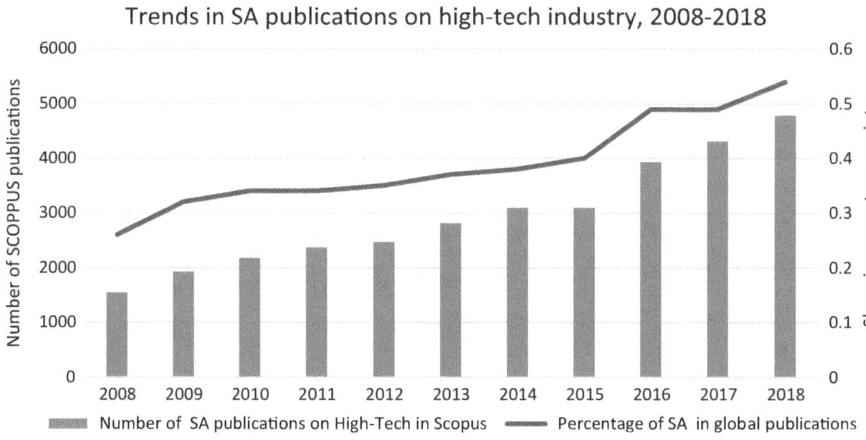

Fig. 9 Trends in scientific publications related to high-tech industrialisation in numbers and SA share of the global total. Source: Authors' plot with data from DSI national foresight exercise interim report

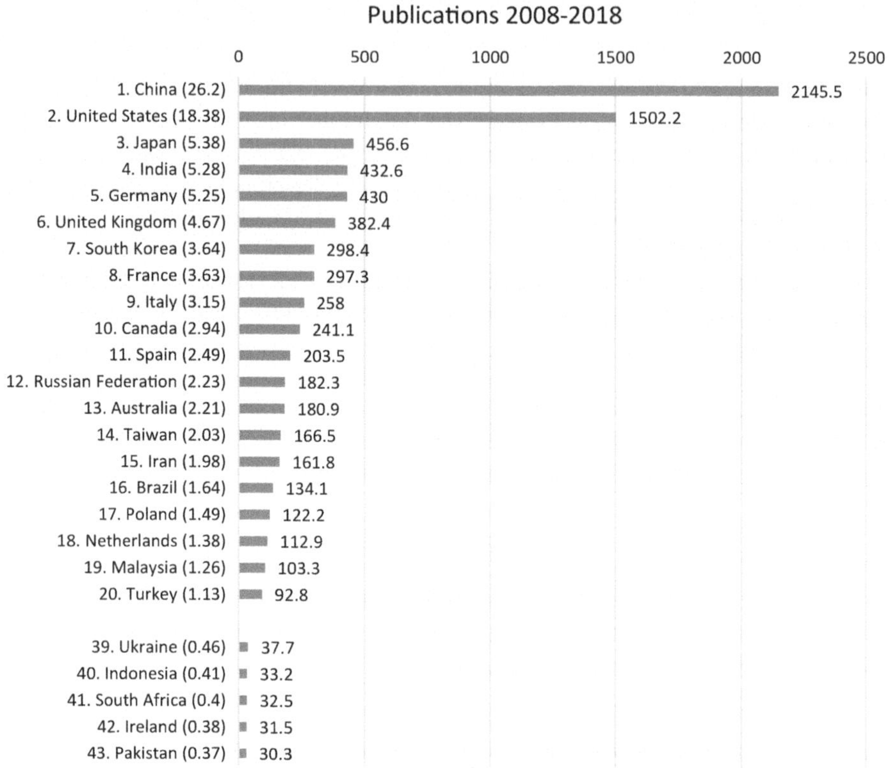

Fig. 10 South Africa's global ranking in cumulative number of publications on high-tech publications indexed in Scopus, 2008–2018. Source: Authors' plot with data from DSI national foresight exercise interim report

Table 1 Top 10 South African institutions by number of publications in "high-tech industry" domain in Scopus for 2008–2018

Institution	Number of publications	Percentage of total
University of Pretoria	4036	12.41%
University of Cape Town	3925	12.07%
University of Stellenbosch	3591	11.04%
University of Kwa Zulu Natal	3493	10.74%
University of Johannesburg	3436	10.57%
University of the Witwatersrand	3166	9.74%
Council for Scientific and Industrial Research	2226	6.85%
Tshwane University of Technology	1756	5.40%
North-West University	1293	3.98%
University of South Africa	1192	3.67%

Table 2 Top 20 partner countries of South Africa in Scopus-indexed international collaboration publications in "high-tech industry" domain, 2008–2018

Country	Number of joint publications with SA scientists	Share of the country in all international collaborative publications of SA (Scopus indexed)
United States	2665	19.10%
United Kingdom	1982	14.20%
India	1724	12.36%
Germany	1451	10.40%
China	1330	9.53%
France	1246	8.93%
Australia	10,771	7.68%
Nigeria	932	6.68%
Canada	893	6.40%
Netherlands	879	6.30%
Italy	717	5.14%
Sweden	681	4.88%
Spain	611	4.38%
Switzerland	559	4.01%
Turkey	505	3.62%
Japan	503	3.60%
Saudi Arabia	503	3.60%
Brazil	467	3.35%
Poland	435	3.12%
Iran	421	3.02%

Source: DSI national foresight exercise, interim report

Fig. 11 IT adoption cluster

4.3 Key Thrusts of High-Tech Industrialisation in South Africa

4.3.1 Enabling Small Businesses to Adopt New Technologies

The first of the three key thrusts identified by the foresight exercise for the future of high-tech industrialisation is the imperative to enable small businesses to adopt new technologies for them to thrive in a changing global technological landscape. This thrust consists of three clusters of emerging topics, namely, "IT adoption", "high-tech startups", and "skills development". Each cluster has been crystallised from the connections between related topics of greater centrality: disruptive innovations and the advantages of their adoption that yielded the IT adoption cluster (Fig. 11), startup funding, entrepreneurship and venture capital investment for the "high-tech

Fig. 12 High-tech startups cluster

Fig. 13 Skills development cluster

startups" cluster (Fig. 12), and employment, retirement, and workforce for the "skills development" cluster (Fig. 13).

As economies worldwide continue to embrace the Fourth Industrial Revolution, connectivity is going to play an increasingly crucial role in enabling businesses to invest and take advantage of new possibilities previously unavailable to them. The potential offered by ultra-fast connectivity to exploit new technologies will undoubtedly be tied to the quality and reliability of the available connectivity infrastructure. The success of strategies aimed to enhance the operating environment of high-tech startups and other small businesses will hinge on ensuring their increased capacity to interact and collaborate wherever possible and reap the benefits of synergy and knowledge spillovers. In this respect, the foresight exercise identified the first-mover advantage as one of the key drivers of success, whereby early adopters of the leading technologies gain a competitive advantage. Disruptive innovations, in the sense of Schumpeterian creative destructions, are most likely to offer such a competitive advantage by displacing obsolete technologies and establishing niches of profitability before technologies are widely adopted by their competitors (Fig. 11).

The "skills development" cluster is dominated by the concerns about unemployment and the preoccupation with retirement and economic downturn but also refers to employment, job creation, and career development as related issues (Fig. 13). This cluster also emphasises the importance of linking these concepts to the promotion of technical talents and the deployment of a skilled workforce in supporting the widespread adoption of new technologies among small businesses.

4.3.2 New Thinking for New Industries

This thrust consists of nine clusters of related concepts, each of which can potentially contribute to charting new thinking in the emerging high-tech industries: artificial intelligence, virtual reality, gaming, cryptocurrency, biometry, cleantech, new technologies, aerospace, and biotechnology.

Artificial intelligence (AI), along with the associated development in machine learning and complex problem-solving, is expected to revolutionise strategic thinking and operations of the industries of the future, in which the application of virtual reality and gaming simulation will play an important role in supporting decision-making (see Figs. 14, 15, and 16).

Blockchain technology, which forms the basis for secure transactions in cryptocurrencies, is equally expected to play an important role in the industries of the future (Fig. 17). Beyond its application to cryptocurrencies, development in blockchain technology has the potential to power such diverse industries as energy

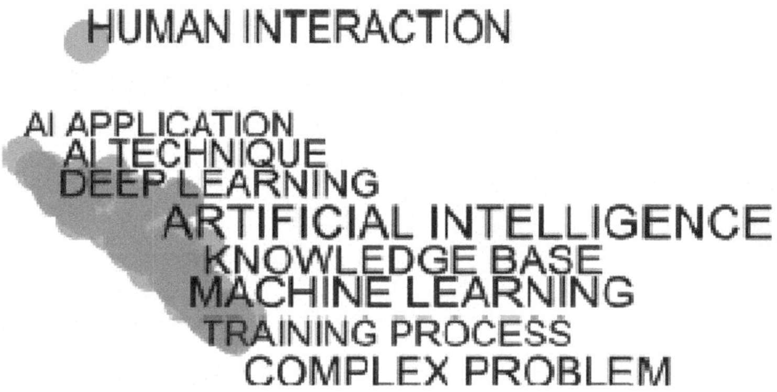

Fig. 14 Artificial intelligence

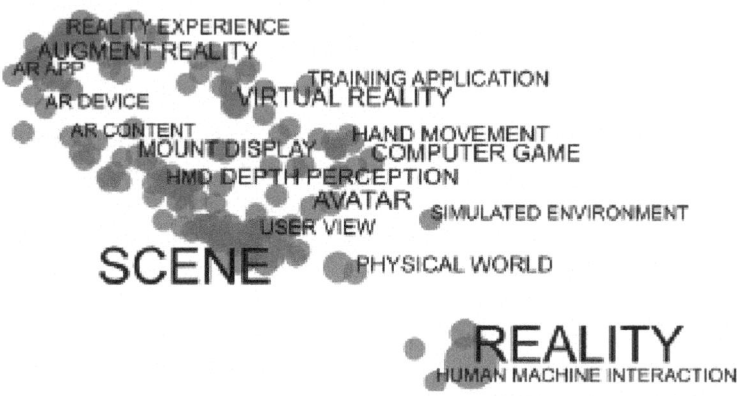

Fig. 15 Virtual reality

Fig. 16 Gaming

Fig. 17 Blockchain technology and cryptocurrency

(generation and distribution) and agriculture. Security of transactions in the emerging industries also requires further development in authentication systems, whereby biometric-based security systems will occupy a central place (Fig. 18).

Within the context of the ongoing threat of global warming, the security and sustainability of new industries are also directly related to the use of clean technologies, putting environmental protection and energy security at the centre of their development. This cluster is particularly concentrated as "cleantech" (Fig. 19).

Other directions of the emerging industries expected to benefit from new thinking are in the new technologies cluster, which includes additive manufacturing, 3D printing, and digital design (Fig. 20).

Fig. 18 Biometry cluster

Fig. 19 Cleantech

Fig. 20 New technology

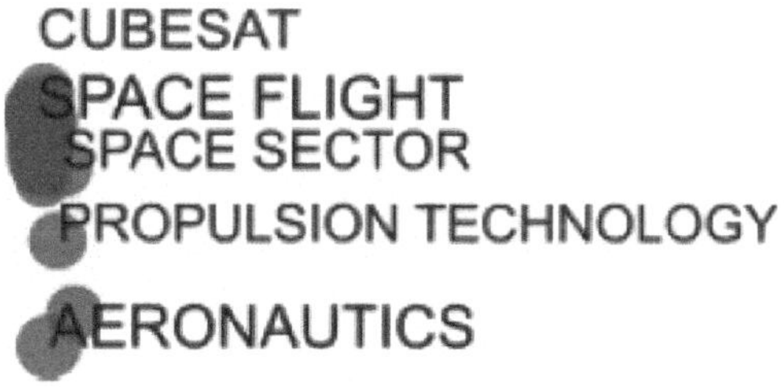

Fig. 21 Aerospace

For South Africa, growth in aerospace technologies and biotechnology are two other areas in which new thinking could generate new opportunities for technological progress—the final clusters encapsulating the coalescence of emerging ideas for new industries in the high-tech sector (Figs. 20, 21, and 22). While the aerospace and biotechnology industries in South Africa are by no means new, potential innovations could contribute to significant societal change if they are harnessed though strategically designed industrial policy.

Fig. 22 Biotechnology cluster

Fig. 23 Economy cluster

4.3.3 New Thinking for Old Industries

The final thrust—new thinking for old industries—consists of eight clusters structured around leading concepts that inspire changes in existing industries in each one of them: economy, productivity, climate change, environmental policy, sustainability, clean transport, energy efficiency, and chemistry (Figs. 23, 24, 25, 26 and 27). The economy cluster combines a multitude of concepts with relatively low centrality but which are linked through their pertinence to making existing industries more competitive (Fig. 23). This can be achieved by supporting innovation and entrepreneurship for the development of the manufacturing sector, by upgrading production technologies, and by expanding export markets. The related productivity cluster suggests that improvements in the productivity of existing industry should be pursued through new capital investments and the reduction of cycle time, among other strategies (Fig. 24).

Related to the "productivity improvement" cluster is the "energy efficiency" cluster, which centres around new energy technologies that improve both the generation and use of energy (Fig. 25). As a corollary of the "energy efficiency" cluster, the "clean transport" and "sustainability" clusters bring together the main ideas around making transport systems greener and less energy-intensive, and therefore more sustainable. Both an increase in the use of renewable energy and the reduction

Fig. 24 Productivity cluster

Fig. 25 Climate change cluster

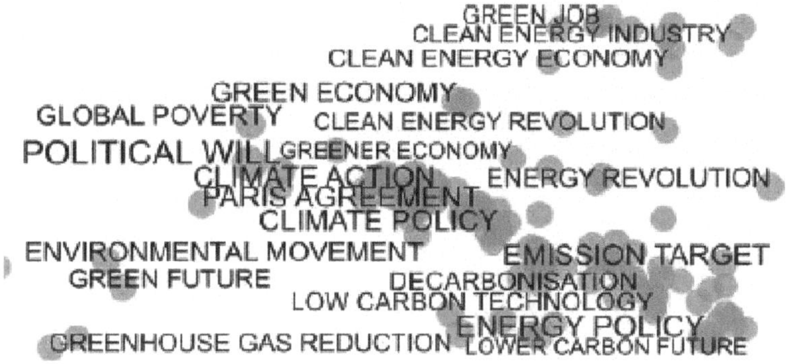

Fig. 26 Environmental policy cluster

Fig. 27 Sustainability cluster

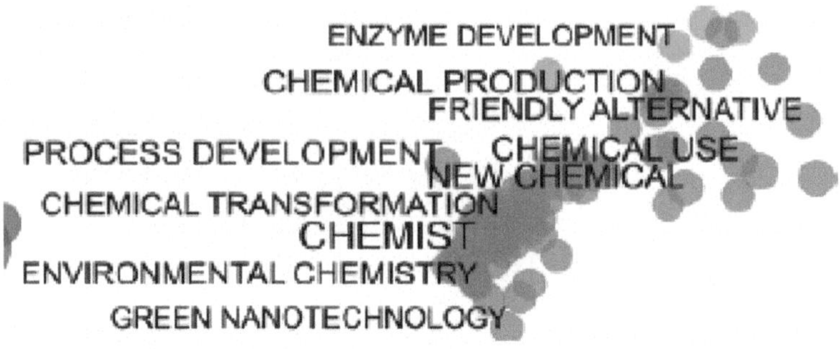

Fig. 28 Chemistry cluster

of fossil fuels in transport and other industries are required to build sustainable production systems that are adapted to the need to face the threat of climate change through appropriate agriculture and water management policies (Fig. 25). Related environmental policies targeting the decarbonisation of production systems and the reduction of greenhouse gas emissions (Fig. 26) can all be linked to building a more sustainable economy (Fig. 27).

The final cluster points to the importance of transforming the chemical industry with new discoveries and using it to produce more friendly alternatives for environmental protection and green materials for emerging technologies such as nanotechnology (Fig. 28).

4.4 Priority Areas for Innovation Strategies

In relation to innovation strategies for high-tech industrialisation relevant to South Africa, the semantic analysis identified seven priority clusters:

1. Blockchain and digital currency.
2. Cybersecurity.
3. Industry 4.0.
4. Technology adoption and industrial upgrade.
5. E-finance.
6. Personal information security.
7. Renewable energy.

The clusters were analysed and evaluated on the basis of their connection to the following 12 indicators of desirable socio-economic outcomes. These are (1) access to and supply of clean water, (2) affordable food, (3) reduction of carbon emissions, (4) economic growth, (5) export growth and competitiveness, (6) high living standards, (7) high-quality health care service, (8) job creation, (9) low greenhouse gas emission, (10) poverty alleviation, (11) renewable energy growth, and (12) skills development.

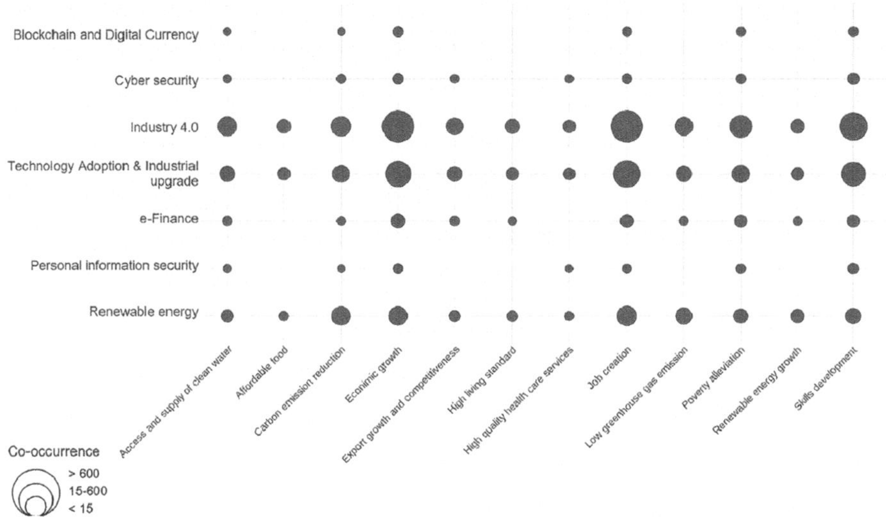

Fig. 29 Connection between the 7 Industry 4.0 priority areas and the 12 leading indicators of socio-economic outcomes. Source: DSI South Africa 2030 STI Foresight final report

The outcome of a bibliometric analysis of all high-tech-related documents mentioning South Africa brings to the fore the following main connections between the main priority areas of emerging technologies and the associated indicators of socio-economic outcomes: "Industry 4.0" and "technology adoption and industrial upgrade" are the clusters with the strongest connections to the largest number of indicators: both clusters connect most strongly to the indicators of skills development, job creation, and economic growth, which are also very relevant for the South African context (see Fig. 29).

In the bibliometric analysis, Industry 4.0 tends to be strongly associated with the "economic growth" indicator. This suggests that emerging technologies associated with Industry 4.0 are perceived as having the potential to be the new drivers of economic growth.

Industry 4.0 is equally strongly associated with the indicators of job creation and skills development. This suggests that intensive strategies need to be put in place to develop the skills required by Industry 4.0 technologies and to ensure that when these technologies are adopted and begin to diffuse widely, there will be enough human capacity to fill the corresponding jobs as soon as they arise from industry demand.

The strong connection between "technology adoption and industrial upgrade" and the indicators of job creation, skills development, and economic growth point to the important role that the application of new technologies as well as the improvement on existing production methods can be expected to play in enhancing firm productivity and, through this, boost aggregate economic growth. Successful adoption and absorption of externally developed technological knowledge for industrial upgrade is intimately linked to pre-emptively building adequate capabilities and supporting the corresponding efforts through an industrial policy with commensurate incentives.

Less intense in the frequency of connections but no less important is the link between the "renewable energy" cluster and the indicators of carbon emission reduction, job creation, and economic growth, which underscore the economic benefits and employment potential of an increased transition to renewable energy, on top of its direct effect on reducing carbon emissions for more sustainable production within the context of a changing climate. It should be evident that a strategy combining all these connections is desirable for the optimisation of socio-economic benefits under the constraints of available inputs.

5 Conclusion

Drawing on longstanding literature and on evidence on South Africa in particular, this chapter has taken a future-oriented perspective on high-tech industrialisation in the South African context. Through a review of the literature, we discussed the importance of high-tech industrialisation, including reflecting briefly on industrial development, innovation, learning and capabilities, technological upgrading and technology intensity, and industrial exports. We provided an empirical overview of high-tech industrialisation in South Africa. Through a novel semantic analysis, the chapter offers a unique and futuristic perspective.

The overview of the South African high-tech industry landscape shows stagnant performance within a manufacturing sector characterised by premature deindustrialisation. The three main thrusts identified by the semantic analysis, which represent potential pillars of a suitable high-tech industrialisation strategy based on current trends in research, highlight the importance of enabling small businesses to adopt new technologies, adopt new thinking to drive emerging industries, and apply new thinking to old industries to upgrade them and make them fit for a sustainable future. The analysis also discussed seven priority areas of high-tech industrialisation strategy in relation to how they connect to using Industry 4.0 technologies to yield desirable socio-economic outcomes for the future of South African economy.

While South Africa shares many characteristics and challenges with other developing countries, it also has unique features (Andreoni et al., 2021; Tregenna et al., 2021). One is its extremely high rate of unemployment, well known to be among the highest in the world. Unemployment is especially high among the low- and semi-skilled. Employment creation thus needs to be a central concern in any policy choices, including where there are possible tensions and trade-offs between competing policy objectives.

In the domain of high-tech industrialisation, the uptake of technologies associated with the Fourth Industrial Revolution (4IR) by manufacturing firms will come not only with challenges but also with multiple opportunities—for employment creation, technological catch-up, and for steering the economy towards a green and sustainable future (Tregenna et al., 2021).

There are ways in which high-tech industrialisation is compatible with employment creation and can, directly and indirectly, contribute to this and more broadly to addressing South Africa's "triple challenges". As discussed earlier, high-tech industrialisation is important to structural transformation and "catching up". Higher-productivity manufacturing activities associated with high-tech industrialisation can be linked to growth in demand and hence in output and employment. Furthermore, there can be strong linkages between some parts of high-tech manufacturing and other parts of the economy—including less technology-intensive and manufacturing as well as other more labour-intensive sectors such as services and agriculture. This suggests that the broader labour-absorbing impact of high-tech manufacturing can be significant through indirect effects and employment multipliers, even when high-tech manufacturing itself is not highly directly labour-intensive. It is also important to recognise that there is not necessarily a direct negative correlation between labour intensity and technology intensity, and even more so when indirect labour intensity is taken into account.

Yet, it is important to also acknowledge possible tensions, trade-offs, and associated policy choices. A path focused exclusively or overwhelmingly on high-tech industrialisation would tend to be intensive in capital, technology, and high-skilled labour, and not directly employment-intensive or specifically intensive in low- and semi-skilled labour. Taking indirect employment creation (through linkages and multipliers) into account mitigates this tension to some degree but does not resolve it. An industrialisation path that is excessively focused on high-tech manufacturing—even if feasible—would not yield the employment generation that South Africa desperately needs.

This suggests that high-tech industrialisation should be thought of as one component of South Africa's industrialisation, and indeed of its broader growth path. This needs to be blended and balanced with efforts to expand parts of manufacturing that are highly labour absorbing, even if they are not especially high productivity, technologically advanced, or growth pulling.

This sort of mixed industrialisation strategy implies that industrial policy needs to support a range of manufacturing sectors. Different industries can play different roles in the economy, some contributing more strongly to employment creation, others to the balance of payments, others to growth pulling through intersectoral linkages, others through innovation and technological advancement, and so on. Naturally, even within the more labour-intensive sectors that are classified as low-tech manufacturing, there is considerable scope for technological upgrading, including through 4IR technologies.

A shift to high-tech industrialisation—even as part of a balanced industrialisation path discussed above—is neither automatic nor easy. Many countries may aim at high-tech industrialisation, but not all will succeed. A developing country such as South Africa is deficient in the types of productive capabilities that are needed for success in high-tech industrialisation. Furthermore, there are both domestic and international constraints and challenges, including barriers to entry.

This points to the importance of robust policy efforts to promote and support high-tech industrialisation. Left to "market forces", we would not see the types of investments and upgrading effort that are needed for high-tech industrialisation.

With the rapid technological changes associated with the 4IR, South Africa needs to move quickly to take advantage of emerging possibilities and mitigate exposure to threats. This underscores the need for a foresight-driven approach that looks towards future challenges and opportunities, identifies the actions that are needed in the current period, and ensures timely implementation of the necessary changes.

References

Abramovitz, M. (1986). Catching up, forging ahead, and falling behind. *Journal of Economic History, 46*, 386–406.

Alcorta, L., Haraguchi, N., & Rezonja, G. (2013). Industrial structural change, growth patterns, and industrial policy. In J. E. Stiglitz, J. Y. Lin, & E. Patel (Eds.), *The industrial policy revolution II* (pp. 457–492). Palgrave Macmillan.

Andreoni, A., Mondliwa, P., Roberts, S., & Tregenna, F. (2021). Framing structural transformation in South Africa and beyond. In A. Andreoni, P. Mondliwa, S. Roberts, & F. Tregenna (Eds.), *Structural transformation in South Africa: The challenges of inclusive industrial development in a middle-income country*. Oxford University Press.

Andreoni, A., & Tregenna, F. (2020). Escaping the middle-income technology trap: A comparative analysis of industrial policies in China, Brazil and South Africa. *Structural Change and Economic Dynamics, 54*, 324–340.

Andreoni, A., & Tregenna, F. (2021). The middle-income technology trap and premature deindustrialisation in South Africa. In A. Andreoni, P. Mondliwa, S. Roberts, & F. Tregenna (Eds.), *Structural transformation in South Africa: The challenges of inclusive industrial development in a middle-income country*. Oxford University Press.

Avenyo, E., & Tregenna, F. (2022). Greening manufacturing: Technology intensity and carbon dioxide emissions in developing countries', *Applied Energy, 324*, 119726.

Avenyo, E., Tregenna, F., & Kraemer-Mbula, E. (2021). Do productive capabilities affect export performance? Evidence from African firms. *European Journal of Development Research, 33*, 304–329.

Avenyo, E., Tregenna, F., & Ngwadleka, K. (2021) "Learning to export" and "learning to innovate": Revisiting the relationship between innovation and exports in African firms', Economic Research Southern Africa Working Papers, No. 861.

Bell, M., & Pavitt, K. (1993). Technological accumulation and industrial growth: Contrasts between developed and developing countries. *Industrial and Corporate Change, 2*(2), 157–210.

Blankenburg, S., Palma, G., & Tregenna, F. (2008). Structuralism. In L. Blume & S. Durlauf (Eds.), *The new Palgrave: A dictionary of economics* (2nd ed., pp. 69–74). Palgrave Macmillan.

Chenery, H. (1955). The role of industrialisation in development programs. *The American Economic Review, 45*(2), 40–57.

Cieślik, A., & Michałek, J. J. (2017). Innovation forms and firm export performance: Empirical evidence from ECA countries. *Entrepreneurial Business and Economics Review, 5*(2), 85–99.

Cieślik, A., Qu, Y., & Qu, T. (2018). Innovations and export performance: Firm-level evidence from Chinese firms. *Entrepreneurial Business and Economics Review, 6*(4), 27–47.

Cohen, W. M., & Levinthal, D. A. (1990). Absorptive capacity: A new perspective on learning and innovation. *Administrative Science Quarterly, 35*, 128–152.

D'Aurio, R. (2021). International business innovation, IPRs, Tech. Transfer and diffusion. Unpublished paper.

Dahlman, C. J., & Westphal, L. (1982). Technological effort in industrial development—An interpretative survey of recent research. In F. Stewart & J. James (Eds.), *The economics of new technology in developing countries* (pp. 105–137). Pinter.

De Fuentes, D., Niosi, J., & Peerally, J. A. (2020). Exploring innovation and export interplay in Canadian firms. *Economics of Innovation and New Technology, 30*(1), 1–21.

Department of Science and Innovations (DSI). (2021). South Africa 2030 science, technology and innovation Foresight. Final report, Moscow.

Department of Science and Technology (DST). (2019). Big data analytics for South African National STI Foresight 2030. Interim report, Pretoria.

Doms, M. E., & McGuckin, R. H. (1992). Trade in high technology products. *Science and Public Policy, 19*(6), 343–346.

Dosi, G., Freeman, C., Nelson, R., Silverberg, G., & Soete, L. (Eds.). (1988). *Technical change and economic theory*. Pinter.

Driver, A. & Hodge, J., (2000). Understanding the high-tech sector in the Cape Metropolitan Region: a contribution to the development of a regional strategy for high tech industry. Development Policy Research Unit Working Paper No. 00/37, University of Cape Town.

ECLAC (United Nations Economic Commission for Latin America and the Caribbean). (1964). *El desarrollo económico de América Latina en la post-guerra*. UN (CEPAL).

Erten, B., Leight, J., & Tregenna, F. (2019). Trade liberalization and local labor market adjustment in South Africa. *Journal of International Economics, 118*, 448–467.

European Academies Science Advisory Council [EASAC]. (2004). Towards 3%: attainment of the Barcelona target. https://royalsociety.org/-/media/Royal_Society_Content/policy/publications/2004/9718.pdf

Fagerberg, J. (2000). Technological progress, structural change and productivity growth: A comparative study. *Structural Change and Economic Dynamics, 11*(4), 393–411.

Fagerberg, J., & Srholec, M. (2009). Innovation, technology and development: Unpacking the relationship(s). In B.-A. Lundvall, K. J. Joseph, C. Chaminade, & J. Vang (Eds.), *Handbook of innovation systems and developing countries. Building domestic capabilities in a global context*. Edward Elgar.

Falkowski, K. (2018). Competitiveness of the Baltic States in international high-technology goods trade. *Comparative Economic Research Central and Eastern Europe, 21*(1), 25–43.

Figueiredo, P. N. (2001). *Technological learning and competitive performance*. Edward Elgar.

Furtado, C. (1964). *Development and underdevelopment: A structural view of the problems of developed and underdeveloped countries*. University of California Press.

Godin, B. (2004). The obsession for competitiveness and its impact on statistics: the construction of high-technology indicators. *Research Policy, 33*(8), 1217–1229.

Gottardi, G. (1996). Technology strategies, innovation without R&D: and the creation of knowledge within industrial districts. *Journal of Industry Studies, 3*(2), 119–134.

Guarascio, D., & Pianta, M. (2017). The Gains from technology: New products, exports and profits. *Economics of Innovation and New Technology, 26*(8), 779–804.

Haraguchi, N., Martorano, B., & Sanfilippo, M. (2019). What factors drive successful industrialisation? Evidence and implications for developing countries. *Structural Change and Economic Dynamics, 49*(C), 266–276.

Hirschman, A. (1958). *The strategy of economic development*. Yale University Press.

Hobday, M. (2003). Innovation in Asian industrialisation: A Gerschenkronian perspective. *Oxford Development Studies, 31*(3), 293–314.

Hobday, M. (2013). Learning from Asia's success: Beyond simplistic 'lesson-making. In A. Szirmai, W. Naudé, & L. Alcorta (Eds.), *Pathways to industrialisation in the twenty-first century: New challenges and emerging paradigms* (pp. 133–153). Oxford University Press.

Hunt, J., & Tybout, J. (1999). *Does promoting high-tech products spur development?. FEEM Working Paper REG 42*. Fondazione Eni EnricoMattei.

Kaldor, N. (1966). *Causes of the slow rate of economic growth of the United Kingdom: An inaugural lecture*. Cambridge University Press.

Khan, H. (2004). Technology and economic development: The case of Taiwan. *Journal of Contemporary China, 13*(40), 507–521.

Kim, L. (1980). Stages of development of industrial technology in a developing country: A model. *Research Policy, 9*, 254–277.
Lall, S. (1992). Technological capabilities and industrialisation. *World Development, 20*(2), 165–186.
Lall, S. (1996). *Learning from the Asian tigers: Studies in technology and industrial policy*. Macmillan.
Lall, S. (2000). The technological structure and performance of developing country manufactured exports, 1985–98. *Oxford Development Studies, 28*(3), 337–369.
Lee, K. (2013). *Schumpeterian analysis of economic catch-up: Knowledge, path-creation and middle-income trap*. Cambridge University Press.
Lo Turco, A., & Maggioni, D. (2015). Dissecting the impact of innovation on exporting in Turkey. *Economics of Innovation and New Technology, 24*(4), 309–338.
Mabugu, R., & Chitiga-Mabugu, M. (2007). *South Africa trade liberalization and poverty in a dynamic microsimulation CGE model*. University of Pretoria Working Paper WP 2007–38.
Mani, S. (2004). Exports of high technology products from developing countries; are the figures real or are they statistical artifacts? In S. Mani & H. Romijn (Eds.), *Innovation, learning, and technological dynamism of developing countries* (pp. 12–47). United Nations University Press.
Montobbio, F., & Rampa, F. (2005). The impact of technology and structural change on export performance in nine developing countries. *World Development, 33*(4), 527–547.
Mustapha, N., & Kondlo, L. (2019). A new approach to R&D intensity classes illustrated on manufacturing industries in South Africa. *International Journal of Technological Learning, Innovation and Development, 11*, 97–118.
OECD. (1995). Classification of high technology sectors and products Doc.DSTI/EAS/IND/STPP(95)1.
Pavitt, K. (1984). Sectoral patterns of technical change: Towards a taxonomy and a theory. *Research Policy, 13*(6), 343–373.
Peña-Vinces, J., & Audretsch, D. (2021). Tertiary education and science as drivers of high-technology exporting firms growth in developing countries. *Journal of Technology Transfer, 46*, 1734–1757.
Porter, A. L., Roessner, J. D., Newman, N., & Cauffiel, D. (1996). Indicators of high technology competitiveness of 28 countries. *International Journal of Technology Management, 12*(1), 1–32.
Prebisch, R. (1950). *The economic development of Latin America and its principal problems. Economic Bulletin for Latin America, No. 7*. United Nations Department of Economic Affairs.
Prebisch, R. (1963). Towards a dynamic development policy for Latin America, United Nations.
Saritas, O., Sokolov, A., & Cele, M. (2024) Introduction. In O. Saritas, A. Sokolov, & M. Cele (Eds.), *21st century foresight: Shaping the future for sustainable social, economic and environmental development in South Africa*. Springer.
Seyoum, B. (2004). The role of factor conditions in high-technology exports: An empirical examination. *The Journal of High Technology Management Research, 15*(1), 145–162.
Seyoum, B. (2005). Determinants of levels of high technology exports an empirical investigation. *Journal of Competitiveness Studies, 13*(1), 64.
Som, O. (2012). *Innovation without R&D: Heterogeneous innovation patterns of non-R&D-performing firms in the German manufacturing industry*. Springer Science & Business Media.
Stockey, N. (1991). Human capital, product quality and growth. *Quarterly Journal of Economics, 106*(2), 587–616.
Stokey, N. (1988). Learning by doing and the introduction of new goods. *Journal of Political Economy, 96*, 701–717.
Szirmai, A. (2012). Industrialisation as an engine of growth in developing countries, 1950–2005. *Structural Change and Economic Dynamics, 23*(4), 406–420.
Szirmai, A., Naudé, W., & Alcorta, L. (2013). Introduction and overview: The past, present, and future of industrialisation. In A. Szirmai, W. Naudé, & L. Alcorta (Eds.), *Pathways to indus-*

trialisation in the twenty-first century: New challenges and emerging paradigms (pp. 1–61). Oxford University Press.

TIPS. (2021). Export Tracker-Q1 2021.

TIPS. (n.d.). South African trade in high-technology products. https://www.tips.org.za/research-archive/trade-and-industry/item/download/17_410930f1c4b3b21d33cb485c6af936eb

Trade and Industrial Policy Strategies [TIPS]. (2017). Manufacturing subsectors: Electronics and appliances.

Tregenna, F. (2008). The contributions of manufacturing and of services to employment creation and growth in South Africa. *South African Journal of Economics, 76*(s2), 175–204.

Tregenna, F. (2009). Characterising deindustrialisation: An analysis of changes in manufacturing employment and output internationally. *Cambridge Journal of Economics, 33*(3), 433–466.

Tregenna, F. (2015). *Deindustrialisation, structural change and sustainable economic growth. MERIT Working Papers 032*. Maastricht Economic and Social Research Institute on Innovation and Technology (MERIT), United Nations University.

Tregenna, F. (2016). Deindustrialisation and premature deindustrialisation. In J. Ghosh, R. Kattel, & E. Reinert (Eds.), *Elgar handbook of alternative theories of economic development* (pp. 710–728). Edward Elgar.

Tregenna, F., & Andreoni, A. (2020). Deindustrialisation reconsidered: Structural shifts and sectoral heterogeneity. Working Paper Series (IIPP WP 2020-06), UCL Institute for Innovation and Public Purpose, University College London, London. https://www.ucl.ac.uk/bartlett/public-purpose/wp2020-06

Tregenna, F., Ewinyu, A., Oqubay, A., & Valodia, I. (2021). The South African economy: challenges and complexities. In A. Oqubay, F. Tregenna, & I. Valodia (Eds.), *The Oxford handbook of the South African economy*. Oxford University Press.

US Department of Commerce. (2017). *High-Tech industries: The role of FDI in driving innovation and growth*. International Trade Administration.

Verdoorn, P. (1993[1949]). On the factors determining the growth of labour productivity. In L. Pasinetti (Ed.), *Italian economic papers* (Vol. II, pp. 59–68). Oxford University Press.

ICTs and Smart Systems

Judy Backhouse

Abstract Four thrusts were identified as important for South Africa to optimally benefit from information and communication technologies (ICTs) in the future. In identifying these thrusts, the team that participated in the Foresight Exercise recognised two lines of thought. On the one hand, South Africa needs to address a backlog in the use of ICTs that results from the digital divide and poor implementation of basic information systems. On the other hand, there is a need to anticipate and prepare for exploiting new technologies, such as advanced data analytics, and understanding their impact on society. Two of the thrusts, **ICT infrastructure and Internet access** and **smart and sustainable municipal service delivery**, address the first of these, while the other two, **big data, data analytics and decision support** and **checks and balances for a digital future**, address the second. This chapter argues for the importance and expected benefits of these four thrusts, reviews some of the research already done in the areas and suggests areas where research is lacking. In addition to research, the nature of ICTs demands that attention be paid to innovation and to the practice of ICTs, including the use of technologies by individuals and their application in organisations. Thus, this chapter also examines these two aspects arguing that ongoing support for and research into innovation with ICTs and into the professional practice and use of ICTs are integral to understanding these thrusts and reaping their full benefits. The chapter ends with a set of recommendations as to activities that will support and develop each of the four thrusts.

J. Backhouse (✉)
United Nations University Operating Unit on Policy-Driven Electronic Governance (UNU-EGOV), Guimarães, Portugal
e-mail: backhouse@unu.egov

O. Saritas et al. (eds.), *21st Century Foresight*, Science, Technology and Innovation Studies, https://doi.org/10.1007/978-3-031-69557-5_14

1 Introduction

The domain of ICT is often associated with new technologies such as artificial intelligence (AI), the Internet of Things (IoT), big data analytics, blockchain, robotics and quantum computing. The deliberations during the Foresight Exercise however revealed two directions of thought. The first was to address the backlog in the use of ICT in South Africa that results from the digital divide, a lack of ICT skills and the poor implementation of basic information systems, particularly in the public sector. These concerns were considered to be prerequisites to the second direction, which was how best to harness the new technologies, and data analytics in particular, to the benefit of South African society.

These two thought directions resolved into four thematic thrusts, the first two addressing the backlog and the second two addressing the future uses of ICT. The first thrust is the need to improve ICT infrastructure and Internet access in order to enable wider use of ICTs which will lead (among other benefits) to more widespread technology skills. The second is the application of ICTs to municipal service delivery, which is considered an appropriate way to spread the use and benefits of ICTs more broadly, building skills and reducing the digital divide. At the same time, increasing the focus on ICT in local government could create local demand for ICT services and stimulate the local ICT industry. The third thematic thrust examines the use of data, data analytics and machine learning for decision-making. Of all the new technologies on the horizon, this was considered the most fundamental and promising area for research and development as it underpins the application of many other technologies. Finally, there is a need to think through the social, environmental and economic impacts of ICTs and to consider what checks and balances are needed to safeguard South African society in the digital future. A fourth thrust is proposed around this theme.

This chapter provides deeper reflections on these four thrusts, arguing for their importance and selection. There are of course many other aspects of ICT that are going to be important for the future as well as other technologies that should be investigated, understood and exploited. However, the participants in the Foresight Exercise deemed these four focus areas to be priorities for South Africa. Addressing these would put the country on a good footing to scale new heights in the innovative use of ICTs.

The analyses conducted show that while research in the ICT domains is increasing in South Africa, output is very low in terms of the global position and the quality and specialisation of research could be improved. Indeed, there are many areas relating to ICT in South Africa that are under-researched and great opportunities to grow our knowledge by focused research attention on the four thematic thrusts. However, one of the challenges of assessing research into ICT is that the wide impacts and application of ICT result in research being published in outlets across many disciplines, as can be seen in the extensive range of resources consulted for this chapter. Assessing the true volume and value of local research into ICT needs to be carefully considered as these thrusts are evaluated in the future. It also becomes apparent that research alone is not going to have the desired impacts without support for innovation and improvements in practice, so this chapter addresses as well the opportunities to boost innovation and improve practice in these four thrusts, with the potential for developing the local ICT industry and realising widespread benefits.

2 ICT Infrastructure and Internet Access

Before government, business and individuals in South Africa can take advantage of information and communication technologies and smart systems, they need to be connected to the Internet. This connection facilitates social networks, access to information and knowledge, visibility and participation in the economy (Backhouse & Chauke, 2020; Buthelezi & Dalvit, 2019). South Africa has an Internet penetration rate (defined as the percentage of the population that uses the Internet) of 55%. This is lower than 16 other countries in Africa including the leader Kenya (87.2%), Nigeria (61.2%) and even neighbouring Zimbabwe (56.5%). South Africa's Internet penetration rate is only slightly higher than the average for Africa (47.1%) and lower than the average in all other world regions (Internet World Stats, n.d.).

Internet access is a function of technology, having the devices and networks to connect and economics and being able to pay to use them. It also depends on people knowing how to use the technology and having the capabilities (in the sense of Sen, 1997) to do so. Moreover, access requires a supportive environment which includes institutions, discourses, programmes and social norms that support the use of the Internet (Kleine, 2010). The lack of these enabling conditions leads to information poverty, defined as not having "the required skills, abilities and/or material means to access and use information in a meaningful way to address (one's) needs" (Strand & Britz, 2018, 364).

Increasing access to the Internet in South Africa is arguably the first and most important task to ensure that the country can take full advantage of ICTs and smart systems. This section examines the current state of these dimensions of access, as well as the obstacles and opportunities for improving access. There is a discussion of what research is needed to fill current knowledge gaps in our understanding of the challenges of access. This section also suggests policy and oversight interventions that can assist in improving access.

2.1 *Infrastructure and the Economics of Access*

2.1.1 Undersea Cables and Terrestrial Backbone Fibre

Internet connectivity depends on undersea cables and terrestrial backbone fibre. South Africa is currently well served by five (some sources say six) undersea cables with a combined capacity of more than 70 terabits. Cables planned for the next 2 years will boost that capacity by an additional 360 terabits (Song, 2020a). In 2019, both Google and Facebook initiated undersea cables to serve Africa using the new generation of fibre-optic technology, with greatly increased capacity (Song, 2020b). The ownership of the cables and the power of the owners to dictate prices and conditions of use may need to be monitored in the future.

There have also been investments in terrestrial backbone fibre across South Africa and with neighbouring countries, although some argue that this is still inadequate (Delport, 2020). Others argue that terrestrial fibre networks remain underutilised due to high prices and a failure to establish an enabling environment. Companies contribute where there is a business case for investing (e.g. see Moyo, 2019 and Rabinovitsj, 2019), but this means that developing terrestrial infrastructure depends on the value that companies attach to it. Government can create an enabling environment for business, reduce the barriers for companies and ensure that policies, incentives and penalties support the use of shared infrastructure.

The greater challenge for South Africa is to make it attractive for companies to extend infrastructure to underserved parts of the country and incentives for this have not been entirely effective. Arguments have been made to consider different technologies to achieve this. For example, satellite might be an option to connect sparsely populated areas where it is difficult to install or maintain cable or telecommunication towers, and the recent launch of the high-throughput Amos 17 satellite should facilitate this. There have also been investigations into the delivery of rural broadband via balloon (Song, 2020b). Further research into such technology alternatives is needed.

2.1.2 Mobile Broadband

The growth of mobile broadband in South Africa has been hampered by limited available International Mobile Telecommunications (IMT) (or high-demand) spectrum and a delayed digital migration process which means that much of the 700 MHz and 800 MHz spectrum is still being used for terrestrial broadcasting. South Africa missed the deadline agreed with the International Telecommunication Union (ITU) in June 2015 to switch from analogue to digital terrestrial television and is still not compliant. The 2015 amendment of the 2008 Broadcasting Digital Migration Policy left the switch-on date for digital broadcasting and the switch-off date for analogue broadcasting open-ended (ICASA, 2015), and the Department of Communications and Digital Technologies is projecting that the process will complete by 2022 at the earliest (Mcleod, 2020).

The Independent Communications Authority of South Africa (ICASA) is the official regulator of the South African telecommunications, broadcasting and postal services sectors. In September 2020 ICASA announced plans to license additional spectrum with the process expected to be concluded by March 2021 (ICASA, 2020). This development is expected to incentivise investment in broadband infrastructure and thus stimulate economic growth, as well as to improve the quality of service for users of mobile services. The successful bidders will have obligations for coverage to address poorly served areas of the country as well as empowerment obligations to address transformation in the sector. There are differences of opinion between the Department of Communications and Digital Technologies and ICASA as to the feasibility of this process (Mcleod, 2020) which could still slow the process and continue to hamper improvements to mobile broadband.

Also announced in 2020 were plans to license spectrum for the wireless open-access network (WOAN), intended to be a public-private sector-owned and managed consortium that will lower barriers to entry for smaller players. The intention is to increase ownership in the ICT sector by historically disadvantaged individuals and to promote competition.

2.1.3 Small-Scale Local Alternatives to Telecom Companies

The cost of data in South Africa considerably exceeds the cost in other African countries and other parts of the world. According to the Data Service Market Inquiry conducted in 2018 by the Competition Commission of South Africa, the cost of the cheapest 1 GB of data in South Africa was seven times higher than the cost of the same amount of data in Egypt and nearly three times that in Ghana, Kenya and Nigeria. The cost of data was lower for the high-end products which are only accessible to a few. Investigations such as that by the Competition Commission as well as ICASA's 2018 Priority Markets Inquiry and the 2020 Mobile Broadband Services Inquiry continue to throw light on the structural conditions in the telecom market that raise prices and potential mechanisms to address them.

Telecom companies argue that high prices are necessary to cover the cost of infrastructure, in particular when it comes to serving areas where population densities are low and the market too small to give a reasonable return. Spectrum auction requirements that operators service rural areas do not seem to have been a successful strategy with critics arguing that companies have been willing to pay penalties rather than meet these conditions (Song, 2019).

For the South African context, it might be worthwhile to explore other models. One example is the community owned network infrastructure provided by Telecomunicaciones Indígenas Comunitarias (TIC) in Mexico, a cell phone network that provides access to more than 3500 people in some of the harshest conditions, at far lower prices than telecom companies can provide in neighbouring cities. This initiative has succeeded because of the dedication of activists and community members who fought legal battles to gain access to the radio frequency spectrum. Costs are kept down because the network is codeveloped by community *asambleas* (assemblies) and infrastructure is maintained by local people so that money is channelled back into the community (Srinivasan, 2020).

A South African example is that of farmer Chris Isted who built up a network company on his farm in the Eastern Cape (Burgess, 2017). Initiated to provide for his own need for reliable network connectivity, he has grown the business to provide for other farmers and companies in the area and now provides infrastructure for telecommunications company Neotel and is contributing to the rollout of connectivity to schools and municipalities.

These models of network provision, initiated by local people, seem to better suit rural conditions and may provide more successful ways to address connectivity in underserved areas. They will also contribute to building a local ICT industry. Key to their success is access to spectrum or terrestrial networks. There have been arguments for new policies on spectrum licensing in the light of findings that some frequency bands are unused, particularly in rural areas, and could be assigned to such service providers (Hadzic, 2019). Further research is needed into what the supportive conditions are that will encourage such innovative initiatives.

2.2 Human Capabilities and a Supportive Environment for Access

It is widely recognised that, after technology, people need skills to be able to use the technology to gain access to the Internet. These skills include basic computer literacy, the ability to connect a device to a network, how to use e-mail and other tools to communicate, knowledge of search engines, information literacy, online etiquette and how to stay safe online, among others (Backhouse & Chauke, 2020, Gcaza & von Solms, 2017, Nyahodza & Higgs, 2017).

The extent to which these skills exist in South Africa is not yet clear. Researchers into ICT literacy in South Africa have investigated the skills of rural women (Tawiah & Setlhodi, 2020), community development workers (Fasasi & Heukelman, 2017), university students (Schlebusch, 2018, Tewari et al., 2018), teachers (Adegbenro & Olugbara, 2019, Kafu-Quvane & Chikoko, 2019) and school leaders (Musgrave & de Wet, 2017). There is however, little systemic data on ICT literacy. Recognising this, researchers at the Vaal University of Technology, which is mandated to provide such training to communities in the Northern Cape and Southern Gauteng, have been working to define a consistent methodology for assessing these skills (Lombard & Jordaan, 2019). There has also been research into the best way to deliver such training, for example, the evaluation of technical and vocational education and training (TVET) colleges in the Eastern Cape for this task (Ngqulu et al., 2019).

Research needs to continue into the question of skills and skill development. However, there are also indications that simply having access gives people the opportunity to develop the necessary skills themselves, with minimal intervention. When the city of Johannesburg provided free public Wi-Fi in libraries, a programme of digital ambassadors was provided to help library staff and the public to connect to the service. In time, this was not necessary as library users simply ask each other how to connect (Backhouse & Chauke, 2020).

In addition to having skills, there are other resources that give individuals the agency to be able to use the Internet. These include psychological resources like self-confidence, cultural resources like a belief that it is appropriate for them to use the Internet and social resources like friends and family who encourage them to do so (Kleine, 2010). For example, research into the use of the Digital Doorway in Zandspruit, an informal settlement outside Johannesburg, showed that girls were being pushed away by boys and told that the equipment was not for them (Walton, 2014). So, a context that encourages use and supportive social attitudes towards the Internet as a resource for everyone is also important for access.

Where Internet access has been provided in public libraries, use is still hampered by power cuts, overcrowding, slow maintenance and repairs, malicious interference with computers and security concerns (Backhouse & Chauke, 2020, Mojapelo, 2020). In addition, gender and education level affects Internet use (Bornman, 2016). For students who gain access through schools or universities, similar problems relating to network speed, facilities and security are reported, but also language problems and reluctance to engage with technology (Nyahodza & Higgs, 2017).

From this it is clear that increasing the use of the Internet among South Africans is a multifaceted problem that needs to be addressed from many directions.

2.3 *Policy and Research for Increasing Access*

There are many sources of data on Internet access and the diffusion of technology across South African society. There is scope for ongoing research that unpacks this data by region, type of access and class, as well as more research into the obstacles and facilitators of access.

South African researchers have examined the benefits of Internet access and mobile connectivity. Studies have shown that access to mobile media expanded and strengthened social capital in a rural area (Buthelezi & Dalvit, 2019). Greyling (2018) showed that in Gauteng, Internet access is positively related to subjective well-being regardless of race and age. A range of different economic benefits were observed to result from the provision of free public Wi-Fi in Johannesburg (Backhouse & Chauke, 2020). More studies of this nature are needed to expand the contexts studied, to better understand these relationships and how to foster them.

In the past decade, South Africa has enacted legislation and established policies to support the development and deployment of the infrastructure necessary for a rich and productive ICT sector. Legislation includes the National Infrastructure Plan (2012), the Infrastructure Development Act (2014), the Integrated ICT Policy (2013), the National Broadband Policy (2013), the Independent Communications

Authority of South Africa Amendment Act (2014) and the Electronic Communications Amendment Act (2014). Much of this regulation has focused on addressing the convergence of telecommunications and broadcasting, clarifying the roles of government departments and agencies, as well as removing regulatory bottlenecks and improving mandatory reporting.

Research is needed that examines and critiques the existing policies and regulations. A search on the Web of Science for studies on South Africa relating to policy and broadband returned only five papers published in the last 5 years. The research conducted has examined the barriers to policy implementation (Manda & Backhouse, 2016a), compared the capabilities of Wi-Fi and television white space (TVWS) for broadband deployment in rural areas (Lysko et al., 2019), developed a framework for providing broadband to South African schools (Ford & Herselman, 2017), analysed the deployment of TVWS in six SADC countries (David et al., 2018) and examined the regulatory frameworks for dynamic spectrum allocation in emerging economies (Somdyala et al., 2017).

The International Telecommunications Union (ITU) has published several best practice guidelines for regulation to support the deployment of broadband, policy and regulatory incentives, encouraging innovation and enabling digital inclusion. These recommendations have been made with goals of supporting economic development, enhancing digital inclusion, encouraging innovation, facilitating the uptake and use of mobile applications and services and catalysing sustainable development. What is not clear is whether South Africa is following these guidelines or, indeed, how well suited these guidelines are to the South African context.

This means that there is significant scope for research into the policy frameworks in place and how effective they are. Research that examines the extent to which South African policy follows international guidelines and critically evaluates the appropriateness of the guidelines for the South African context would be valuable, as would comparison studies of regulations across similar contexts, like the work done by David et al. (2018) and Somdyala et al. (2017).

3 Smart and Sustainable Municipal Service Delivery

The second thrust identified for ICT during the 2018 Foresight Exercise was the use of ICT to improve municipal service delivery. The effective functioning of municipalities is particularly important for South Africa because it has a direct impact on the life experiences and well-being of individuals across the country. Effective use of ICTs in municipalities is likely to have a wide and direct impact on how South Africans live and secondary economic benefits by building a strong local IT sector and so is considered a priority.

Municipalities are in close contact with communities and individuals and constitute the coalface of government interaction with the people (Reddy, 2016). Municipalities are responsible for the provision of public goods and benefits including housing, roads, water and public transport. They also include less tangible services such as public drainage, sewerage and public safety standards (Ndudula, 2013).

There are shortcomings in municipal service delivery in South Africa at present. Challenges include onerous levels of responsibility and insufficient funding, including unfunded mandates (Reddy, 2016). While municipalities have several funding sources at their disposal, the demarcation process has resulted in some municipalities having no viable economic base (Zybrands, 2012). In addition, a lack of skills and political appointments have resulted in local governments that lack management skills and the technical expertise to mobilise different forms of funding (Reddy, 2016). Despite major reforms in local government procurement, supply chains continue to be used for corrupt ends (Reddy & Wallis, 2015; Tshishonga, 2015). According to the auditor general of South Africa, the combination of misspending, a culture of non-payment for services and the need for outsourcing to compensate for non-contributing employees has led to "an inevitable downward spiral to a financial cul-de-sac" (AGSA, 2020:2) for local municipalities and districts.

Information and communications technologies have been effective at improving municipal service delivery in many different contexts (Lopez-Lopez et al., 2018; Schmidthuber et al., 2019; Sevin, 2016). Given the importance of effective municipalities and the problems they currently face, a focus on how ICT can be used to enhance their functioning is an obvious national priority.

3.1 Information Systems for Municipal Administration

While the use of information systems for administrative functions is widespread in the commercial sector, their application in government, and particularly at the local level, has been less effective. However fundamental information systems such as e-procurement, e-payment systems, document tracking and workflow management systems that allow for accurate record-keeping and transaction processing could improve government processes.

Financial information systems have been shown to improve revenue collection and improved internal financial controls (Misilingi, 2013). E-procurement systems can reduce corruption through improved inventory management and record-keeping and through improved accountability provided by audit trails (Aduwo et al., 2020; Mamokhere, 2020). Billing and collection systems, used with or without smart meters, can ensure accurate and timely invoicing of customers and proper tracking of payments and can be used to enforce rules such as discounting, differential pricing and credit arrangements. Where smart metering is used, there can be reductions in the staff needed to collect data. Having effective financial and collection systems in place increases the chance that municipalities can better manage their finances and cash flows.

Record-keeping is a key part of municipal service provision. Many municipalities still rely on paper-based forms for much of their service provision. Such processes are time-consuming and require support staff to interact physically with residents, as well as costly filing and storage facilities. They are also prone to errors and physical files can easily be lost. An electronic document and records management system can increase municipal efficiency and address these issues. Such systems can be provided in a low-cost manner. In Spain, more than 6000 small- and medium-sized cities have implemented a cloud-based electronic document and records management system that is tailored to their needs.[1] The system comes with a web interface for employees and another for the public to use to access municipal services. This system had the added benefit that during the 2020 COVID19 lockdowns, municipal employees could immediately work from home with no interruption in service provision in those municipalities that had implemented the system (U4SSC, forthcoming).

There are many other information systems that could improve the functioning of municipalities. Ticketing and call centre management systems provide the means to track issues raised and their resolution. Such systems result in better management of problems and queries and collect data that facilitate innovation by highlighting where services or processes need to change. Fleet management systems can effectively manage the range of vehicles that cities own and operate including emergency response vehicles, public transport, waste collection and security patrols, to reduce costs, improve maintenance and reduce emissions. Fleet management systems are well established, readily available and inexpensive.

Well-functioning administrative information systems can address many of the challenges that municipalities face by enabling them to be more responsive and reduce delays, improving perceptions of reliability and lowering the cost of providing services. Information systems increase fairness and consistency in service provision and improve transparency and accountability. They also enable services to be delivered conveniently, and the remote delivery of services has become particularly important with the advent of COVID19.

Municipalities have little in the way of support for their ICT function (Sibanda & von Solms, 2019). For example, the website of the South African Local Government Association (SALGA) is sorely lacking in resources for ICT. Municipalities are expected to adhere to the Corporate Governance of ICT Policy Framework (CGICTPF) for effective ICT governance, but many have struggled to implement it in practice (Sibanda & von Solms, 2019). A greater national focus on both innovation and research in this area would bring the energy of technologists, software developers and researchers to bear on the challenges and potential solutions. The development of municipal management information systems for South Africa would go a long way to improving municipal service delivery at low cost and, with thoughtful procurement processes, has the potential to grow the IT sector.

[1] https://info.gestiona.espublico.com/clientes/

3.2 Communication and Trust

Apart from the information capabilities of ICTs, they provide ways of communicating that can greatly enhance service provision, improve communication and build trust between municipalities and the people they serve. Simply having access to basic office automation tools like email, messaging systems and shared files and calendars within municipal offices can improve internal communication and increase efficiency while developing skills in ICTs. Improved communication can also facilitate knowledge sharing and coordination across different departments and agencies.

An obvious way for cities to start providing electronic services is through a city website, and websites have come to be regarded as a standard measure of local e-government service provision (UNDESA, 2020). City websites can provide information for local businesses and residents about services and about the performance and plans of the city. More sophisticated websites can offer online requests for services, monitor progress, facilitate feedback and engage in consultative processes. The development and hosting of websites can easily be outsourced, and many administrative systems come with built-in web interfaces. Research has investigated what makes government websites effective (Asamoah, 2019; Langer & Gottel, 2020), but few papers address the South African context.

ICTs also offer tools for communicating with the public. Municipalities can take advantage of websites and social media tools to publicise their plans and achievements, to communicate changes in processes to the public, to solicit feedback on projects and service delivery and to brand the city (Sevin, 2016). Providing e-government services increases the transparency and improves reputation of local governments (Lopez-Lopez et al., 2018). Many South African municipalities use these communication channels (Sevin, 2016), but not always to good effect (Asamoah, 2019; Chatfield, 2017; Mawela, 2017). People who actively engage with government platforms experience improved information flow, increased trust and greater satisfaction with local government (Schmidthuber et al., 2019).

There is limited research into the use and effectiveness of electronic communication channels by local government in South Africa. This is an area where more studies are needed in order to inform efforts within local government and to better understand provincial and national interventions that could improve their use. Research should be encouraged into these matters.

3.3 ICTs and Municipal Capacity, Culture and Change

While ICTs have the potential to improve municipalities and how they are run, implementing them can be difficult. Municipalities face basic challenges like having reliable electricity, computers and the skills to be able to use them effectively (Asamoah, 2019). Internal processes may need to change, and regulations may need to be modified to allow for online processes, electronic documents and digital signatures (Twum-Darko, 2007).

The increased use of information systems in municipalities would build ICT skills among employees. At the same time, it would give employees greater awareness of the capabilities of ICTs, how they can improve processes and what their limits are as well as the risks and challenges of using such technologies. These basic skills and technology awareness are important if municipal employees are going to use ICTs effectively and be able to innovate, experimenting with new ways to use technology to improve services. Finally, there is potential for local job creation and upskilling as people are needed to help with the rollout of infrastructure and systems and to provide support for employees.

Information systems can change the culture and values of institutions. For example, where a fleet management system was implemented in the Municipality of Ezeiza, near Buenos Aires, the outcomes included an improved culture of compliance as rules for replacement, maintenance and vehicle allocation were enforced (U4SSC, forthcoming). They can also be used as "a powerful and crucial tool to support legislation led reform" (Twum-Darko, 2007: 240) as was the case in the reform of municipal property valuation practices in South Africa. Information systems can thus be used to change practices and perceptions across local government.

3.4 Smart Cities

Globally, cities are being encouraged to become "smart" through the implementation of smart technologies. Smart technologies have the potential to improve the economy, mobility, environment, people, quality of life and governance of cities. In South Africa many smart city projects have been piloted and implemented (Backhouse & Chauke, 2020; Booysen et al., 2019; City of Tshwane, n.d.; CPI, 2016; Xulu, 2013). However, research shows that many of the claimed benefits of smart cities are hypothetical with little evidence of results (Lim et al., 2019). There is also research that shows that smart city initiatives can increase inequality (Brannon, 2017; Datta, 2015) which is clearly unacceptable in the South African context. What this means is that cities need to approach the use of smart technologies with caution and careful monitoring of results as well as ongoing research is needed to be sure that the outcomes are desirable.

Smart technologies that are likely to benefit municipal service delivery include things like smart lighting which can reduce the cost of street lighting, improve the maintenance of city infrastructure, and improve the safety of cities (Tang et al., 2019). Such projects are easy to justify as they reduce costs for the city and make it possible for cities to extend street lighting more widely. The technology for smart street lighting is also well established and simple to manufacture so that increasing demand from cities could have economic benefits by developing a supply chain around smart lighting.

The smart use of data by cities promises great benefits for cities in improving planning, monitoring and decision-making through the collection and analysis of data and particularly with the application of machine learning. However, such projects depend on good data, and so cities need to start by developing basic data management capabilities. Once cities have digitised their processes and are collecting data and properly managing data, many interesting opportunities will open up for the use of that data. The use of data and data analytics for decision support is discussed further in the next section.

On the other hand, technologies like blockchain should be approached circumspectly by cities. Recent work by U4SSC examined block chain projects and found that most were still in pilot phases. The report emphasises that the technology is not yet fully understood and that risks, legal status of contracts, regulatory requirements, data privacy aspects and ethical concerns are still to be addressed (U4SSC, 2020). This is a technology to be avoided by all but the most well-resourced municipalities.

Cities in South Africa need guidelines, supported by research and evidence, of beneficial outcomes, as to which smart technologies are simple to implement and have proven benefits. This is not to say that they should not aspire to using more advanced technologies but that any such use should be driven by local needs, rather than an attempt to appear smart.

Municipal service delivery is perhaps not the first thing that springs to mind when considering the potential of new information and communications technology for the country. Making municipalities work effectively is a lot less sexy than shiny high-tech. Yet, by applying technology to municipal service delivery, there is the potential for the benefits of ICT to reach the widest possible number of people in the country and make a meaningful impact on many lives. There is much that can be done by researchers to support this project both in investigating the technologies and information systems and in studies of implementation effectiveness in order to strengthen our knowledge of how to successfully deploy these systems in the South African context. There is also a role for innovators and companies to develop solutions and provide support to municipalities and for regulators to deepen their understanding of what is needed to support municipalities to function more effectively.

4 Big Data, Data Analytics and Decision Support

A third ICT thrust that was identified during the Foresight Exercise as of particular importance to South Africa was the use of data and advanced data analytical tools to improve decision-making. The growth of techniques and tools that make use of large data sets, advanced statistical techniques and machine learning has made it possible to gain understanding, more quickly and accurately, of relationships and dynamics, particularly in complex systems such as social systems, economic systems and biological systems.

Technologies offer new ways to collect data through sensors, to use data that has been difficult to process (like audio and video data), to make sense of data through natural language processing and machine learning and to act on the knowledge that is created through expert systems. Such tools replace costly and time-consuming manual data collection processes and enable more frequent or even real-time monitoring. They offer more accurate and comprehensive data as well as the possibility for new insights into some of the challenges that face the country and new ways of addressing those challenges.

There is already considerable research being carried out in South Africa into the application of advanced statistical and machine learning techniques, with a growing research base, and the tools have already found widespread application in business. This section looks at the kind of research that is being undertaken, the applications of advanced analytical tools in business, and discusses the data that is being used. It also considers the support and conditions that are needed to develop and exploit this line of research and innovation more fully.

4.1 Applications of Data Analytics in South Africa

A search of the Web of Science for papers on the application of artificial intelligence in the South African context in the last 5 years revealed more than 50 papers. These cover a wide range of fields of application and indicate the usefulness of advanced data analytic techniques. The breadth of research and wide adoption of advanced data analysis tools and techniques suggest that the methods are accessible to researchers and that capacity exists among the research community to take advantage of them.

With climate change concerns in mind, researchers have developed models for the use of firewood as an energy source in South Africa (Willcock et al., 2018). They argue that South Africa provides a useful case because there is good data available on firewood use and models developed in South Africa can be applied to other sub-Saharan countries where the data to build such models is lacking. Being able to model such energy sources is important in developing appropriate policies to counter climate change. There has also been research done on forecasting loads on power distribution networks (Motepe et al., 2019) and forecasting energy demand of major industrial sectors in South Africa (Olanrewaju, 2019).

Machine learning can be used to identify patterns which enable more effective medical diagnoses. Among the work done in South Africa, applications in the medical domain include improving diagnostic tests for tuberculosis (Beccaria et al., 2019) and for kidney disease (Ogunleye & Qing-Guo, 2017). Researchers have also worked to improve the recruitment and placing of healthcare staff by building models to predict how long healthcare workers will stay in underserved communities (Moyo et al., 2018).

A study of factors that impact the performance of schools in South Africa used data mining and machine learning techniques to show that the availability of clean water, toilets, nearby hospitals, electricity, household goods, cell phone Internet access and safety in communities are all important variables (Wandera et al., 2019). This kind of study will be particularly useful in policy decisions to address the long-standing underperformance of education in South Africa.

Land cover maps are used to model climate change, environmental impact assessments, spatial development planning, policy development, greenhouse gas inventories and biodiversity conservation planning. The costly and time-consuming process of updating such maps means that they are typically only updated every 4 to 5 years, which is inadequate to monitor change. Researchers have developed a "locally tuned" system that updates these maps from satellite images and is accurate enough for many of the applications of land cover maps (Wessels et al., 2016). Satellite images have also been used to map the spread of invasive species in order to plan for managing them (Mureriwa et al., 2019), to detect the encroachment of bush into grasslands (Graw et al., 2017), to identify water reservoirs (Fang et al., 2019) and to monitor salt accumulation in the Vaalharts and Breede River irrigation schemes, replacing costly and time-consuming collection of soil samples (Vermeulen & Van Niekerk, 2017).

In the domain of agriculture, researchers have used machine learning to accurately map both smallholder and commercial farms, identifying crops from high-resolution images (Debats et al., 2016). Beyond establishing what is growing where, researchers were able to predict levels of maize production (Adisa et al., 2019), to detect and map the nutrients in grasslands in KwaZulu-Natal (Singh et al., 2018) and to map gullies in grazing lands (Makaya et al., 2019).

Data analytic tools support many aspects of transportation. They have been used to quickly identify congestion caused by failed traffic lights, in order to prioritise responses (Mathibela, 2017). Social media posts were used to analyse movement patterns of commuters in Gauteng to better understand where they were coming from and going to (Malema & Musakwa, 2016). This information can be used to plan improvements to transportation systems. And when it comes to improving transport systems, researchers have even developed a machine learning technique to assess how uncertainty might delay highway construction projects (Moghayedi & Windapo, 2019). These tools can contribute to long-term planning, management and immediate responses to traffic conditions.

There are other applications that are useful for city management such as systems to automate the appraisal of property values (Yacim & Boshoff, 2020) and to estimate the duration of protests (Paul et al., 2018). Researchers have looked at improving the design of municipal solid waste management systems (Monzambe et al., 2019) and assessing bacteria on waste-water sludge (Deepnarain et al., 2020).

Research is complemented by widespread use of artificial intelligence in South African businesses. A survey that interviewed senior staff in South African companies learned that businesses are using AI applications in professional services, financial services, infrastructure and transport, ICT and media, retail and health sectors (Microsoft, 2019). Applications include the analysis of sales data to profile customers and predict behaviour, operational improvements by providing staff with expert decision support, evaluating and managing risk, developing new products better aligned with customer needs and new ways to interact with customer such as chatbots. South African companies concluded deals worth some US$ 1658 million for artificial intelligence-related technology in the past 10 years, and investment continues to grow.

Big data and the applications that arise from analysis also hold promise for developing new businesses with many startup companies basing their businesses on these technologies. South African examples include DataProphet (dataprophet.com) that works in the area of autonomous manufacturing, Aerobotics (aerobotics.com) that uses satellite and drone imagery to improve agricultural management and Xineoh (xineoh.com) that provides customer insights for retail companies.

Ultimately, all of these applications of data analytics have the aim of supporting better decision-making. Whether the decisions are how to combat climate change, how to improve medical care or education, how to manage land or how to improve traffic in cities, the insights provided by data analytics help to ensure that resources and effort are directed effectively. What is heartening is that this wide range of application makes it clear that the skills to apply advanced data analytic tools do exist in South Africa and are being applied by researchers and in the business domain.

4.2 *The State of Data in South Africa*

Advanced data analytic techniques depend on reliable, high-quality sources of data. Part of the promise of the Fourth Industrial Revolution is that ubiquitous systems produce large amounts of data that is collected automatically as a consequence of ongoing service and commercial activity (Schwab, 2017). Data is routinely collected by satellites, by telecommunications companies, from transactions conducted with business and government, from social media activities and from sensors that track light, sound and air quality among others. The management of data entails the collection of data using appropriate techniques, correct labelling with metadata, data cleaning and secure storage. Each of these steps requires knowledge and expertise in the use of appropriate technology tools. In addition, it is necessary to keep track of the appropriate uses of the data, including what has been consented to by individuals providing the data.

4.2.1 Data Sources

Recent research that uses advanced statistical and machine learning techniques has drawn on data from a variety of sources. Examples include using images from Google Earth to assess bush encroachment (Graw et al., 2017), satellite images from the European Space Agency to map gullies in grazing land (Makaya et al., 2019), government agricultural land cover maps and satellite images to identify crops (Debats et al., 2016) and climate data and agricultural production data to predict maize crop production (Adisa et al., 2019). Data from Twitter and Facebook helped to understand movement patterns of commuters (Malema & Musakwa, 2016), while a traffic congestion model used data from Twitter and TomTom as well as traffic flow and incident data (Mathibela, 2017). The retention of healthcare staff was examined using recruitment and retention data from Africa Health Placements (Moyo et al., 2018), while researchers predicting the duration of protests made use of textual descriptions of the protests from news and other media (Paul et al., 2019).

South African companies make use of data that they collect as part of their ongoing business. For example, Medscheme has built a data warehouse based on medical aid claims submitted for all the schemes that it administers and is able to use this to detect optimal treatment paths for conditions, based on the actual outcomes of previous cases. Such internal, structured data is the primary basis of most business-based applications of advanced analytical tools. However, companies also make use of external data such as economic (structured) data and news reports and social media (unstructured) to better understand the business environment and their customers (Microsoft, 2019).

Ownership of and access to data restrict who can use and benefit from it but also create business opportunities. Researchers who make use of advanced data analytic techniques draw their data from freely available open sources as well as more restricted government databases and from the private sector. Some public and private sector data are made available on platforms for public use, but much of it is not available, and access depends on being able to buy the data or on developing relationships with data owners. To take full advantage of the potential of big data and data analytics, it's important to understand what data sources exist and to address complex issues of data ownership and access to data.

4.2.2 South African Data Institutions

South Africa has a solid network of institutions and organisations that collect data, including government departments and institutes, non-profit organisations and private companies. Not only do these institutions have a wealth of historical data, but they also have the capabilities to understand data management and use. A number of portals have been established making data available online and, in the process, developing skills around creating and maintaining meta data, presenting and sharing data online, as well as issues of data security and privacy.

The South African Environmental Observation Network (SAEON) (www.saeon.ac.za) offers a wide range of data on all aspects of the environment. It provides a repository where such data can be accessed and where researchers can upload their own data sets. The South African Weather Service (https://www.weathersa.co.za) collects meteorological and climatological data and related information which is made available on request.

Statistics South Africa (https://www.statssa.gov.za/) conducts the national census every 5 years and makes available economic and demographic data on their website, including information on how to use it. The South Africa Data Portal (https://southafrica.opendataforafrica.org/) provides a large number of data sets from Statistics South Africa in an interactive format. The South African Reserve Bank provides historic macroeconomic data. The Companies and Intellectual Property Commission (CIPC) tracks data on businesses and intellectual property in South Africa.

The Gauteng City-Region Observatory (GCRO), an initiative of provincial government, makes available high-quality data collected from the province. There is data available about South Africa's cities with municipal open data repositories, such as Cape Town's Open Data Portal (https://odp.capetown.gov.za/) and those provided by non-profit organisations like the Open Cities Lab (https://opencitieslab.org/odd/about) which evolved from Open Data Durban. The national treasury provides information on the financial performance of municipalities at Municipal Money (https://municipalmoney.gov.za/).

The National Advisory Council on Innovation (NACI) supports a portal on innovation in the country and provides data on research areas, outputs and funding as well as links to other data sources. The Higher Education Management Information System (HEMIS) provides details of all tertiary education institutions, the courses they offer, student demographics, staffing and funding.

The National Health Laboratory Service (NHLS) is a diagnostic pathology service with a network of laboratories serving 80% of the population in South Africa. The NHLS collects medical and genetic data and works closely with a network of researchers in local universities. The National Department of Health provides a data dictionary to standardise terms used in measuring health (https://dd.dhmis.org/).

Developing open data policies that clarify ownership of and access to data that is generated by public services and private sector transactions would be an important first step to provide clarity and increase access to data.

Machine learning requires baseline data in order to verify that conclusions reached are meaningful. In this regard South Africa is well placed because there are many repositories of national data available to researchers with good historical records in areas like land management, medicine, demography, education and business. This historic data enables researchers to track changes and can be used in the training of algorithms. Researchers have pointed out that many countries in sub-Saharan Africa lack such reference data and that South Africa provides a useful test case for developing models which can then be applied in similar countries (Willcock et al., 2018).

4.3 Supporting Effective Decision-Making

There is a solid base of data and many data collection initiatives active in the country, as well as many examples of the use of big data and data analytics both in business and within the research community. However, for South Africa to take full advantage of the benefits of these technologies, there is work to be done. In particular, data-generating institutions need to be strengthened, databases expanded and improved, and much of the work that is done by researchers needs to be operationalised and more widely implemented. For this to happen a culture of evidence-based decision-making also needs to be developed.

A good starting point would be to evaluate the public institutions that are already collecting data to understand their strengths, weaknesses and resources. From this analysis it would be possible to devise a plan for developing one or more centres of excellence around data science in the country. This may be in the form of one central national data science centre, or centres based around specific types of data such as spatial data, medical data, economic data and demographic data. Such centres of excellence could serve to bring together researchers and practitioners in both the public and private sectors to share knowledge and data and to develop good practices, learning programmes and analytic techniques. National data science centres could be tasked with developing data sources, including identifying necessary sensor networks or other data collection opportunities, as well as managing and taking ownership of the data.

Among researchers and the private sector, there is clearly a solid base of skills in the management data and the application of advanced data analysis tools. However, this base needs to be developed in anticipation of a rapidly growing demand for such skills. In addition, there is a need for less technical skills in the application of data and analysis among government officials and business more widely. Raising awareness of what can be achieved will ensure that both the public and private sectors are able to take advantage of the potential of these technologies. To this end it would be useful to analyse what learning opportunities exist for both deep technical and more application-oriented skills and where the gaps are. While universities will probably be the primary site for learning deep technical skills, it would be advantageous to consider programmes that are more widely accessible, to develop broad understandings of evidence-based decision-making, the applications of data analysis and the management of data.

Finally, there is a need to develop a culture of evidence-based decision-making, particularly in the public sector. Widespread training in the public sector that raises awareness of the need for good data, and what can be learned from it, as described above, is one element. Another element is to increase this awareness among the public and non-profit institutions, since pressure from these sources will help to ensure that public institutions adopt evidence-based approaches. Internal processes that require evidence to support decisions can also be used to strengthen the culture, along with existing efforts at ongoing monitoring as are in use by the Department of Planning, Monitoring and Evaluation (DPME).

In short, South Africa is well placed to take advantage of big data and advanced data analytic tools to strengthen decision-making in the country, to improve the effectiveness of public and private sector organisations with the prospect of social and economic benefits. Consolidating efforts and strengthening existing pockets of excellence will make this a reality.

5 Checks and Balances for a Digital Future

Although the development of new information and communication technologies hold great promise for social and economic benefits, they can also pose threats to accustomed economic, political and social structures and behaviours. Data stored and shared online is vulnerable to abuse by companies and criminals. Social media has changed the way in which people think and relate. Platform companies have disrupted industries with new models of operation and come into conflict with local regulations and needs. Targeted advertising, based on machine learning algorithms, has been used to influence the outcome of voting.

The changes that are wrought by these technologies are complex and systemic and thus can be far-reaching and difficult to predict. We lack effective mechanisms for thoughtfully examining technologies and anticipating their impacts, and, as changes happen rapidly, it is often not possible to consider the impacts before they have happened. Appropriate measures to contain undesirable outcomes can sometimes only be identified after the consequences become apparent. Part of the complexity of the changes that arise from new information and communication technologies is that they present moral and ethical dilemmas that have not yet been encountered. Navigating these requires a well-developed set of common societal values. It is also necessary to have mechanisms in place to debate moral and ethical conundrums widely.

Monitoring the ongoing and evolving social, environmental and economic impacts that arise from new technologies will require established structures, centres, research agendas and mechanisms. Understanding technology developments and their observed and potential impacts, as well as creatively developing mitigating interventions, will be necessary to ensure that the economic system, government, businesses and individuals are able to adapt proactively and in the best interests of all.

In this section we examine some of the risks and impacts that have arisen from new technologies as well as the mechanisms and tools that are available to mitigate undesirable effects. We propose that these matters need to form the fourth thrust for ICTs. An ongoing programme of research, as well as structures and processes that deal with risks, is an important adjunct to activities to harness the benefits of ICTs for South African society.

5.1 Risks and Impacts

First, there are risks in terms of the **reliability and security** of technology. As individuals, businesses and government departments depend more heavily on technology to conduct their lives, businesses and service provision, interruptions to the functioning of hardware, networks and systems will have greater impacts. The more we turn to digital solutions, the more difficult it is to maintain alternatives that can be used when the technology fails. Indeed, for many digital services, there is no acceptable analogue equivalent. As a result, ensuring the uninterrupted functioning of these systems is important.

Problems can arise at the level of hardware and software (physical failure, errors, malicious attacks), at the level of data (corruption, loss, theft) or within networks (reliability, speed, interception etc.). With the growth of the Internet of Things, reliability and security concerns extend to networked physical infrastructure. Phishing, malware and botnet attacks are designed to gain access to devices and hence to data which is then used for criminal ends. Attacks are not only external though, there is a real threat in organisations of internal attacks by disgruntled employees (Rapanyane & Sethole, 2020).

Increasing digitalisation and the rapid spread of computing devices, ubiquitous networks and the increasing demand for processing power create high energy demands as well as ongoing demand for electronics that have negative **environmental impacts**. These impacts arise during manufacture with the demand for rare resources, during use with high energy demands and at the end of life with devices that are difficult to recycle, mostly going to landfills (Plepys, 2002). The lifetime impacts of new technologies need to be taken into account and managed appropriately.

Data has become a valuable resource which can be used by companies for legitimate business ends, but this value means that it is also vulnerable to misuse for profit and for criminal activities. Services such as e-mail, social media and cloud storage create privacy concerns as companies are able to mine data that they store on behalf of customers to better profile individuals, to understand broad trends, for surveillance or for criminal ends. Identity theft is a significant problem in South Africa, and difficult for those affected to recover from (Cassim, 2015). Threats to data privacy can be external with well-orchestrated attacks on networks or data storage to steal data or direct attempts to get individuals to reveal personal data. Risks also arise from careless behaviour by employees with access to personal data (Cassim, 2015). However, even legitimate business activities need to be questioned when companies exploit user ignorance, opting for covert data collection and one-sided user agreements which give them unnecessary access to personal data.

Emerging technologies present **legal and regulatory challenges** as they present never-before-imagined scenarios that are not catered for in local laws and regulations. Examples include platform companies like Uber that has necessitated changes to public transport regulations and technologies like blockchain with inbuilt characteristics of anonymity and immutability that challenge legal practices and principles (U4SSC, 2020). Blockchain has been proposed as a tool for implementing smart contracts, but while such contracts can be used to codify agreements, they do not necessarily have legal value, depending on the applicable law. Understanding the requirement of such new technologies and modifying laws and regulations appropriately takes time.

Another concern is the **power of large, global technology companies** which presents challenges as they operate across national borders and thus may not be subject to local laws and regulations (Nead, 2020). Some platform companies are perceived to negatively affect local conditions, for example, the impact of Airbnb on the availability of residential property (Brauckman, 2017). Where companies provide services that are very widely used, or used extensively by their peers, people may have no option but to agree to one-sided contracts in order to access the services (Iaconesi, 2017).

Algorithms form the basis of automated decision-making and are created and refined through machine learning. Such algorithms learn from the data that is presented to them and as a result have been shown to replicate existing biases, including race and gender biases (Ntoutsi et al., 2019). Algorithms can also be opaque (Iaconesi, 2017), with the basis for their construction difficult to assess by those who are not experts. There are concerns that decisions made by algorithms may be rigid, not allowing the consideration of exceptional circumstances and conditions. These concerns have led to calls in legal circles for the right to a human decision (Huq, 2020). Where algorithms are responsible for operating robots or driving cars or are used to give advice, dilemmas arise around who would take the blame in the case of a failure or for poor advice (Xuying & Young, 2020). There are also moral dilemmas to be resolved so that ethical guidelines can be coded into algorithms, such as how to make trade-offs where both alternatives cause harm. In addition, where algorithms are based on simple optimising criteria such as maximising profit, they can have undesirable impacts on people as discussed below.

Mining of data on human behaviour and the use of optimising algorithms has enabled efficient **exploitation of human frailties**. Systems, including online games, social media, pornography sites and advertising have evolved algorithms that tap into human fears, desires and motivations, effectively using biochemical reward systems to re-enforce behaviours that can be destructive and addictive. Impacts include lack of sleep, loss of daytime functioning, addiction and mental health problems, including stress, anxiety, depression and increased suicide attempts (Tandon et al., 2020; Wang et al., 2020).

Search engines, news and content providers and social media employ algorithms to identify and serve content appropriate to individual's interests and tastes. While this can be useful and help people to navigate the wealth of online information, such algorithms have been shown to **limit exposure to divergent positions**, so that people only see content that confirms positions and ideas that they already hold (Iaconesi, 2017). The danger is that people are not exposed to new ideas and may come to view complex issues in binary terms and without nuance, which leads to polarisation.

Online media, including social media, has been abused for political ends, with misinformation being spread to carefully targeted groups of people. **Fake news** has been shown to change people's behaviour (Bastick, 2021) and has increased support for right-wing populist parties in Europe (Van Kessel et al., 2021). Machine learning and artificial intelligence have been used to create "deepfakes", images or videos where faces have been replaced with those of well-known people (Kietzmann et al., 2020). While this can be used to delightful effect in movie-making, the images produced are highly realistic and so lend themselves to deception. They have been used in pornography, to misrepresent politicians, to create sock puppets (fake personas who are active in the media), for propaganda and for financial fraud (Bezmalinovic, 2018; Damiani, 2019; Dickson, 2019).

New technologies are expected to have wide-ranging **impacts on labour markets** as they remove jobs, transform jobs and create new jobs. In South Africa, an estimated one in three jobs are likely to be totally automated, exacerbating already low employment rates (Rapanyane & Sethole, 2020). Robotics threatens many low-skilled jobs as it replaces physical labour so that sectors like mining and agriculture can anticipate job losses (Leeuw & Mtegha, 2018). In addition, many low-level office workers who do routine tasks, such as call centre agents, can also be replaced. Moreover, the rise of data analytics and artificial intelligence also threatens skilled jobs as analytical and even creative tasks that were once thought to be the exclusive domain of humans can now be carried out by machines. Computers have become proficient at tasks such as translation, legal research and writing. Even the work of professionals will change through automation, as some tasks are de-skilled and can be transferred to lower cost workers, other tasks are automated, transferred to customers through self-service opportunities or eliminated entirely (Sampson, 2020).

Researchers in South Africa are already grappling with these impacts. They have analysed the impact of artificial intelligence and the Internet of Things on employment in South Africa (Malapane, 2019), developed frameworks to understand the process of introducing such automation into the workplace (Mamphiswana & Sinha, 2019) and examined the future skills needs of the public sector (Nhede, 2018).

5.2 Mechanisms to Mitigate Risks

Having spelt out the types of risks and the potential for unintended consequences that exist, this section turns to the question of how these risks can be mitigated. Elements include being aware of the risks, monitoring how they play out and identifying and implementing measures to deal effectively with them. While many risk mitigating measures need to be implemented by individuals and companies, much work can be done at a national level to identify and monitor risks as well as to identify ways to address them and to implement those that need to be coordinated nationally. It is necessary as well to ensure that responsible understandings of the potential risks are communicated widely and effectively.

Responses to the risks fall into three broad categories: the first of these are technological solutions, then there are solutions focused on raising awareness and developing capacity, and finally there are legal and regulatory mechanisms that can be put in place. All three of these necessitate ongoing research and development to understand their effectiveness and impact.

5.2.1 Technology

Secure and reliable systems require physical measures to protect them, including premises that are protected against fire, water damage, infestation, power failures and physical access. Where it is impractical to provide such physical measures, such as in rural municipalities, the use of remote and cloud storage options provides an alternative. Systems can be protected through built-in redundancy, monitoring to identify problems before they escalate and software designed to protect them from direct attack. Databases are designed with sophisticated protection based on limited access and profiling of different types of users. Their contents can also be protected by encryption. Networks rely on ever more sophisticated protocols to ensure accurate transmission of data and layers of encryption to hide the contents being transmitted.

How to identify the people gaining access to information systems is an ongoing challenge. Most systems require passwords, but these can be obtained for fraudulent use. Biometric systems have been developed that offer alternatives based on physical characteristics of individuals such as fingerprints, retinal patterns and voice and facial recognition. The feasibility and effectiveness of such systems, particularly in practice, needs ongoing research. Some responses to mitigate the risks described can be made by employing the same technologies. For example, machine learning algorithms have been used to detect fake news (Utkarsh et al., 2021).

Much of this technology is available and is continually being improved; however its effectiveness depends on how well it is implemented. While risks can, to some extent, be mitigated through technology solutions, they also depend on awareness and knowledge as well as changes in the behaviour of individuals, in organisational processes and in regulations.

5.2.2 Regulations and Legislation

Ensuring the security of systems and networks and the appropriate use of data and new technologies is a complex task that needs to be supported by appropriate regulations and legislation. An appropriate policy framework should comprise both "proactive" and "reactive" elements and should aim to build institutional, economic and regulatory confidence (Ayentimi & Burgess, 2019). Many of the challenges of new technologies include a moral and ethical component, and these are perhaps best dealt with through the established consultative processes that precede legislation.

South Africa is not short on legislation and regulations to address aspects of ICT use. However, particularly in the ICT domain, legislation is disjointed and lacks coherence. The sector has suffered from long delays in implementing regulations (Manda & Backhouse, 2016a).

The National Cyber Security Policy Framework (NCPF) was approved in 2012 with the aim of creating a "secure, dependable, reliable and trustworthy cyber environment that facilitates the protection of critical information infrastructure whilst strengthening shared human values" (NCPF, 2012: 14). It provided measures to address national security to combat cybercrimes and to review, update and align relevant laws and measures to build confidence and trust in the use of ICT involving both the South African Accreditation Authority and the South African Post Office. An important element of the NCPF is developing a national cybersecurity research agenda and capacity. However, implementation has been slow with little activity on the part of the South African Accreditation Authority (www.saaa.gov.za, accessed 11 December 2020) or the Trust Centre launched in 2013 by the South African Post Office. In the 2019/2020–2021/2022 strategic plan, the Trust Centre is mentioned only in terms of what it has yet to do (SAPO, 2020).

Discussions to establish an integrated national ICT policy framework stalled after the publication of the white paper in 2016. The white paper proposed establishing new regulators and re-arranging the functions between the Department of Telecommunications and Postal Services (DTPS) and the Department of Communications (DoC) which were merged (in 2019) to form the Department of Communications and Digital Technologies (DCDT), possibly making the proposals defunct.

Other relevant legislation includes the Protection of Personal Information Act (POPI, Act 4 of 2013) that governs the use of personal data; the Electronic Communication and Transactions Act, 2002, that regulates the collection, use and protection of personal information obtained through electronic transactions; the Consumer Protection Act of 2008 which sets out an overarching framework for consumer protection in South Africa including citizen's right to privacy; and the Cybercrime and Cybersecurity Act of 2015 which regulates offences and penalties relating to cybercrime (Manda & Backhouse, 2016b). The Public Administration Management Act (2014) provides for the use of information and communication technologies in public administration, and in 2019 the Electronic Deeds Registration Systems Act provided explicitly for electronic deeds registration.

5.2.3 Raising Awareness and Building Capacity

Ensuring that the average person in South Africa is aware of the risks posed by new technologies and has sufficient knowledge to protect themselves, as far as possible, is an important first step towards mitigating risks. To this end activities and interventions are needed that raise awareness of the security challenges of sharing personal information and transacting online and the ways in which information may be manipulated and misrepresented. The social impacts of artificial intelligence and algorithm-driven exposure to information mean that people have to develop sophisticated understandings of what information is reliable and how to judge that. Information literacy ensures that people approach information sources with a critical eye and are able to make judgements about the reliability of sources. In addition, people need to be aware of the addictive nature of information products and develop ways to manage their own use constructively. There are many avenues for spreading such information including advertising, awareness campaigns and targeted training.

The effectiveness of these various avenues needs to be monitored as well as overall levels of awareness among the populace.

One obvious avenue is through schools. Researchers have examined the need for developing a healthy cyber-safety culture among schoolchildren in South Africa and have recommended greater involvement by teachers and the Department of Basic Education (Kritzinger, 2017). In particular they have recommended including cyber-safety awareness in the school curriculum, training teachers to improve their knowledge and assisting schools and implementing cyber-safety policies. Researchers have argued that the new technologies of the Fourth Industrial Revolution demand a more cosmopolitan education that encourages "engaging critically and self-reflexively with one's own values, beliefs and worldviews" (Waghid et al., 2019:7).

The technology tools available to increase systems security are not always used by information technology professionals. A study carried out in Portugal showed that few municipalities were implementing even basic security measures on their websites (Gomes et al., 2019). Monitoring technology developments and translating research into practical advice which can then be used to develop the capabilities of first line IT staff will help to ensure that basic technology security measures are effectively implemented. This could be supported by a stronger professional practice for IT workers.

The ICT sector in South Africa lacks a level of professionalism that would raise the skills of front line IT workers, despite ongoing efforts of professional bodies such as the Institute of Information Technology Professionals South Africa (IITPSA) and the Institute of Chartered IT Professionals (ICITP). Strengthening these bodies to provide and standardise ICT training is important. They could also play a role in developing and enforcing the use of standardised protocols for practice, in the manner of other professions. As the use of ICT grows in the public sector, there is an opportunity to raise the level of professionalism by insisting on accredited qualifications for IT staff. Again, research into the impact and effectiveness of ICT training is needed.

At the highest levels, South Africa needs to nurture and grow university training and research programmes focused on ICT. This includes strengthening existing departments of computer science and information systems and supporting ambitious research endeavours to learn more about new information and communication technologies, as well as their impacts on society.

6 Conclusion and Recommendations

This chapter has reported on the four thrusts that were identified during the 2018 Foresight Exercise by the group that examined the potential of ICTs. It presented a rationale for selecting these thrusts in terms of the challenges that the country faces and the promise of the new technologies. Two of these thrusts, ICT infrastructure and Internet access and smart and sustainable municipal service delivery, are needed to bring the country up to speed in the use of well-established ICTs and lay the groundwork for more advanced use. The remaining two, big data, data analytics and decision support and checks and balances for a digital future, will ensure that the country is well positioned to take advantages of future developments in ICTs.

Recommendations to support the thrust **ICT infrastructure and Internet access** include the following:

- Support research into ICT infrastructure provision, regulation and legislation that impact on the provision of ICT infrastructure though targeted research funding and postgraduate support.
- Establish enabling policy and a regulatory framework to ensure digital interoperability and quality of service, and a level playing field for healthy competition.
- Seek out, encourage and support innovations in technology and modes of access provision to increase access to devices and networks.
- Support research into all aspects of Internet access, from skills to the enabling conditions though targeted research funding and postgraduate support.

Recommendations to support the thrust **smart and sustainable municipal service delivery** include the following:

- Create a national focus on e-government in municipalities in order to strengthen service delivery, grow ICT skills and stimulate the local ICT sector.
- Strengthen the institutions that provide support for municipalities in their ICT deployment, and harness the skills in the private sector through partnerships.
- Make use of existing technology innovation structures and the private sector to develop ICT tools, systems and support structures for municipal service delivery.
- Support research into the use and effectiveness of ICTs in local government through targeted research funding and postgraduate support.

Recommendations to support the thrust **big data, data analytics and decision support** include the following:

- Develop a national open data policy to clarify ownership of and access to data generated by public and private sectors, to identify where open access and data sharing are feasible and desirable and to stimulate the use of open-access data.
- Review the data management and analysis experience and skills of institutions in South Africa that collect national data to evaluate their resources, needs, capacities and networks with a view to strengthening existing institutions.
- Examine the feasibility of establishing one or more national data science centres to bring together researchers and practitioners in the public and private sectors to develop data sources, share data and knowledge and develop good practices, learning programmes and analytic techniques.
- Review learning programmes that are available for data management, understanding of the application of data science and advanced data science technical skills and support the development of suitable programmes to fill gaps, including online open-access awareness training aimed at public servants and the public.

Recommendations to support the thrust **checks and balances for a digital future** include the following:

- Support research into the impact of new ICTs on society, including, for example, social interactions, business behaviour, political systems and trust, though targeted research funding and postgraduate support.
- Support research into the South African policy landscape for ICTs including the effectiveness of existing policies and alternatives though targeted research funding and postgraduate support.
- Support ICT professional bodies to develop practice guidelines for the implementation of ICTs and existing technology solutions that can be used to mitigate risks, including materials to improve dissemination of these practices widely.
- Develop information literacy and cyber-safety awareness tools for the public including school-level curriculum materials, freely available online resources and public awareness programmes.

References

Adegbenro, J. B., & Olugbara, O. O. (2019). Investigating computer application technology teachers' procedural knowledge and pedagogical practices in ICT enhanced classrooms. *Africa Education Review, 16*(1), 1–18.

Adisa, O. M., Botai, J. O., Hassen, A., Abiodun, A., Darkey, D., Botai, C., & Tesfamariam, E. (2019). Application of Artificial Neural Network for Predicting Maize Production in South Africa. *Sustainability, 11*(4), Article 1145.

Aduwo, E. B., Ibem, E. O., Afolabi, A. O., Oluwnmi, A. O., Tunji-Olayeni, P. F., Ayo-Vaughan, E. A., Uwakonye, U. O., & Oni, A. A. (2020). Exploring anti-corruption capabilities of e-procurement in construction project delivery in Nigeria. *Construction Economics and Building, 20*(1), 56–76.

AGSA. (2020). *Auditor General South Africa Media release 1 July 2020 on the publication of the MFMA 2018–2019 general report*. Accessed December 5, 2020, from https://www.gov.za/sites/default/files/gcis_documents/2020MFMA.pdf

Asamoah, K. (2019). E-governance in Africa's local governments: Do district assemblies in Ghana optimize the use of websites and social media? *Electronic Journal of Information Systems in Developing Countries, 85*, e12082. https://onlinelibrary.wiley.com/doi/full/10.1002/isd2.12082

Ayentimi, D. T., & Burgess, J. (2019). Is the fourth industrial revolution relevant to sub-Sahara Africa? *Technology Analysis & Strategic Management, 31*(6), 641–652. https://doi.org/10.1080/09537325.2018.1542129

Backhouse, J., & Chauke, H. (2020). Development impacts of free public Wi-Fi in Johannesburg. In R. Boateng (Ed.), *Handbook of research on managing information systems in developing economies*. IGI Global.

Bastick, Z. (2021). Would you notice if fake news changed your behaviour? An experiment on the unconscious effects of disinformation. *Computers in Human Behaviour, 116*, 106633.

Beccaria, M., Bobak, C., Maitshotlo, B., Mellors, T. R., Purcaro, G., Franchina, F. A., Rees, C. A., Nasir, M., Black, A., & Hill, J. E. (2019). Exhaled human breath analysis in active pulmonary tuberculosis diagnostics by comprehensive gas chromatography-mass spectrometry and chemometric techniques. *Journal of Breath Research, 13*(1), Article 016005.

Bezmalinovic, T. (2018). Wenn Merkel plötzlich Trumps Gesicht trägt: die gefährliche Manipulation von Bildern und Videos. *Aargauer Zeitung*, 3 February 2018. Accessed December 10, 2020, from https://www.aargauerzeitung.ch/leben/digital/wenn-merkel-ploetzlich-trumps-gesicht-traegt-die-gefaehrliche-manipulation-von-bildern-und-videos-132155720#

Booysen, M. J., Visser, M., & Burger, R. (2019). Temporal case study of household behavioural response to Cape Town's "Day Zero" using smart meter data. *Water Research, 149*, 414–420.

Bornman, E. (2016). Information society and digital divide in South Africa: results of longitudinal surveys. *Information Communication and Society, 19*(2), 264–278. https://doi.org/10.1080/1369118X.2015.1065285

Brannon, M. M. (2017). Datafied and divided: Techno-dimensions of inequality in American cities. *City & Community, 16*(1), 20–24.

Brauckman, S. (2017). City tourism and the sharing economy – potential effects of online peer-to-peer marketplaces on urban property markets. *Journal of Tourism Futures, 3*(2), 114–126.

Burgess M (2017) A livestock farmer's thriving rural Wi-Fi network. *Farmer's Weekly*. Accessed November 5, 2020, from https://www.farmersweekly.co.za/agri-business/empowerment/livestock-farmers-thriving-rural-wi-fi-network/

Buthelezi, M., & Dalvit, L. (2019). Exploring how mobile phones mediate bonding, bridging and linking social capital in a South African Rural Area. In B. Mutsvairo & M. Ragnedda (Eds.), *Mapping the digital divide in Africa: A mediated analysis* (pp. 195–213). https://doi.org/10.5117/9789462986855_CH10

Cassim, F. (2015). Protecting personal information in the era of identity theft: Just how safe is our personal information from identity thieves? *Potchefstroomse Elektroniese Regsblad (PER), 18*(2), 69–110.

Chatfield, A. (2017). A social media text analytics framework for double-loop learning for citizen-centric public services: A case study of a local government FAQ. *Government Information Quarterly, 34*(1), 110–125.

City of Tshwane. (n.d.). *City of Tshwane's experience with implementing the University of Pretoria pilot bike share system and capacity developed to roll out future bicycle promotion schemes*. Accessed November 20, 2020, from https://www.cyclecities.co.za/themes/core/views/portal/content/conference_info/presentations/Implementing_the_University_of_Pretoria_Pilot_Bike_Share_System.pdf?v=464361341893

CPI. (2016). *Waste-to-energy innovation in KwaZulu-Natal*. Centre for Public Impact (CPI), April 5, 2016. Accessed November 20, 2020, from https://www.centreforpublicimpact.org/case-study/waste-energy-innovation-south-africa/

Damiani J (2019). A voice Deepfake was used to scam a CEO out of $243,000. *Forbes*, 3 September 2019. Accessed December 10, 2020, from https://www.forbes.com/sites/jessedamiani/2019/09/03/a-voice-deepfake-was-used-to-scam-a-ceo-out-of-243000/

Datta, A. (2015). New urban utopias of postcolonial India: 'Entrepreneurial urbanization' in Dholera smart city, Gujarat. *Dialogues in Human Geography, 5*, 3–22.

David, S., Muchanga, A. & Zennaro, M. (2018). The co-regulation of TV-white spaces: The Southern Africa Development Community Approach. Lecture Notes in Computer Science, *10th international conference on cross-cultural design (CCD) held as part of 20th international conference on human-computer interaction (HCI International)*, Jul 15–20, 2018, Las Vegas, NV.

Debats, S. R., Luo, D., Estes, L. D., Fuchs, T. J., & Caylor, K. K. (2016). A generalized computer vision approach to mapping crop fields in heterogeneous agricultural landscapes. *Remote Sensing of Environment, 179*, 210–221.

Deepnarain, N., Nasr, M., Kumari, S., Stenstrom, P., Pillay, K., & Bux, F. (2020). Artificial intelligence and multivariate statistics for comprehensive assessment of filamentous bacteria in waste water sludge bulking. *Environmental Technology and Innovation, 19*, 100853.

Delport, J. (2020). Is South Africa capable of managing its own telecoms infrastructure? *IT News Africa*. Accessed November 6, 2020, from https://www.itnewsafrica.com/2020/09/is-south-africa-capable-of-managing-its-own-telecoms-infrastructure/

Dickson, E. J. (2019). Deepfake Porn is still a threat, particularly for K-Pop stars. *Rolling Stone*, 7 October 2019. Accessed December 10, 2020, from https://www.rollingstone.com/culture/culture-news/deepfakes-nonconsensual-porn-study-kpop-895605/

Fang, W., Wang, C., Xi, C., Wan, W., Li, H., Zhu, S., Fang, Y., Liu, B., & Hong, Y. (2019). Recognizing global reservoirs from landsat 8 images: A deep learning approach. *IEEE Journal of Selected Topics in Applied Earth Observations and Remote Sensing, 12*(9), 3168–3177. https://doi.org/10.1109/JSTARS.2019.2929601

Fasasi, R. O., & Heukelman, D. (2017). ICT: Performance evaluation of community development worker's in South Africa on e-skills. *Information Technology for Development, 23*(2), 388–402.

Ford, M. & Herselman, M. (2017). Broadband in schools: Towards a definition and model of broadband for South African Schools. *IST-Africa Week Conference (IST-Africa)* MAY 30–JUN 02, 2017, Windhoek, Namibia.

Gcaza, N., & Von Solms, R. (2017). A strategy for a cybersecurity culture: A South African perspective. *Electronic Journal of Information Systems in Developing Countries, 80*(1).

Gomes, H., Zúquete, A., Dias, G. P., & Marques, F. (2019). Usage of HTTPS by municipal websites in Portugal. In Á. Rocha, H. Adeli, L. Reis, & S. Costanzo (Eds.), *New knowledge in information systems and technologies. WorldCIST'19 2019. Advances in intelligent systems and computing* (Vol. 931). Springer. https://doi.org/10.1007/978-3-030-16184-2_16

Graw V, Carsten O, Dubovyk O, & Piroska R (2017). Bush encroachment detection in Africa - A multi-scale approach. *Proceedings of the IEEE international geoscience and remote sensing symposium*, Fort Worth, Jul 23–28, 2017, pp. 4270–4273.

Greyling, T. (2018). Internet access and its relationship to subjective well-being in a developing region. *South African Journal of Economic and Management Sciences, 21*(1). https://doi.org/10.4102/sajems.v21i1.1841

Hadzic, S. A global south perspective on alternative spectrum policy. *Policy Paper Series* 5, Research ICT Africa, Dec 2019., https://researchictafrica.net/2019/12/18/a-global-south-perspective-on-alternative-spectrum-policy/

Huq, A. Z. (2020). A right to a human decision. *Virginia Law Review, 106*(3):611–688. Accessed December 9, 2020, from https://www.virginialawreview.org/volumes/content/right-human-decision

Iaconesi, S. (2017). Interface and data biopolitics in the age of hyperconnectivity. *Implications for Design, The Design Journal, 20*(sup1), S3935–S3944. https://doi.org/10.1080/14606925.2017.1352896

ICASA. (2015). *Amendment of broadcasting digital migration policy 2015*. Staatskoerant, 18 Maart 2015, No. 38583, No. 232. Accessed November 2, 2020, from https://www.icasa.org.za/legislation-and-regulations/amendment-of-broadcasting-digital-migration-policy-2015

ICASA. (2020). *Chairperson of ICASA announcing plans for the licensing of high demand spectrum and the wireless open-access network*. https://www.icasa.org.za/news/2020/plans-for-the-licensing-of-high-demand-spectrum-and-the-woan accessed 6 November 2020

Internet World Stats. (n.d.). Accessed November 2, 2020, from https://www.internetworldstats.com/stats1.htm

Kafu-Quvane, B. & Chikoko, V. (2019). Factors shaping teachers' experiences in managing teaching in the era of information and communication technology (ICT) integration across the curriculum. *Proceedings 11th international conference on education and new learning technologies (EDULEARN)*, JUL 01–03, 2019, Palma, pp. 4041–4047.

Kietzmann, J., Lee, L. W., McCarthy, I. P., & Kietzmann, T. C. (2020). Deepfakes: Trick or treat? *Business Horizons, 63*(2), 135–146. https://doi.org/10.1016/j.bushor.2019.11.006

Kleine, D. (2010). ICT4What? Using the Choice Framework to operationalise the capability approach to development. *Journal of International Development, 22*, 674–692.

Kritzinger, E. (2017). Cultivating a cyber-safety culture among school learners in South Africa. *Africa Education Review, 14*(1), 22–41.

Langer, P. F., & Gottel, V. (2020). Antecedents and consequences of public administration's social media website attractiveness. *International Review of Administrative Sciences, 86*(1), 38–61.

Leeuw, P., & Mtegha, H. (2018). The significance of mining backward and forward linkages in reskilling redundant mine workers in South Africa. *Resources Policy, 56*, 31–37.

Lim, Y., Edelenbos, J., & Gianoli, A. (2019). Identifying the results of smart city development: Findings from systematic literature review. *Cities, 95*(June), 102397. https://doi.org/10.1016/j.cities.2019.102397

Lombard, A. & Jordaan, A. (2019). A pilot E-literacy environmental scan in the Northern Cape Province of South Africa. *IST-Africa Week Conference (IST-Africa)* May 08–10, 2019, Nairobi, Kenya.

Lopez-Lopez, V., Iglesias-Antelo, S., Vazquez-Sanmartin, A., Connoly, R., & Bannister, F. (2018). e-Government, transparency and reputation: An empirical study of Spanish local government. *Information Systems Management, 35*(4), 276–293.

Lysko, A. A., Burger, C. R. & Hagopian, E. (2019). Comparison of the modelled maximum wireless link distance for TVWS and Wi-Fi within South African regulatory constraints: Which technology wins and in which scenarios? *2019 IEEE 2nd Wireless Africa Conference (WAC)*, Aug 18–20, 2019, Pretoria.

Makaya, N. P., Mutanga, O., Kiala, Z., Dube, T., & Seutloali, K. E. (2019). Assessing the potential of Sentinel-2 MSI sensor in detecting and mapping the spatial distribution of gullies in a communal grazing landscape. *Physics and Chemistry of the Earth, 112*(Special issue), 66–74.

Malapane, T. A. "The Impact of Artificial Intelligence and Internet of Things in the Transformation of E-Business Sector," 2019 Systems and Information Engineering Design Symposium (SIEDS), Charlottesville, VA, USA, 2019, pp. 1-5, doi:https://doi.org/10.1109/SIEDS.2019.8735644.

Malema, H. K. & Musakwa, W. (2016). Tweets and Facebook posts, the novelty techniques in the creation of origin-destination models. *The international archives of the photogrammetry, remote sensing and spatial information sciences*, Volume XLI-B2, 2016 XXIII ISPRS Congress, 12–19 July 2016, Prague, 555–562.

Mamokhere, J. (2020). An assessment of reasons behind service delivery protests: a case of greater Tzaneen municipality. Journal of Public Affairs, 20, e2049.

Mamphiswana, R. & Sinha, S. (2019). Machine and human is the new workspace in emerging economies: A phased approach as the strategic framework to reach sustainable economic system readiness. In *Proceedings of the Portland international conference on management of engineering and technology (PICMET)*, Portland, August 25–29, 2019.

Manda, M. I. & Backhouse, J. (2016a). *Towards a "Smart Society" through a connected and smart citizenry in South Africa: A review of the National Broadband Strategy and Policy*. Lecture notes in computer science. CT 15th IFIP Electronic Government (EGOV)/8th IFIP WG 8.5 International conference on electronic participation (ePart), SEP 05-08, 2016, Guimaraes. doi:https://doi.org/10.1007/978-3-319-44421-5_18.

Manda, M. I. & Backhouse, J. (2016b). Addressing trust, security and privacy concerns in the government integration, interoperability and information sharing through policy: A case of South Africa. *Proceedings of the International conference on information resources management (CONF-IRM)*. Paper 67. http://aisel.aisnet.org/confirm2016/67

Mathibela, B. (2017). Another broken traffic light? Reducing traffic congestion in resource constrained environments. *Proceedings of the 28th IEEE Intelligent Vehicles Symposium (IV)*, Redondo Beach, CA, June 11–14, 2017, pp. 1305–1310.

Mawela, T. (2017). Exploring the role of social media in the G2C relationship: A South African perspective. *Information Development, 33*(2), 117–132. https://doi.org/10.1177/0266666916639743

Mcleod, D. (2020). *No digital migration before 2022: Director-general*. Tech Central. Accessed November 6, 2020, from https://techcentral.co.za/no-digital-migration-before-2022-director-general/101264/

Microsoft. (2019). *Artificial Intelligence in Middle East and Africa, South Africa: Outlook for 2019 and beyond*. Report commissioned by Microsoft and conducted by EY. Accessed December 6, 2020, from https://info.microsoft.com/ME-DIGTRNS-CNTNT-FY19-05May-26-AIMaturityStudy-SRGCM1070_01Registration-ForminBody.html

Misilingi, E. M. (2013). Effect of information systems on revenue collection by local authorities in Home Bay County, Kenya. *Universal Journal of Accounting and Finance, 1*(1).

Moghayedi, A., & Windapo, A. (2019). Key uncertainty events impacting on the completion time of highway construction projects. *Frontiers of Engineering Management, 6*(2), 275–298.

Mojapelo, S. M. (2020). The internet access and use in public libraries in Limpopo Province, South Africa. *Public Library Quarterly, 39*(3), 265–282.

Monzambe, G. M., Mpofu, K., & Daniyan, I. A. (2019). Statistical analysis of determinant factors and framework development for the optimal and sustainable design of municipal solid waste management systems in the context of industry 4.0. *Proceedings of the 29th CIRP design conference (CIRP Design)*, Povoa de Varzim, May 08–10, 2019, pp. 245–250.

Motepe, S., Hasan, A. N., & Stopforth, R. (2019). Improving Load forecasting process for a power distribution network using hybrid AI and deep learning algorithms. *IEEE Access, 7*, 82584–82598. https://doi.org/10.1109/ACCESS.2019.2923796

Moyo, A. (2019). *Seacom's 1TB fibre link between Joburg, Cape Town goes live*. ITWeb. Accessed November 2, 2020, from https://www.itweb.co.za/content/KzQenqj8XRd7Zd2r

Moyo, S., Doan, T. N., Yun, J. A., & Tshuma, N. (2018). Application of machine learning models in predicting length of stay among healthcare workers in underserved communities in South Africa. *Human Resources for Health, 16*, Article 68.

Mureriwa, N. F., Adam, E. & Adelabu, S. (2019). Cost effective approach for mapping prosopis invasion in Arid South Africa Using Spot-6 imagery and two machine learning classifiers. *Proceedings of the IEEE International Geoscience and Remote Sensing Symposium (IGARSS)*, Yokohama, JAPAN, Jul 28–Aug 02, 2019, pp. 3724–3727.

Musgrave, S., & de Wet, C. (2017). An evaluative study of an ICT module for a school leadership and management preparation program. *Turkish Online Journal of Distance Education, 18*(2), 43–55.

NCPF. (2012). National Cybersecurity Policy Framework (NCPF). *Government Gazette*, 4 December 2015, No. 39475.

Ndudula, M. R. (2013). *An analysis of the politics–administrative interface and its impact on delivery of municipal services: A case of the Mnquma Local Municipality*. MPA Dissertation, University of Fort Hare, Alice.

Nead, N. (2020). *Do big tech companies have too much power?* Readwrite, 28 September 2020. Accessed December 11, 2020, from https://readwrite.com/2020/09/28/do-big-tech-companies-have-too-much-power/

Ngqulu N, Gumbo S & Nogwina M (2019) Modelling TVET Colleges as Alternative Centres to Deliver eSkills Training in Rural Communities of Eastern Cape. IST-Africa Week Conference (IST-Africa) May 08–10, 2019, Nairobi.

Nhede, N. (2018). The fourth Industrial Revolution and novel skills needed for the public sector. In de Vries MS, van Dijk G & Chitiga-Mabugu M (Eds) *Public administration: Reflection, relevance and readiness. The network of institutes and schools of public administration in Central and Eastern Europe (NISPAcee)*, pp. 201–215.

Ntoutsi, E., Fafalios, P., Gadiraju, U., Iositidis, V., Nejdl, W., Vidal, M. E., et al. (2019). Bias in data driven artificial intelligence systems – An introductory survey. *WIREs Data Mining and Knowledge Discovery, 10*, e1356. https://doi.org/10.1002/widm.1356

Nyahodza, L., & Higgs, R. (2017). Towards bridging the digital divide in post-apartheid South Africa: A case of a historically disadvantaged university in Cape Town. *South African Journal of Libraries and Information Science, 83*(1), 39–48.

Ogunleye, A. A. & Qing-Guo, W. (2017). XGBoost model for chronic kidney disease diagnosis. *IEEE/ACM transactions on computational biology and bioinformatics*, Epub 2019, Apr 17.

Olanrewaju, O. A. (2019). Predicting industrial sector's energy consumption: Application of support vector machine. *Proceedings of the IEEE international conference on industrial engineering and engineering management* (IEEM), Macao, MACAO, Dec 15–18, 2019, pp. 1597–1600.

Paul S, Hasija M, Mangipudi RV, & Marwala T (2018). Early estimation of protest time spans: A novel approach using topic modeling and decision trees. *Proceedings of the 1st international conference on soft computing in data analytics (SCDA)*, Chilakapalem, March 10–11, 2018. Advances in Intelligent Systems and Computing, 758:107–116.

Paul S, Hasija M, Mangipudi RV, Marwala T (2019) Early Estimation of Protest Time Spans: A Novel Approach Using Topic Modeling and Decision Trees. In: Nayak J., Abraham A., Krishna B., Chandra Sekhar G., Das A. (eds) Soft Computing in Data Analytics. Advances in Intelligent Systems and Computing, vol 758. Springer, Singapore.

Plepys, A. (2002). The grey side of ICT. *Environmental Impact Assessment Review, 22*(5), 509–523. https://doi.org/10.1016/S0195-9255(02)00025-2

Rabinovitsj, D. (2019). Enabling better global connectivity through new partnerships and technologies. *Facebook Engineering*. Accessed November 2, 2020, from https://engineering.fb.com/2019/02/25/connectivity/mobile-world-congress-2019/

Rapanyane, M. B., & Sethole, F. R. (2020). The rise of artificial intelligence and robots in the 4th Industrial Revolution: Implications for future South African job creation. *Contemporary Social Science, 15*(4), 489–501. https://doi.org/10.1080/21582041.2020.1806346

Reddy, P. S. (2016). The politics of service delivery in South Africa: The local government sphere in context. *The Journal for Transdisciplinary Research in Southern Africa, 12*(1), a337. https://doi.org/10.4102/td.v12i1.337

Reddy, P. S., & Wallis, M. (2015). Involving SMMEs. In P. S. Reddy & M. De Vries (Eds.), *Quo Vadis?: Local governance and development in South Africa post 1994* (pp. 237–259). Bruylant Publishers.

Sampson, S. E. (2020). A strategic framework for task automation in professional services. *Journal of Service Research*, July 31, 2020 doi:https://doi.org/10.1177/1094670520940407.

SAPO. (2020). *South African Post Office SOC Ltd. Strategic Plan 2019/20–2021/22*. Accessed December 11, 2020, from https://www.postoffice.co.za/About/SAPostOfficestrategicplan1922.pdf

Schlebusch, C. L. (2018). Computer anxiety, computer self-efficacy and attitudes towards the internet of first year students at a South African University of Technology. *Africa Education Review, 15*(3), 72–90.

Schmidthuber, L., Stutz, S., & Hilgers, D. (2019). Outcomes of open government. Does an online platform improve citizens' perception of local government? *International Journal of Public Sector Management, 32*(5), 438–456.

Schwab, K. (2017). *The fourth industrial revolution*. Random House USA Inc.

Sen, A. (1997). Maximisation and the act of choice. *Econometrica, 65*(4), 745–779.

Sevin, E. (2016). Branding cities in the age of social media: A comparative assessment of local government performance. In M. Z. Sobaci (Ed.), *Social media and local governments: theory and practice* (pp. 301–320). Springer.

Sibanda, M., & von Solms, R. (2019). Devising a strategy for IT governance implementation in municipalities. A case study of South Africa. *Electronic Journal of Information Systems in Developing Countries, 85*, e12067.

Singh, L., Mutanga, O., Mafongoya, P., & Peerbhay, K. Y. (2018). Multispectral mapping of key grassland nutrients in KwaZulu-Natal, South Africa. *Journal of Spatial Science, 63*(1), 155–172. https://doi.org/10.1080/14498596.2017.1341352

Somdyala, B., Rananga, S., Mfupe, L., Masonta, M. & Mekuria, F. (2017). Spectrum regulation for future internet networks in developing economies. *IST-Africa Week Conference (IST-Africa)* May 30–Jun 02, 2017, Windhoek.

Song, S. (2019). *Spectrum auctions are killing competition and failing rural access*. Accessed November 10, 2020, from https://manypossibilities.net/2019/04/spectrum-auctions-are-killing-competition-and-failing-rural-access/

Song, S. (2020a). *African undersea cables*. Accessed November 2, 2020, from https://manypossibilities.net/african-undersea-cables/

Song, S. (2020b). *Africa Telecoms Infrastructure in 2019*. Accessed November 10, 2020, from https://manypossibilities.net/2020/01/africa-telecoms-infrastructure-in-2019/

Srinivasan, R. (2020). *In Oaxaxa, an unlikely union between hackers and indigenous peoples*. The MIT Press Reader. Accessed November 10, 2020, from https://thereader.mitpress.mit.edu/oaxaca-hackers-and-indigenous-peoples-network/

Strand, K. J., & Britz, J. (2018). The evolving role of public libraries in South Africa in addressing information poverty: A historical context. *Library Management, 39*(6–7), 364–374.
Tandon, A., Kaur, P., Dhir, A., & Mantymaki, M. (2020). Sleepless due to social media? Investigating problematic sleep due to social media and social media sleep hygiene. *Computers in Human Behavior, 113*, 106487. https://doi.org/10.1016/j.chb.2020.106487
Tang, Z., Jayakar, K., Feng, X., Zhang, H., & Peng, R. X. (2019). Identifying smart city archetypes from the bottom up: A content analysis of municipal plans. *Telecommunications Policy, 43*(10), 101834. https://doi.org/10.1016/j.telpol.2019.101834
Tawiah, S., & Setlhodi, I. I. (2020). Introducing Information and Communication Technology Training for Rural Women in South Africa: Innovative Strategies for the Advancement of Livelihoods. *International Journal of Adult Education and Technology (IJAET), 11*(11), 45–59.
Tewari, D. D., Mtose, X., & Ilesanmi, K. D. (2018). Impact of computer literacy training on the academic performance of first year students in the University of Zululand, South Africa. *International Journal of Educational Sciences, 21*(1–3), 48–56.
Tshishonga, N. (2015). The increased politicisation of human resources recruitment. In P. S. Reddy & M. De Vries (Eds.), *Quo Vadis?: Local governance and development in South Africa post 1994* (pp. 129–146). Bruylant Publishers.
Twum-Darko, M. (2007). *The role of Information Systems in legislation-led reform: A case study in the context of the new municipal rates act of South Africa*. PhD thesis, University of Cape Town, https://open.uct.ac.za/handle/11427/11709
U4SSC. (2020). *Blockchain for Smart Sustainable Cities*. United for Smart Sustainable Cities (U4SSC). Accessed December 8, 2020, from https://www.itu.int/en/publications/Documents/tsb/2020-U4SSC-Blockchain-for-smart-sustainable-cities/index.html
U4SSC. (forthcoming). *Simple ways to be smart*. United for Smart Sustainable Cities (U4SSC).
UNDESA. (2020). *United Nations e-Government Survey 2020: Digital government in the decade of action for sustainable development*. United Nations, 2020.
Utkarsh, S., Azeez, S. N., Darshan, B. C., & Chaya Kumari, H. A. (2021). A study on discernment of fake news using machine learning algorithms. *Lecture notes on Data Engineering and Communications technologies, 53*, 645–657.
Van Kessel, S., Sajuria, J., & Van Hauwaert, S. M. (2021). Informed, uninformed or misinformed? A cross-national analysis of populist party supporters across European democracies. *West European Politics, 44*(3), 585–610.
Vermeulen, D., & Van Niekerk, A. (2017). Machine learning performance for predicting soil salinity using different combinations of geomorphometric covariates. *Geoderma, 299*, 1–12.
Waghid Y, Waghid Z and Waghid F (2019). The Fourth Industrial Revolution Reconsidered: On advancing cosmopolitan education. South African Journal of Higher Education, 33(6), 1-9. DOI: https://doi.org/10.20853/33-6-3777
Walton, S. M. (2014). *Determining the impacts of an information and communications technology intervention: Empowering indigent youths and their community*. PhD Thesis, Monash University. https://monash.hosted.exlibrisgroup.com/permalink/f/1gr7b58/catau21276175310001751
Wandera, H., Marivate, V. & Sengeh, M. D. (2019). Predicting national school performance for policy making in South Africa. *Proceedings of the 6th international conference on soft computing and machine intelligence (ISCMI)*, Johannesburg, South Africa, Nov 19–20, 2019, pp. 23–28.
Wang, L., Liu, X., Liu, Z. Z., & Jia, C. X. (2020). Digital media use and subsequent self-harm during a 1-year follow-up of Chinese adolescents. *Journal of Affective Disorders, 277*, 279–286.
Wessels, K. J., van den Bergh, F., Roy, D. P., Salmon, B. P., Steenkamp, K. C., MacAlister, B., Swanepoel, D., & Jewitt, D. (2016). Rapid land cover map updates using change detection and robust random forest classifiers. *Remote Sensing, 8*(11), Article 888.
Willcock, S., Martínez-López, J., Hooftman, D. A. P., Bagstad, K. J., Balbi, S., Marzo, A., Prato, C., Sciandrello, S., Signorello, G., Voigt, B., Villa, F., Bullock, J. M., & Athanasiadis, I. N. (2018). Machine learning for ecosystem services. *Ecosystem Services, 33 Part B*, 165–174. ISSN 2212-0416,. https://doi.org/10.1016/j.ecoser.2018.04.004

Xulu S (2013). Johannesburg's smart meter implementation programme: from concept to reality. *24th AMEU Technical Convention 2013*. https://dev.ee.co.za/wp-content/uploads/2014/05/AMEU-Convention-2013-p121-128.pdf

Xuying L & Young EH (2020). Who gets the blame for service failures? Attribution of responsibility toward robot versus human service providers and service firms, Computers in Human Behavior, 113, 106520, ISSN 0747-5632, doi:https://doi.org/10.1016/j.chb.2020.106520.

Yacim, J. A., & Boshoff, D. G. B. (2020). Neural networks support vector machine for mass appraisal of properties. *Property Management, 38*(2), 241–272.

Zybrands, W. (2012). Municipal structures and finances – Predicaments and performance challenges. In S. Booysen (Ed.), *Local elections in South Africa: Policies, parties and politics* (pp. 67–88). Sun Press.

Food Security

Jennifer Thomson

Abstract This chapter explores the role of agriculture, particularly precision agriculture, in enhancing food security in South Africa amidst the challenges posed by climate change. It addresses the country's current state of food security, emphasizing the importance of agriculture as the foundation of the economy. Despite being foundational, agriculture tends to be overlooked in priority lists. The chapter highlights a shift in rural households' food production to market purchases, leading to significant expenditures. Smallholder agriculture could mitigate these costs, emphasizing the need to boost their productivity through sustainable intensification and improved farming methods. The chapter underscores the need for regulatory authorities to provide guidelines for genome editing, expressing concerns about the potential stifling of research due to funding constraints and regulatory barriers. Drawing attention to the importance of a "product" rather than a "process" approach, the chapter urges regulators to align decisions with scientific parameters, citing Argentina as an example. The conclusion stresses the pivotal role of agriculture in South Africa's economy and food security, urging the government to prioritize and acknowledge its significance. A call is made for collaboration between the Department of Science and Innovation and provincial departments to promote the agricultural sector, particularly in rural areas, addressing poverty, unemployment and inequality challenges.

1 Introduction

How can agriculture, and specifically precision agriculture, contribute to increasing food security in South Africa? In doing so, the impacts of climate change need to be taken into account, as we only have to look at the increase in devastating weather-related floods, wildfires, storms and droughts of recent years to realize that the

J. Thomson (✉)
University of Cape Town, Rondebosch, Cape Town, South Africa
e-mail: jennifer.thomson@uct.ac.za

O. Saritas et al. (eds.), *21st Century Foresight*, Science, Technology and Innovation Studies, https://doi.org/10.1007/978-3-031-69557-5_15

worse is still to come. This chapter will discuss these issues, but before going into detail, let us look at current levels of food security in the country.

Although agriculture is the foundation of the South African economy and the basis of our food security, it does not feature high on the list of priorities. In the past rural households produced most of their own food, but recent studies have shown an increase in dependence on market purchases by such households. Food expenditure can account for as much as between 60 and 80% of total household income. However, this does not have to continue to be the case if smallholder agriculture were to play more of a role in reducing such costs. There is, therefore, a need to significantly increase the productivity of such farms, and one way to achieve this is by the use of sustainable intensification using improved farming methods (https://www.researchgate.net/publication/46536430_The_Contribution_of_Subsistence_Farming_to_Food_Security_in_South_Africa).

This is not an issue that can sit on the backburner as the South African Foresight Exercise for Science, Technology and Innovation (2030) states that malnutrition and stunting affect 25% of the under-fives in this country, permanently compromising their cognitive and physical development and amplifying current socio-economic inequalities. In addition, there is an upward trend in overweight/obese adults and children, with more than 60% of South African women affected. Precision agriculture has the potential to take South Africa's agricultural sector to the next level, as it can improve yields, reduce waste and pollution, reduce the land area required and reduce water and energy demands. Farms need to be transformed into agribusinesses, both small and large, as these might hold greater appeal for the younger generation to be drawn into farming as a career. The next sections will deal with aspects of precision agriculture, namely, genetically modified crops and genome editing, with the emphasis on improving food security and the importance of developing local research capacity.

2 The Use of Genetically Modified (GM) Crops in South Africa

In a recent book titled *GM Crops and the Global Divide* (Thomson, 2021), I have a chapter on countries that got the use of GM crops right. What do I mean by a country "getting it right"? Such a country needs to have a government which is supportive of innovations and new technologies that improve agricultural production and make the lives of farmers more profitable and less stressful. It should also have a regulatory system in place that is flexible, operates on a case-by-case system and whose decisions are science based. The regulations should not act as barriers to the development in implementation of GM crops. And finally, they should encourage private enterprises to develop and commercialize such crops. One such country is South Africa.

South African farmers started planting GM crops commercially as early as 1998. Thus, even in those early days, the government was supporting these innovative technologies which enabled farmers to become more profitable. By the year 2018, they were growing 2 million hectares of GM maize, 0.7 million of soybeans and 37,000 of cotton. This means that approximately 85% of maize farmers chose to plant GM maize in that year as some will alternate between GM and non-GM seeds depending on personal requirements such as varietal preferences, intended use of the crop, etc. (ISAAA, 2018). Most (66%) of GM maize carry both the *Bt* (for insect resistance) and herbicide tolerance (HT) traits, commonly referred to as stacked traits. The fall armyworm, which has in recent years devastated many of the maize fields in other parts of Africa, had much less of an impact here. This has been attributed to the high percentage of *Bt* maize. In addition, over the years the country has been planting GM maize, higher yields have meant that there is less of a need to plant the crops on marginal land. Indeed, the maize yield has doubled since 1998 (ISAAA, 2017).

Another of the reasons why South Africa "got it right" was because it was an early adopter of GM crops. Why was this the case? One of the reasons is that the country had been dealing with GMOs since 1978 when the Council for Scientific and Industrial Research (CSIR), then the major government research funding agency, started the South African Committee for Genetic Experimentation (SAGENE). This was because they started to receive research grant proposals for work on GM bacteria, and they decided they needed guidelines along the lines of those developed in the United States by the National Institutes of Health, the NIH Guidelines for Research into Recombinant DNA.

SAGENE fulfilled an excellent development role as it required universities to improve laboratory standards in order for their academic staff to obtain research grants for work on GMOs. They also ran training programmes which helped to bring many young scientists into this area of research. After some years SAGENE felt it had accomplished much of what it was set up to achieve and therefore went into abeyance, meeting only once or twice a year. That, however, changed in the 1990s.

In 1990, SAGENE received an application from the multinational company Calgene Inc. for field trials of GM cotton resistant to the herbicide bromoxynil (BXM™). Shortly thereafter they received an application from Clark Cotton to conduct trials of Bollgard® cotton seed, carrying a *Bt* gene. Faced with these applications, the South African government officially reconstituted SAGENE, announcing it in the *Government Gazette* of 15 May 1992 (Thomson, 2013).

SAGENE then drew up procedures for assessing whether to approve applications for importation, trial release or general release of GM crops. However, it was not a statutory body of the government and therefore could not impose penalties for any infringements of its procedures. As a result, the government set about developing a Genetically Modified Organisms Act, which was published on 23 May 1997. However, in true government style, it took a considerable time before the act could be implemented as the regulations had to be approved by parliament. As a result, SAGENE took on an interim role until this took place in December 1999. In this interim, SAGENE handled 27 applications of which 13 were for maize; 4 for cotton; 2 for soybeans; 1 each for canola, strawberries, eucalyptus and apple; and 4 for microorganisms. The traits included insect resistance, fungal and viral resistance and herbicide tolerance.

As a result, by the time the GMO Act came into being, the government was reasonably familiar with GM crops. It also had advisors from the scientific and agricultural sectors to assist them. One of these advisory bodies was, and still is, the National Advisory Committee on Innovation which advises the Department of Science and Innovation. Together they drew up the National Biotechnology Strategy in 2001, which resulted in a number of biotechnology regional innovation centres, including the National Innovation Centre for Plant Biotechnology (PlantBio). In 2013 they produced the bioeconomy strategy which aimed to bring innovative research to commercialization. The encouragement of private enterprise to become involved in the mobilization of GM crops for public use is another feature of countries that got it right.

Another reason why South Africa became an early acceptor of GM crops was the organization AfricaBio. This was started in 1990 and became a registered section 21 company in 2000 (Thomson, 2013). According to their website, they are "an independent, non-profit biotechnology stakeholders' association for the safe, ethical, and responsible research, development, and application of biotechnology and its products" (www.africabio.org). In the early years of its existence, it played a pivotal role in informing farmers, commercial and smallholders alike, about the role that GM crops could play in agriculture. They ran workshops to teach them about the sustainable use of both *Bt* and HT crops to prevent the build-up of resistance in either insects against the *Bt* toxin or in weeds against the relevant herbicides.

A third reason for the early acceptance of GM crops was that the great majority of commodity crop production is carried out on modern, commercial farms where the benefits of the GM input traits were apparent. By making GM seeds available to such farmers as early as 1998, the South African government certainly "got it right". In addition, by facilitating, more recently, the payment of levies to breeders by soybean farmers, it enabled breeders to continue to bring novel varieties to market. This could be used as a model for other countries facing similar problems.

But what about smallholder, mainly subsistence, farmers? Unfortunately, in this respect, the government has not yet "got it right". Why is this? South Africa is the only country where the majority of its citizens eat GM white maize up to three times a day. Yellow maize is used for animal feed. In 2016, Marnus Gouse and colleagues carried out a study to determine what effect growing GM maize had on smallholder farmers in a region where this is grown extensively (Gouse et al., 2016). They stated that "although there is an extensive and growing body of literature on the economic impact of the adoption of GM crops in both developing and developed economies, there is only scant evidence that the technology has had any specific and distinguishable impact among female and male [smallholder] farmers".

In order to provide evidence of the role that GM maize plays in the lives of smallholder farmers in South Africa, Gouse and his colleagues studied gender-specific adoption and performance of *Bt* and HT maize in the KwaZulu-Natal province.

They chose this region as many of the smallholder farmers there are women. Their results, both quantitative and qualitative, showed that women farmers, traditionally responsible (together with their children) for weeding, valued the labour-saving advantages of HT maize. Labour demand is greatest during the crucial period when land is prepared and crops are planted and weeded. Labour-saving advantages are amplified by the trend of migration away from rural areas to cities. Among male farmers, higher yields were the main reason behind adoption (Gouse et al., 2016).

Another study looked specifically at the use of *Bt* maize by smallholder farmers. Kotey et al. (2016) found that there is a problem linked to the possible build-up of insects that become resistant to the *Bt* toxin. Farmers need to plant a buffer zone of non-*Bt* plants around their *Bt* crop. This maintains a population of *Bt*-susceptible insects that can reproduce with any resistant insects that develop and sustain *Bt*-susceptible individuals in the pest population. The problem which they identified was with the lack of good extension and advisory support received by these farmers, especially in the poorer areas of the country. Smallholder *Bt* maize farmers need to be made aware of the stewardship requirements that require the buffer zone, and this information must be transferred to them in a manner that is understandable. If this is not done, the long-term sustainability of this technology is at risk. The authors' study showed that the extension personnel in the regions they studied lacked adequate training to effectively disseminate GM maize technology to those farmers.

The authors (Kotey et al., 2016) end by recommending that the Department of Rural Development and Agrarian Reform (DRDAR) train the extension personnel on good stewardship methods and the consequences of non-compliance. In addition, the DRDAR should make non-*Bt* hybrids available to the farmers in order to allow them to use such plants as buffers to prevent insect resistance from developing. If this is not done, although the country's commercial farmers will have "got it right", their smallholder counterparts will not.

3 Lessons from African Countries Which Delayed Growing Genetically Modified Crops

If South African farmers, albeit mainly commercial one, are growing GM crops so profitably, why are not more African countries growing GM crops? It is not as if South African farmers have had bad experiences, especially when growing GM white maize. In a recent article entitled "Economic and Ecosystem Impacts of GM Maize in South Africa" (Ala-Kokko et al., 2021), the authors state that the key benefits from growing GM white maize were estimated to amount to US$5 million from 2001 to 2018, with lower pesticide requirements compared to conventional white maize. In 2017, South Africa produced approximately 1.1 million hectares of GM maize varieties for direct human consumption, representing an 85% adoption rate. In the light of food insecurity in African countries, which will only become worse with climate change, why do not more grow GM crops?

Indeed, South Africa's use of GM crops stands out as a beacon of forward-looking policies based on scientific evidence, when compared with other countries in Africa. For many years South Africa and Sudan were the only African countries growing such crops commercially, with Eswatini and Nigeria having joined more recently and a few others in the pipeline. To estimate the detrimental effect this has had on some countries in East Africa, Justus Wesseler, an agricultural economist from Wageningen University, and colleagues investigated the cost of delaying the introduction of three GM crops, namely, disease-resistant cooking bananas (matoke) and insect-resistant cowpea and maize. In their 2017 paper (Wesseler et al., 2017), they commented "If Kenya had adopted GE [genetically engineered] corn in 2006, between 440 and 4,000 lives could theoretically have been saved. Similarly, Uganda had the possibility in 2007 to introduce the black sigatoka resistant banana, thereby potentially saving between 500 and 5,500 lives over the past decade".

In addition, the authors expressed concern over the delay in approval of *Bt* cowpea by Nigerian officials, as they estimate that "A one-year delay in approval [of the insect-resistant *Bt* cowpea], would especially harm Nigeria, as malnourishment is widespread there … [and] cost Nigeria about 33 million USD to 46 million USD and between 100 and 3,000 lives". Fortunately, in 2019, Nigeria approved the commercial release of *Bt* cowpea that is resistant to the pod borer, *Maruca vitrata.* Commenting on this announcement, Professor Ishayaku, the project's principal investigator from the Institute of Agricultural Research in Nigeria, urged other African countries to stop raising problems about the use of GM crops when such problems have already been addressed numerous times. He wrote: "Regulators should stop the attitude of re-inventing the wheel: we have global and universal principles. What is applied in Rwanda can be applied in Kenya as long as it is healthy and environmentally friendly … [Regulators] should bear in mind that the financial burden through regulations reduces the profitability of the technology". In other words, why don't African countries learn from each other and not begin afresh each time (http://africenter.isaaa.org/prof-ishayaku-nigerias-approval-bt-cowpea-signifies-africas-capacity-adopt-biotech-crop)?

More recently, in August 2021, the African Agricultural Technology Foundation (AATF) reported that Nigerian trials of GM maize stacked with drought tolerance and insect resistance traits were showing outstanding results. The maize variety, called Tela, produced 9 tonnes per hectare in areas of the country where yields are normally around 3 tonnes per hectare. Again, Prof Ishayaku stated: "The savings farmers will make from this maize variety are estimated to be over 3 billion nairas from insecticide spray of 500 hectares land and over 6 billion nairas from drought effects" (Transgenic Maize Trials Show Outstanding Results in Nigeria- Crop Biotech Update (August 4, 2021) | ISAAA.org).

Also in 2021, Kenya approved national performance trials of GM cassava which is resistant to the highly destructive cassava brown streak disease (CBSD). This was developed by the Kenya Agricultural and Livestock Research Organization (KALRO) and could prove to be the answer to cassava farmers who can sometimes lose up to 98% in their crops. Even in good years, the yield might only be 10 tonnes per hectare compared with 50 tonnes per hectare commonly achieved by farmers in Asia and South America. Cassava is the world's most widely grown starch storage root crop and accounts for a third of the total production of staple food crops in sub-Saharan Africa (Kenya approves disease-resistant GMO cassava - Alliance for Science (cornell.edu)).

These few examples of African countries only now starting to grow GM crops shows the wisdom of South African regulators. The advantages might become even more obvious in the future as climate change leads to increasing problems due to drought and concomitant decreased areas of land suitable for cultivation.

4 The Importance of Drought-Tolerant Maize

Almost all countries in Africa are projected to experience increased drying during the twenty-first century and could become more vulnerable to the impacts of drought. The daily maximum temperature is projected to increase by up to 8^0C; precipitation could decrease, while warm spells increase (Gan et al., 2016). With this in mind, a number of groups are investigating ways to improve the drought tolerance of crops. One such initiative is the Water Efficient Maize for Africa (WEMA) project under the management of the AATF (mentioned above). Monsanto donated, royalty free, the gene, *csp*B, which encodes a cold shock protein from the bacterium *Bacillus subtilis*. This protein has been shown to confer tolerance to dehydration in transgenic maize (Chang et al., 2014). The gene has been introduced into African maize varieties by the International Maize and Wheat Improvement Center (CIMMYT), which, although based in Mexico, has centres in Africa.

Through the WEMA project (now known as the TELA Maize Project as TELA is derived from the Latin word *tutela* which means protection), a variety of genes are available to farmers. These include the drought-tolerant transgene (*csp*B, DroughtGard®, MON87460), as well as three *Bt* genes encoding insect resistance (MON810 (*Cry1Ab*) and MON89034 (*Cry1A.105* and *Cry2Ab2*)). All these genes were accessed royalty free for smallholder farmers in Africa by the AATF. The *Bt* genes have been stacked with the DroughtGard® gene to provide several variety options for the farmers who operate in drought-prone environments.

The fall armyworm (*Spodoptera frugiperda*; FAW) is an insect pest, native to the Americas, which has recently been reported in Africa, where it is ravaging staple crops, particularly maize, causing serious crop destruction. It was first reported in Nigeria in 2016, but it has since spread to over 40 countries in Africa, thus posing a major threat to food and nutrition security in Africa. Confined field trials (CFTs) in various African countries have shown that the *Bt* MON810 gene confers significant control of FAW as well as stemborers.

In South Africa, which has a long history of commercializing biotech crops, five TELA® hybrids have been commercialized to smallholder farmers since 2016. Farmers are currently growing these with good protection against stemborer and

FAW (https://maize.org/new-maize-hybrid-shows-resistance-to-stem-borers-in-south-africa/).

5 The Development of Genome-Edited Crops

This reasonably new technology for crop and animal improvement could make a difference to the lives of both the poor and the not-so-poor worldwide, but already the EU is threatening to stigmatize it as unsafe. The first, and still widely used, method used for genome editing is called CRISPR. It allows scientists to change DNA sequences in a very precise manner and thus change the function of a gene. It has the capacity to improve crops by, for instance, changing a gene that might code for sensitivity to a harmful pathogen.

How does it work? CRISPR is an acronym for *c*lustered *r*egularly *i*nterspaced *s*hort *p*alindromic *r*epeats. These short palindromic repeat sequences of ~30 base pairs are found clustered at regular intervals in a genome. CRISPR is itself shorthand for CRISPR/Cas9 (*C*RISPR *as*sociated). The CRISPR part is a sequence of DNA that binds to the repeat sequence in a particular gene, for example, one that codes for sensitivity to a plant virus. CRISPR is bound to Cas9, which is an enzyme that cuts the plant's DNA at that particular spot like a pair of scissors. The plant's own DNA repair system then comes into play to mend the cut by introducing random bases. Thus, a change or mutation is introduced at that spot, and the gene no longer codes for virus sensitivity.

The technology, which was discovered and developed in 2012, has already been used for crop improvement in Africa, one example being resistance in cassava to its brown streak virus, a major problem in East and Central Africa (Gomez et al., 2019). But how will regulators react to this new tool? One problem they face is that, unlike "old-fashioned" genetic engineering, the plant is left without any signature DNA sequences that can be used to detect standard GMOs. The only way to detect the difference between a conventional plant and its gene-edited counterpart is by fine-scale DNA sequencing to determine whether a particular gene has been mutated or not, not a straightforward exercise.

The first country to take the plunge was the United States, which gave the green light to the CRISPR-edited white mushroom that is resistant to browning. It was developed by a plant pathologist, Yinong Yang, at Pennsylvania State University, who deleted a few base pairs in a family of genes coding for the enzyme polyphenol oxidase that is responsible for browning (Waltz, 2016).

In response to this, a rather interesting dialogue has been taking place in the organic farming sector. Most of them, exemplified by the pro-organic Cornucopia Institute in Wisconsin, USA, express their abhorrence for genetic engineering, including gene-edited crops. However, some organic growers see this technology as a boon for their industry and are calling for a revision in the rules so that they can benefit from gene editing. Among these is the influential Klaas Martens, who farms in New York's Finger Lakes region and runs Lakeview Organic Grain. At a conference in 2019 he said, "If it could be used in a way that enhances the natural system, and mimicked it, then I would want to use it. But it would definitely have to be case by case". Others noted that CRISPR-like techniques could help them to reduce pesticide use and grow more disease-resistant crops (Cerier, 2019). If this approach were to gain momentum in this farming sector, the use of this powerful technology could become much more widely accepted.

What is the situation in the EU, because, as I mentioned at the start of this section, they were threatening to brand this technology as unsafe? On 25 July 2018, the Court of Justice of the European Union determined that "Organisms obtained by mutagenesis are GMOs and are, in principle, subject to the obligations laid down by the GMO directive". However, "organisms obtained by mutagenesis techniques which have conventionally been used in a number of applications and have a long safety record are exempt from these obligations" (Court of Justice of the European Union, 2018).

The next day, in an article entitled "Scientific Community Defeated by Green Groups in European Court Ruling on Gene Edited Crops", Mark Lynas commented that Europe now faced the bizarre situation of regulating new plant varieties that had been produced by precise genome editing but allowing the "hit-and-miss" mutagenesis by chemicals or radiation (Lynas, 2018).

In a response to this ruling, scientists representing 93 European plant and life sciences research centres and institutes endorsed a position paper calling upon the court to alter the legislation and to thoroughly revise the GMO regulations to "correctly reflect scientific progress in biotechnology". This is because "organisms that have undergone simple and targeted genome edits by means of precision breeding and which do not contain foreign genes are at least as safe as if they were derived from classical breeding techniques" (VIB, 2018).

The scientists expressed great concern that this ruling would put European agricultural industries at a competitive disadvantage because it would make precision breeding hyper-expensive. Farmers too would miss out on new crops that are more nutritious and better suited to respond to climate change (VIB, 2018). Subsequently, in May 2019, 14 EU countries called for a "unified approach" to gene editing in plants. In response the European Commission promised to come up with a "robust response" to the EU court ruling and draft a legislative proposal "in due time"—whenever that may be (Fortuna, 2019).

As far as developing countries are concerned, many of them could follow the lead of the EU, as they did with GM crops, thereby denying their farmers and seed companies access to these new crops to be bred with highly precise, modern technology. Time will tell whether science or politics holds sway.

But what about South Africa, which proved to be way ahead of most African countries when it came to approving safe GM crops? Unfortunately, the regulatory authorities have been slow to accommodate requirements regarding genome-edited crops in its governance frameworks (Entine et al., 2021). In 2017, the Academy of Sciences of South Africa (ASSAf), under the auspices of the Department of Science and Innovation (DSI), published a consensus study on "The Regulatory Implications of New Breeding Techniques (NBTs)" (ASSAf, 2017), but no formal guidelines were published. This is despite the fact that the study stated that such new breeding techniques hold great potential and are relevant to South Africa. Moreover, although the study recommended a science-based risk analysis approach, authorities appear to have become bogged down in the legal interpretation of the existing GMO Act regulations between "product"- or "process"-based approaches. Another problem is that South Africa continues to use the European-based definition for a GMO using the value laden, capricious concept of "naturally occurring genetic variation" as a threshold for GMO regulation. According to the 2021 Entine et al. paper, "These new technological developments and consequent need for updated governance guidelines should be used as an opportunity to realign regulatory frameworks with scientific risk analysis principles. Despite the enduring public debate on genetics-based innovations there is wide consensus among the scientific community that these technologies can be applied in a useful and safe manner. The well-established science-based risk analysis frameworks for GMOs have evolved over more than three decades into robust tools that can be effectively applied to any genetically engineered organism, to ensure its sustainability, if the principle of a case-by-case, comparative risk analysis is applied". South Africa therefore has an opportunity not only to address the shortcoming of past regulations but can also, as the paper states, "future-proof" them. This, however, will only be possible if the regulations are based on well-established, scientific risk analysis principles.

Another problem with South Africa's non-comital stance on genome-edited crops is that such an attitude can result in the research community hesitating to embark on the research necessary to bring such crops to commercialization. Argentina, a country in the forefront of regulatory systems in the field of agricultural biotechnology (Vicien & Trigo, 2017), has adopted an approach that is already stimulating local innovation processes (Whelan et al., 2020). Indeed, they comment that genome-edited products constitute a window of opportunity for developing countries. South African regulators would do well to take heed of this advice, especially as the Argentine experience has shown that gene-edited products seem to follow a much faster development rate from bench to market. Part of this accelerated development has been led in that country by small and medium enterprises (SMEs) and public research institutions. They also note that, in addition, product profiles are more diversified in terms of traits and organisms. Such advances are totally dependent upon a "product"-based approach to regulations as opposed to a "process" one.

As yet, no South African scientist or organization has applied for permission to conduct field trials of a genome-edited crop. However, as this could well occur in the near future, the regulatory authorities should urgently consider how such applications will be considered.

6 Conclusions

In the introduction the question posed was how agriculture, and specifically precision agriculture, could contribute to increasing food security in South Africa. The point was also made that although agriculture is the foundation of the South African economy and the basis of our food security, it does not feature high on the list of priorities. Two aspects of precision agriculture, namely, genetically modified (GM) crops and genome editing, were considered.

Regarding the former, South Africa was in the forefront of accepting GM crops as early as 1998 and recently ranked eighth in the list of countries growing such crops (ISAAA, 2019). One of the outcomes of this approach to the technology was that scientists were encouraged to undertake research in the development of GM crops. As a result, many students from other parts of Africa came to universities and research institutes in South Africa to gain experience in this field of expertise. Sadly, for those other African countries, the majority of these students, including those in my own lab and many of my colleagues' labs, has chosen to remain in South Africa in order to pursue their careers. This pattern has been seen in many other countries where scientists have realized that the work which they were carrying out would never have the opportunity to be used by farmers. They have therefore chosen to work elsewhere or changed their field of research.

This was highlighted in an article entitled "The Perils of GMO Research: A Scientist Speaks Out". The author stated that "as many as a quarter of European universities have shut down their GMO research programs, some due to a loss of funding and others because scientists are leaving the GMO sphere, tired of the backlash and criticism" (https://allianceforscience.cornell.edu/blog/2018/05/perils-gmo-research-scientist-speaks/).

Turning to the newer technology of genome editing, how will South Africa fare in embracing this area of precision agriculture? As mentioned above, regulatory authorities here have not yet published formal guidelines regarding this technology, despite acknowledging that it holds great potential for improving crop production. Will South Africa end up, like so many other African countries did with GM crops, by stifling research in this area due to lack of funding and/or opportunities for such research to come to fruition in the hands of farmers? It is time for regulators to take action and allow scientific parameters to guide their decisions, as have their counterparts in Argentina. Furthermore, it is imperative that South Africa's decisions be "product" based rather than "process" based.

Finally, it is also imperative that the importance of agriculture, as the foundation of South Africa's economy and its peoples' food security, be acknowledged by the government and given its rightful place on the list of the country's priorities. Indeed, in the 28 January 2021 audit of the South African Bioeconomy Sector prepared by the National Advisory Committee on Innovation (NACI), one of the recommendations stated:

> the Department of Science and Innovation (DSI) should work with selected provincial departments in developing the agricultural sector of the bioeconomy countrywide, espe-

cially in rural provinces. Such an approach would increase the DSI footprint in provinces and contribute toward addressing the triple challenges of poverty, unemployment and inequality, and in developing indigenous knowledge applications of technology.

Acknowledgements The author wishes to thank Dr Hennie Groenewald of Biosafety SA on information and discussions regarding genome-edited crops. (Still to come) Kingston Mashingere ARC on TELA in South Africa

References

Ala-Kokko, K., Nalley, L. L., Shew, A. M., Tack, J. B., Chaminuka, P., Matlock, M. D., & D'Haese, M. (2021). Economic and ecosystem impacts of GM maize in South Africa. *ScienceDirect, 2021*, 79100544. https://doi.org/10.1016/j.gfs.2021.100544)

ASSAf (Academy of Science of South Africa). (2017). *The regulatory implications of new breeding techniques*. ASSAf. doi: https://doi.org/10.17159/assaf.2016/0011

Cerier, S. (2019). 'Is there a place for CRISPR gene editing in organic farming? Many farmers say 'yes''. *Genetic Literacy Project*, December 3 December, 2019. https://geneticliteracyproject.org/2019/12/03/is-there-a-place-for-crispr-gene-editing-in-organic-farming-many-farmers-say-yes

Chang, J., Clay, D. E., Hansen, S. A., Clay, S. A., & Schumacher, T. E. (2014). Water stress impacts on transgenic drought-tolerant corn in the northern great plains. *Agronomy Journal, 106*, 125–130. https://doi.org/10.2134/agronj2013.0076

Court of Justice of the European Union. (2018). *Judgment in Case C-528/16*. https://curia.europa.eu/jcms/upload/docs/application/pdf/2018-07/cp180111en.pdf

Entine, J., Felipe, M. S. S., Groenewald, J.-H., Kershen, D. L., Lema, M. H. A., Nepomuceno, A. L., Ohsawa, R., Reynante, L. O., Ordonio, R. L., Parrott, W. A., Quemada, H., Ramage, C., Slamet-Loedin, I., Smyth, S. J., & Wray-Cahen, D. (2021). Regulatory approaches for genome edited agricultural plants in select countries and jurisdictions around the world. *Transgenic Research*. https://doi.org/10.1007/s11248-021-00257-8

Fortuna, G. (2019). 14 EU countries call for 'unified approach' to gene editing in plants. *EURACTIV*, 24 May 24, 2019. https://www.euractiv.com/section/agriculture-food/news/14-eu-countries-call-for-unified-approach-to-gene-editing-in-plants

Gan, T. Y., Ito, M., Hülsmann, S., Qin, X., Lu, X. X., Liong, S. Y., Rutschman, P., Disse, M., & Koivusalo, H. (2016). Possible climate change/variability and human impacts, vulnerability of drought-prone regions, water resources and capacity building in Africa. *Hydrological Sciences Journal, 61*, 1209–1226. https://doi.org/10.1080/02626667.2015.1057143

Gomez, M. A., Lin, Z. D., Moll, T., Chauhan, R. D., Hayden, L., Renninger, K., Beyene, G., Taylor, N. G., Carrington, J. C., Staskawicz, B. J., & Bart, R. S. (2019). Simultaneous CRISPR/Cas9-mediated editing of cassava eIF4E isoforms nCBP-1 and nCBP-2 reduces cassava brown streak disease symptom severity and incidence. *Plant Biotechnology Journal, 17*, 421–434. https://doi.org/10.1111/pbi.12987

Gouse, M., Sengupta, D., Zambrano, P., & Zepeda, J. F. (2016). Genetically Modified Maize: Less Drudgery for Her, More Maize for Him? Evidence from Smallholder Maize Farmers in South Africa. *World Development*, pp. 27–38.

ISAAA. (2017). *Global status of commercialized biotech/GM crops in 2017: Biotech crop adoption surges as economic benefits accumulate in 22 years*. ISAAA Brief No. 53. ISAAA, Ithaca, NY.

ISAAA. (2018). *Global status of commercialized biotech/GM crops in 2018: Biotech crops continue to help meet the challenges of increased population and climate change*. ISAAA Brief No. 54. ISAAA, Ithaca, NY.

ISAAA. (2019). *Global Status of Commercialized Biotech/GM Crops in 2019: Biotech crops drive socio-economic development and sustainable environment in the new frontier.* ISAAA Brief No. 55. ISAAA, Ithaca, NY

Kotey, D. A., Assefa, Y., Obi, A., & Van den Berg, J. (2016). Disseminating genetically modified (GM) maize technology to smallholder farmers in the Eastern Cape province of South Africa: Extension personnel's awareness of stewardship requirements and dissemination practices. *South African Journal of Agricultural Extension, 44*, 59–74.

Lynas, M. (2018). Scientific community defeated by green groups in European court ruling on gene edited crops. *Alliance for Science*, 26 July 26, 2018. https://allianceforscience.cornell.edu/blog/2018/07/scientific-community-defeated-green-groups-european-court-ruling-gene-edited-crops

Thomson, J. A. (2013). *Food for Africa: The life and work of a scientist in GM crops.* UCT Press.

Thomson, J. (2021). *GM crops and the global divide.* CSIRO Press.

VIB. (2018). *Regulating genome edited organisms as GMOs has negative consequences for agriculture, society and economy.* Position Paper on the ECJ ruling on CRISPR, 12 Nov 2018. http://www.vib.be/en/news/Documents/Position%20paper%20on%20the%20ECJ%20ruling%20on%20CRISPR%2012%20Nov%202018.pd

Vicien, C., & Trigo, E. (2017). The Argentinian GMO biosafety system: An evolving perspective. In A. A. Adenle, E. J. Morris, & D. J. Murphy (Eds.), *Genetically modified organisms in developing countries: Risk analysis and governance* (pp. 247–257). Cambridge University Press. https://doi.org/10.1017/9781316585269.022

Waltz, E. (2016). Gene-edited CRISPR mushroom escapes US regulation. *Nature, 532*, 293–303. https://doi.org/10.1038/nature.2016.19754

Wesseler, J., Smart, R. D., Thomson, J., & Zilberman, D. (2017). Foregone benefits of important food crop improvements in sub-Saharan Africa. *PLoS One, 12*, e0181353. https://doi.org/10.1371/journal.pone.0181353

Whelan, A. I., Gutti, P., & Lema, M. A. (2020). Gene editing regulation and innovation economics. *Frontiers in Bioengineering and Biotechnology, 8*, 303. https://doi.org/10.3389/fbioe.2020.00303

Water Security

Ozcan Saritas

Abstract The water security domain in SAForSTI 2030 aims to address water wastage, pollution, and usage challenges, focusing on rural areas. Despite the importance of water recycling, the country lags, especially in tailoring technologies to its unique environment. Although efforts to deliver breakthrough technologies increases in small- and medium-sized enterprises (SMMEs) in the water sector and enhanced rural water access exists, South Africa still needs to improve potency and management challenges. Neglecting water issues poses severe risks to economies, livelihoods, and populations, potentially leading to catastrophic consequences. Weak hydrological monitoring systems, data storage and sharing issues, reliance on outdated data, and managing complexity in growing cities hinder progress. Key policy considerations in water security include the need to synergize the National Water RDI Roadmap, the national water and sanitation master plan, and the water components of the industrial policy action plan. Improved coordination among institutions engaged in water research, innovation, and deployment is crucial. Supporting co-learning platforms and developing a clear framework delineating the roles of relevant parties in the water sector are also recommended to address the complex challenges faced by South Africa in water security.

1 Introduction

The water security domain is dedicated to addressing critical issues related to water resources, primarily focusing on mitigating water wastage, curbing water pollution, and optimizing water usage, particularly in rural areas. An essential facet of water security is the implementation of efficient water recycling practices, a vital element for sustainable water management in South Africa. However, the country faces

O. Saritas (✉)
Rochester Institute of Technology Dubai, Dubai Silicon Oasis, UAE

National Research University, Higher School of Economics, Moscow, Russia
e-mail: oxscad1@rit.edu; osaritas@hse.ru

O. Saritas et al. (eds.), *21st Century Foresight*, Science, Technology and Innovation Studies, https://doi.org/10.1007/978-3-031-69557-5_16

challenges in this domain, needing to catch up with water recycling technologies. Notably, the existing expertise is predominantly concentrated in industries such as iron, steel, and mining, necessitating a shift toward broader and more diverse applications in other sectors.

> Water recycling is very important for South Africa, but we lag behind. Most of our expertise is in the iron and steel, and mining industries … The water-saving technologies that we need are different from those in the developing world and should be tailored to the African environment. –Stakeholder interview (2 August 2018)

As emphasized in a stakeholder interview conducted on 2 August 2018, it is crucial to recognize that the water-saving technologies required for South Africa differ significantly from those applicable in the developing world. Tailoring these technologies to the African environment is imperative to address unique challenges and optimize their effectiveness.

To guide and structure efforts in enhancing water security, South Africa has developed the Water Research, Development, and Innovation (RDI) Roadmap for 2015–2025. This strategic framework aims to align RDI activities with national policies, strategies, and plans for water resource management. The National Water Research, Development, and Deployment Program, a component of this road map, focuses on achieving breakthrough technologies at regular intervals; fostering the growth of small-, medium-, and micro-sized enterprises (SMMEs) in the water sector; improving water accessibility in rural communities; and ensuring sustainable sanitation practices for the entire population.

Despite these initiatives, South Africa continues to grapple with substantial water efficiency and management challenges. The consequences of neglecting water-related issues pose severe threats to economies, livelihoods, and populations. Such negligence can lead to catastrophic outcomes, including extensive financial burdens and potentially irreparable ecosystem damage. Unsustainable water management further jeopardizes hard-earned development gains, reverses poverty reduction efforts, and undermines job creation endeavors.

Recent years have seen significant efforts and notable achievements. However, the nation's water sector still needs to overcome hurdles such as weakly maintained hydrological monitoring systems, data storage and sharing challenges, reliance on outdated data for decision-making, and difficulties in managing complexity as urban areas expand and resources become more integrated. To address these issues, specific policy considerations are proposed. These include synergizing and harmonizing various water-related initiatives such as the National Water RDI Roadmap, the National Water and Sanitation Master Plan, and the water components of the Industrial Policy Action Plan. Additionally, there is a call for improved coordination among institutions engaged in water research, innovation, and solution deployment. Establishing co-learning platforms to bring stakeholders involved in water-related activities together is suggested, along with developing a clear framework outlining the roles of relevant parties in the water sector. This multifaceted approach is essential for fostering sustainable water security and ensuring the prudent management of this critical natural resource in South Africa.

2 Issues and Drivers in Water Security

Equitable access to water is a basic human right, yet marginalized and rural communities often do not have access to clean water or related technologies. With climate models predicting that South Africa will become drier, increasing water shortages will be a growing trend over the coming decades. Water will become more expensive. During the SAForSTI 2030 exercise, the following issues and drivers were considered to be significant for water security in South Africa.

2.1 People Using at Least Basic Drinking Water Services (% of Population)

The percentage of people using at least basic water services. This indicator encompasses both people using basic water services and those using safely managed water services. Basic drinking water service is defined as drinking water from an improved source, provided collection time is not more than 30 min for a round trip. Improved water sources include piped water, boreholes or tubewells, protected dug wells, protected springs, and packaged or delivered water.

Data on drinking water, sanitation, and hygiene are produced by the Joint Monitoring Programme of the World Health Organization (WHO) and United Nations Children's Fund (UNICEF) based on administrative sources, national censuses, and nationally representative household surveys. WHO/UNICEF defines a basic drinking water service as drinking water from an improved source, provided collection time is not more than 30 min for a round trip. Improved water sources include piped water, boreholes or tubewells, protected dug wells, protected springs, and packaged or delivered water.

Figure 1 illustrates that the number of people accessing clean water services in South Africa is increasing on a continuous basis in the last couple of decades. There is still approximately 10 percent of the population lacking access to clean water services. According to Fig. 2, South Africa is leading in terms of providing clean water services to its population among the sub-Saharan African countries. However, the country is still behind the other benchmark countries.

2.2 People Using at Least Basic Drinking Water Services, Rural (% of Rural Population)

The percentage of people using at least basic water services. This indicator encompasses both people using basic water services and those using safely managed water services. Basic drinking water service is defined as drinking water from an improved source, provided collection time is not more than 30 min for a round trip. Improved

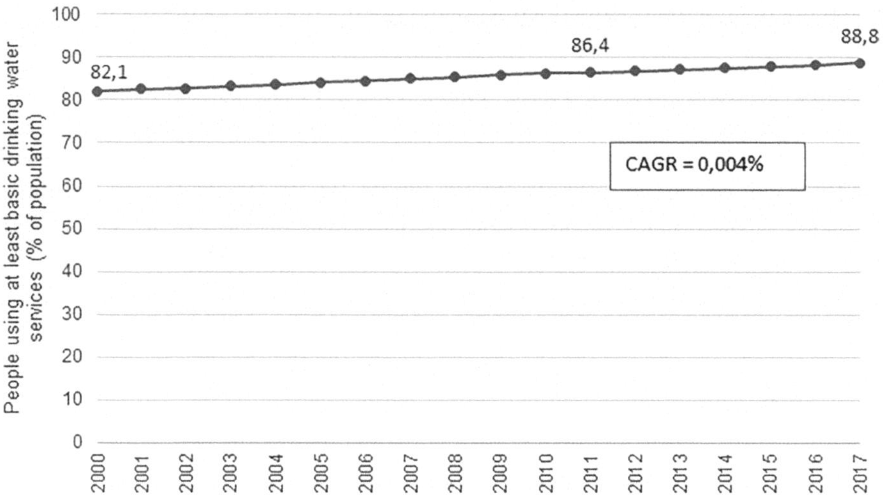

Fig. 1 People using at least basic drinking water services, South Africa

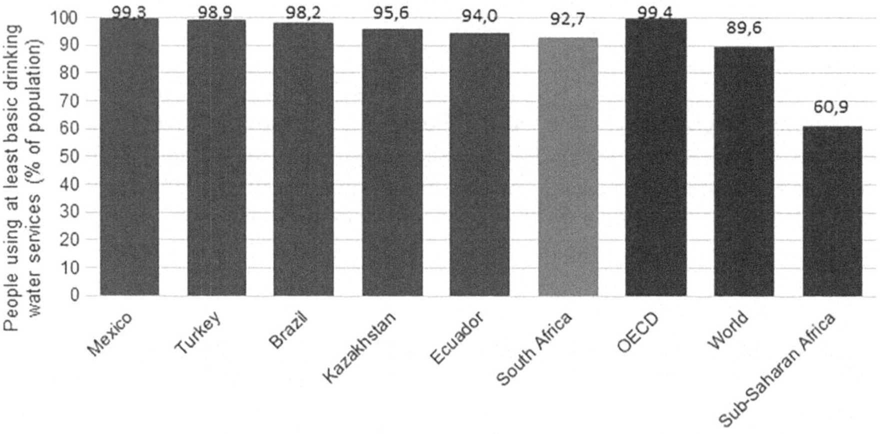

Fig. 2 People using at least basic drinking water services, 2017. Source: WHO/UNICEF Joint Monitoring Programme (JMP) for Water Supply, Sanitation and Hygiene (washdata.org)

water sources include piped water, boreholes or tubewells, protected dug wells, protected springs, and packaged or delivered water.

Data on drinking water, sanitation, and hygiene are produced by the Joint Monitoring Programme of the World Health Organization (WHO) and United Nations Children's Fund (UNICEF) based on administrative sources, national censuses, and nationally representative household surveys. WHO/UNICEF defines a basic drinking water service as drinking water from an improved source, provided collection time is not more than 30 min for a round trip. Improved water sources include piped water, boreholes or tubewells, protected dug wells, protected springs, and packaged or delivered water.

Access to water services in rural areas is relatively lower compared to urban areas. In South Africa, approximately 80% of the rural population has access to clean water services (Fig. 3). This figure is again much higher compared to the other sub-Saharan African countries but is still behind the other benchmark countries (Fig. 4).

2.3 People with Basic Handwashing Facilities Including Soap and Water (% of Population)

The percentage of people living in households that have a handwashing facility with soap and water available on the premises. Handwashing facilities may be fixed or mobile and include a sink with tap water, buckets with taps, tippy-taps, and jugs or basins designated for handwashing. Soap includes bar soap, liquid soap, powder detergent, and soapy water but does not include ash, soil, sand, or other handwashing agents.

Data on drinking water, sanitation, and hygiene are produced by the Joint Monitoring Programme of the World Health Organization (WHO) and United Nations Children's Fund (UNICEF) based on administrative sources, national censuses, and nationally representative household surveys. WHO/UNICEF defines a

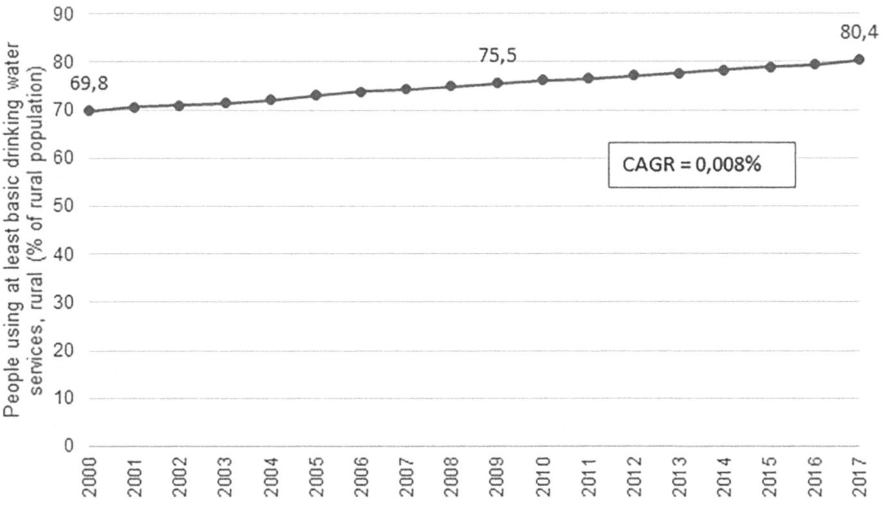

Fig. 3 People using at least basic drinking water services, rural areas of South Africa. Source: WHO/UNICEF Joint Monitoring Programme (JMP) for Water Supply, Sanitation and Hygiene (washdata.org)

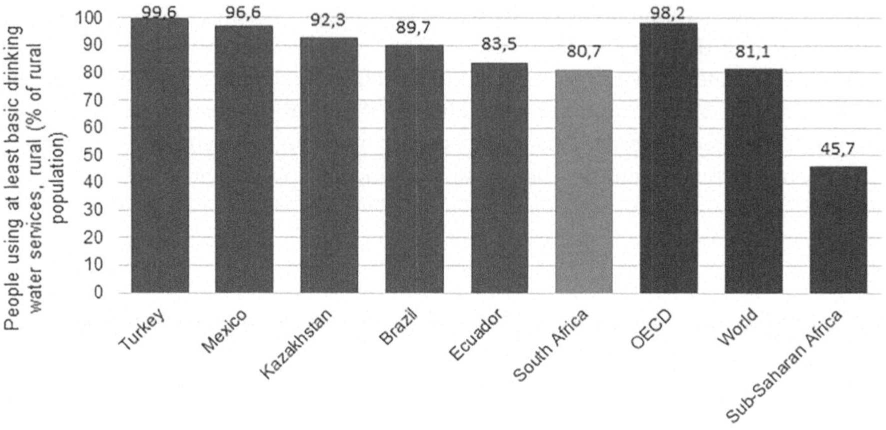

Fig. 4 People using at least basic drinking water services, rural, 2017

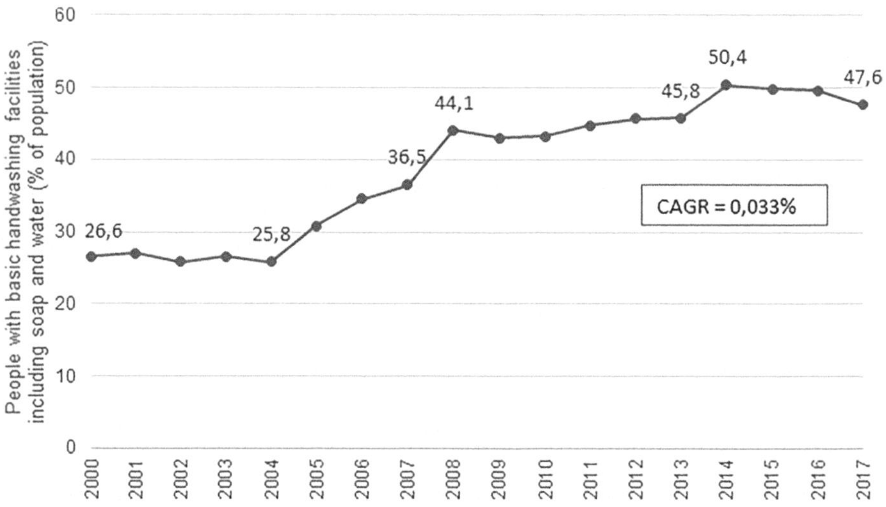

Fig. 5 People with basic handwashing facilities including soap and water, South Africa. Source: WHO/UNICEF Joint Monitoring Programme (JMP) for Water Supply, Sanitation and Hygiene (washdata.org)

basic handwashing facility as a device to contain, transport, or regulate the flow of water to facilitate handwashing with soap and water in the household.

The availability of basic handwashing functions in South African households is no more than 50% (Fig. 5). This creates serious sanitation issues, especially considering the ongoing COVID-19 pandemic, for which water and soap are essential not to spread the virus. With these figures South Africa is well behind other benchmark countries (Fig. 6). This is one of the areas where immediate action is required.

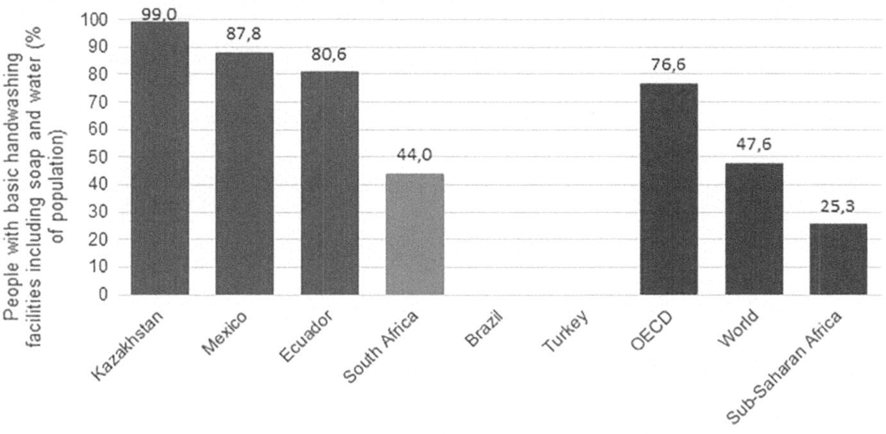

Fig. 6 People with basic handwashing facilities including soap and water, 2017

2.4 People with Basic Handwashing Facilities Including Soap and Water, Rural (% of Rural Population)

The percentage of people living in households that have a handwashing facility with soap and water available on the premises. Handwashing facilities may be fixed or mobile and include a sink with tap water, buckets with taps, tippy-taps, and jugs or basins designated for handwashing. Soap includes bar soap, liquid soap, powder detergent, and soapy water but does not include ash, soil, sand, or other handwashing agents.

Data on drinking water, sanitation, and hygiene are produced by the Joint Monitoring Programme of the World Health Organization (WHO) and United Nations Children's Fund (UNICEF) based on administrative sources, national censuses, and nationally representative household surveys. WHO/UNICEF defines a basic handwashing facility as a device to contain, transport, or regulate the flow of water to facilitate handwashing with soap and water in the household.

Considering the rural areas, the situation is more dramatic in terms of accessing basic hygiene facilities. Only around 40% of the rural population has access to soap and water (Fig. 7). These figures are well behind the other countries for comparison as well as the OECD and the world average (Fig. 8).

2.5 People with Basic Handwashing Facilities Including Soap and Water, Urban (% of Urban Population)

The percentage of people living in households that have a handwashing facility with soap and water available on the premises. Handwashing facilities may be fixed or mobile and include a sink with tap water, buckets with taps, tippy-taps, and jugs or

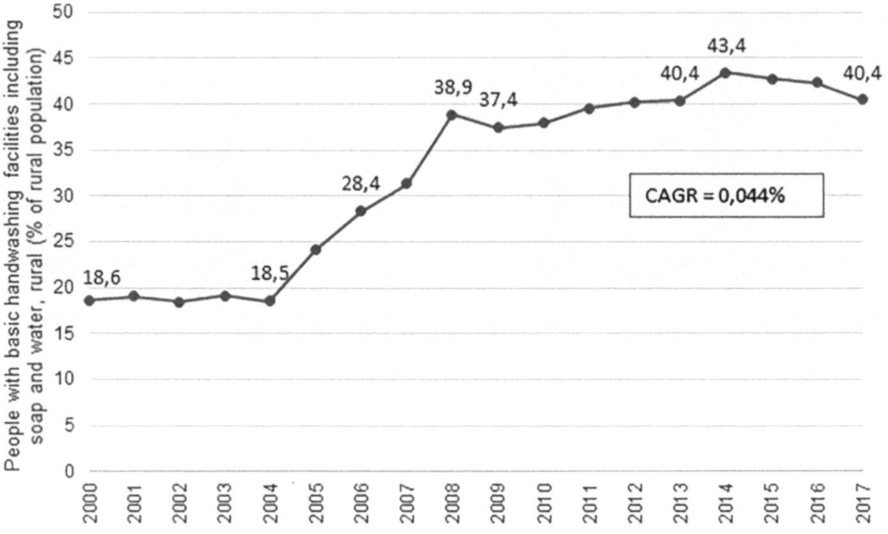

Fig. 7 People with basic handwashing facilities including soap and water, rural, South Africa. Source: WHO/UNICEF Joint Monitoring Programme (JMP) for Water Supply, Sanitation and Hygiene (washdata.org)

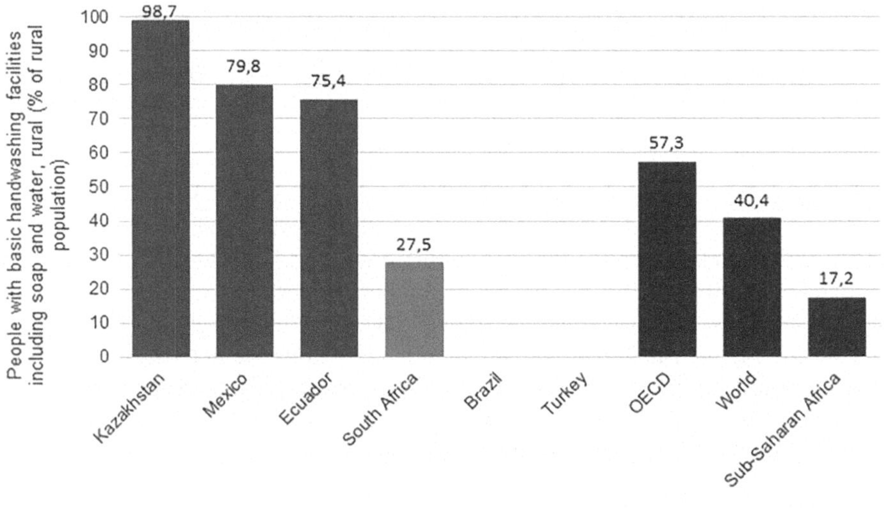

Fig. 8 People with basic handwashing facilities including soap and water (rural), 2017

basins designated for handwashing. Soap includes bar soap, liquid soap, powder detergent, and soapy water but does not include ash, soil, sand, or other handwashing agents.

Data on drinking water, sanitation, and hygiene are produced by the Joint Monitoring Programme of the World Health Organization (WHO) and United

Nations Children's Fund (UNICEF) based on administrative sources, national censuses, and nationally representative household surveys. WHO/UNICEF defines a basic handwashing facility as a device to contain, transport, or regulate the flow of water to facilitate handwashing with soap and water in the household.

Urban population of South Africa is relatively better positioned compared to the rural populations in terms of availability of basic handwashing facilities with 56.2% in 2017 (Fig. 9). However, this is still very low compared to other benchmark countries, and below the world average, which is also very low (Fig. 10).

2.6 *Normalized Indicator Values and Composite Index for the "Access and Supply of Clean Water" Factor*

Figure 11 illustrates that South Africa is relatively better positioned in terms of accessing basic drinking water and well behind in terms of the availability of basic handwashing facilities. The overall performance of the country in access and supply of clean water is just above average.

Aging and poorly maintained water infrastructure and the failure to invest in water management not only represent missed opportunities but may also impede economic growth and job creation. This is particularly relevant given that the largest portion of South Africa's energy supply relies on water for generation.

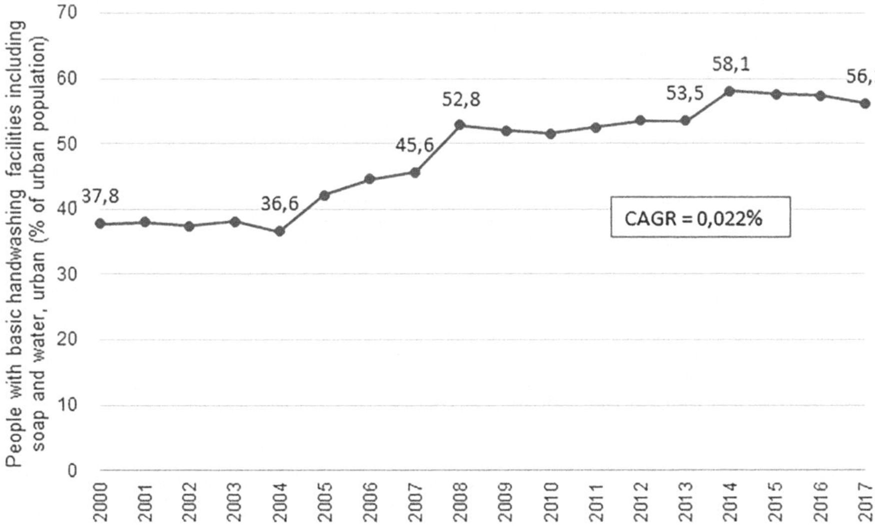

Fig. 9 People with basic handwashing facilities including soap and water (urban), South Africa. Source: WHO/UNICEF Joint Monitoring Programme (JMP) for Water Supply, Sanitation and Hygiene (washdata.org)

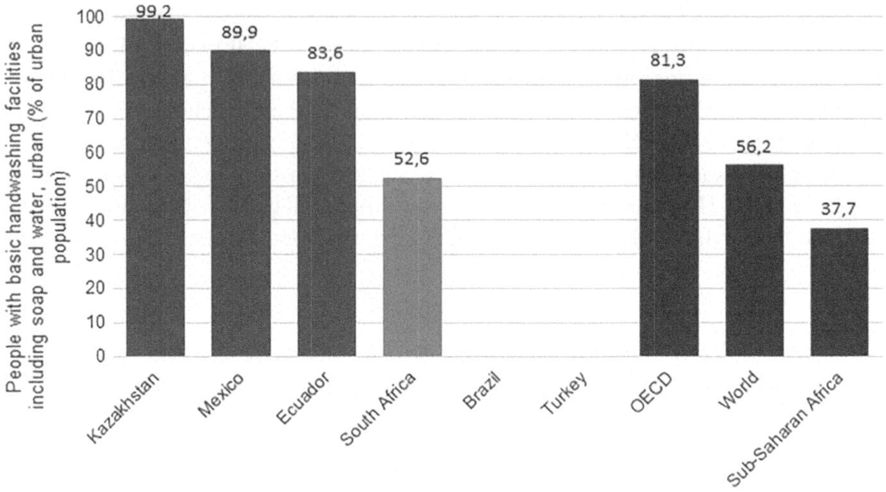

Fig. 10 People with basic handwashing facilities including soap and water (urban), 2017

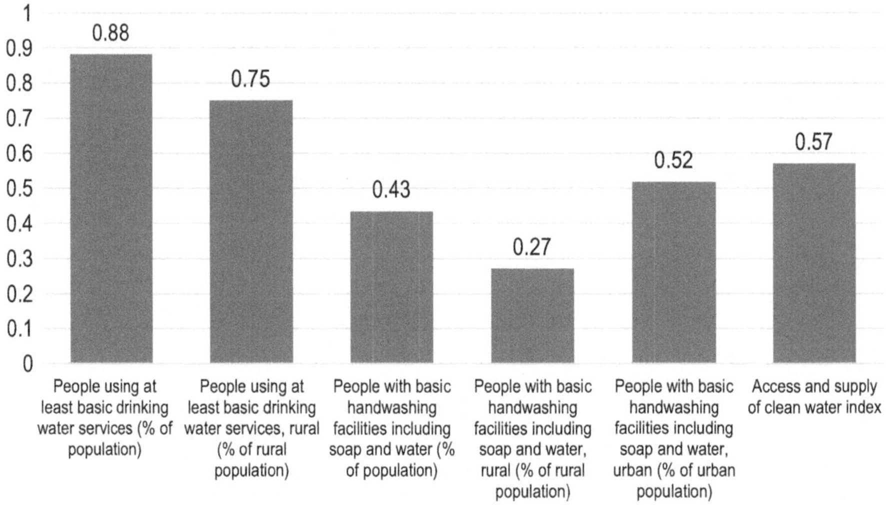

Fig. 11 Normalized indicator values and a composite index for the "access and supply of clean water" factor (0 = worst value among the analyzed countries; 1 = best value among the analyzed countries)

Proactive, innovative, interdisciplinary, environmentally sustainable solutions are needed, giving attention to, among other aspects, demand control, land management, wastewater treatment, local-level approaches to supply and management, the role of data science, and social change management to effect sustainable behavioral changes.

3 Research Capacity and Knowledge Generation on Water Security

Figure 12 illustrates South African researchers' contribution to the global knowledge generation concerning water security from 2008 to 2018. Notably, there has been a consistent increase in scientific focus within South Africa on topics pertaining to water security, especially evident from 2016 onward. As depicted in the figure, the number of publications has been gradually increasing, along with South Africa's proportionate share in the global scientific output in this field—ranking 23rd with 6600 publications (NACI, 2019).

Figure 13 illustrates publication activity in comparison to other countries from 2008 to 2018. The number of publications has been adjusted relative to both GDP and population. The BRICS nations are specifically highlighted. Notably, among the BRICS countries, South Africa has the highest number of publications pertaining to the water security domain, both in terms of GDP and population size normalization (NACI, 2019).

Figure 14 depicts the global ranking of South Africa.

South Africa boasts considerable specialized research capacity that yields top-notch outputs in fields like the environment, earth sciences, and agriculture. Additionally, notable research outputs of high quality are evident in disciplines such as biochemistry, energy, and chemistry, albeit with relatively lower degrees of specialization. Furthermore, significant specialized capacity exists in the social

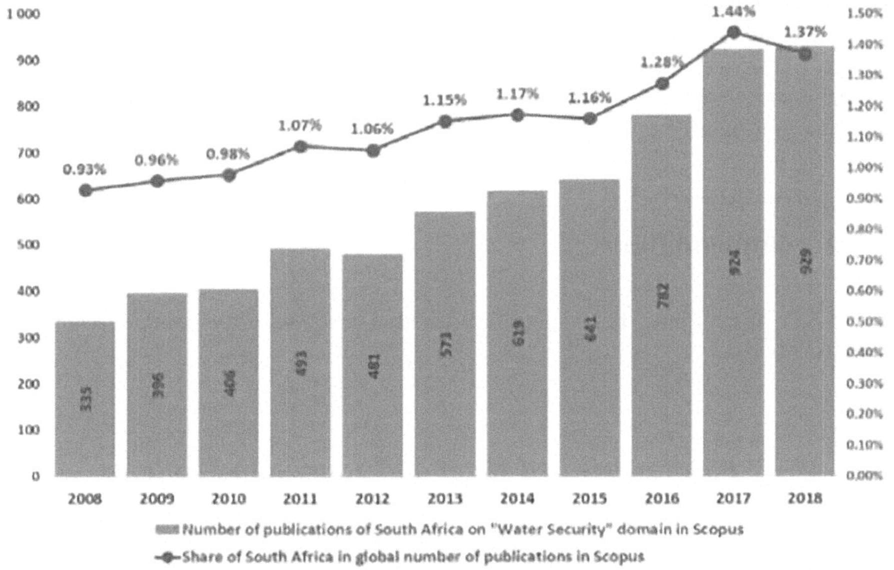

Fig. 12 Publication activity related to the water security domain

Country	Publ ('000)	GDP (USDbn)	Publ/GDP
Iran	15.6	454.0	34.4
Portugal	6.1	219.3	27.8
Malaysia	7.7	314.7	24.5
South Africa	6.6	348.9	18.9
Australia	24.9	1323.4	18.8
Spain	21.3	1314.3	16.2
Poland	8.4	526.5	16.0
Canada	23.7	1647.1	14.4
Netherlands	11.9	830.6	14.3
Sweden	7.0	535.6	13.1
India	33.2	2650.7	12.5
Switzerland	8.3	679.0	12.2
Turkey	9.3	851.5	10.9
UK	27.8	2637.9	10.5
China	120.4	12237.7	9.8
Italy	17.6	1943.8	9.1
South Korea	12.9	1530.8	8.4
Brazil	16.9	2053.6	8.2
France	20.7	2582.5	8.0
Germany	28.0	3693.2	7.6
US	120.5	19485.4	6.2
Mexico	6.4	1150.9	5.6
Russia	8.3	1578.4	5.3
Japan	17.3	4872.4	3.6

GDP (2017)

Country	Publ ('000)	Pop (m)	Publ/Pop
Australia	24,9	24,6	101,2
Switzerland	8,3	8,5	97,6
Netherlands	11,9	17,1	69,6
Sweden	7,0	10,1	69,3
Canada	23,7	36,7	64,6
Portugal	6,1	10,3	59,2
Spain	21,3	46,6	45,7
UK	27,8	66,0	42,1
US	120,5	325,1	37,1
Germany	28,0	82,7	33,9
France	20,7	67,1	30,8
Italy	17,6	60,5	29,1
South Korea	12,9	51,5	25,0
Malaysia	7,7	31,6	24,3
Poland	8,4	38,0	22,1
Iran	15,6	81,2	19,2
Japan	17,3	126,8	13,6
South Africa	6,6	56,7	11,6
Turkey	9,3	80,7	11,5
China	120,4	1386,4	8,7
Brazil	16,9	209,3	8,1
Russia	8,3	144,5	5,7
Mexico	6,4	129,2	5,0
India	33,2	1339,2	2,5

Population (2017)

Fig. 13 Number of publications relevant to the water security domain (normalized according to GDP and population size)

sciences and economics domains, albeit these findings typically do not get published in first-quartile journals (NACI, 2019).

4 Recommendations

Three STI thrusts have been identified for the water security domain. They are described below.

4.1 WA1: Future-Oriented Water and Sanitation Solutions

This thrust tackles the imperative of developing and implementing water and sanitation solutions that can adapt to evolving challenges and opportunities, aiming to bolster water security as a foundation for societal and economic prosperity. Water supply strategies will necessitate a holistic blend of context-specific water sources

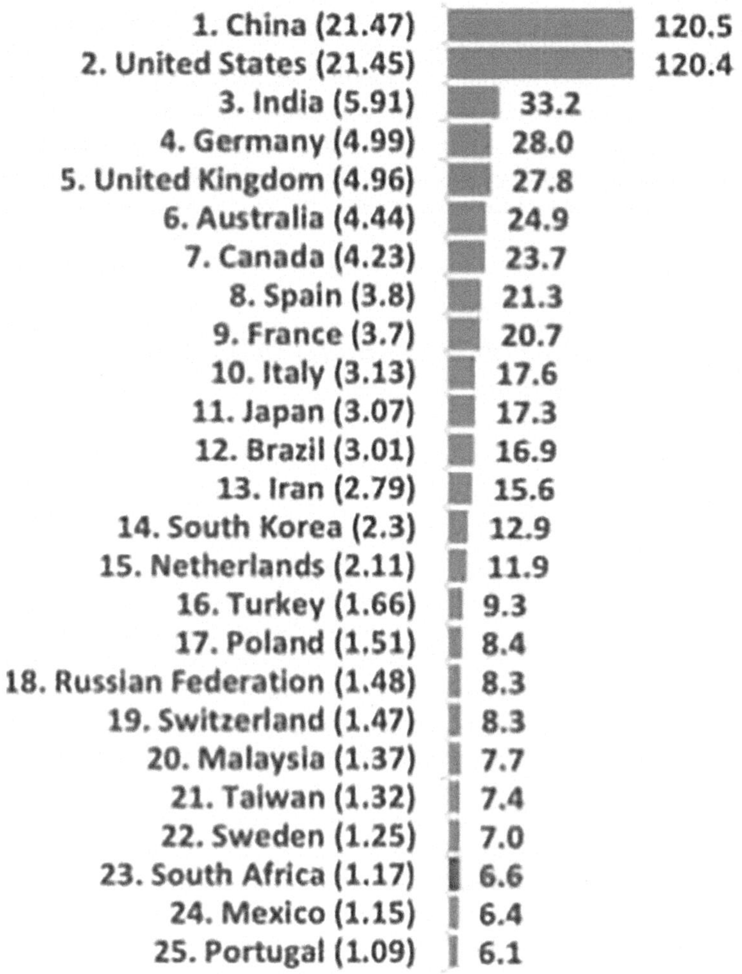

Fig. 14 South Africa's global ranking on water security publications

at various levels, encompassing bulk, regional, and local scales. Moreover, it advocates for the introduction, comprehension, and mainstreaming of innovative sanitation and wastewater solutions, including low- or zero-water usage toilets and technology that enhances energy and water efficiency. Incorporating water-sensitive designs into urban, peri-urban, and rural planning processes is essential, encompassing water and sanitation considerations from the outset (NACI, 2019).

The thrust aims to steer municipalities toward embracing a more diverse array of water sources, incorporating alternative options to enhance resilience, particularly during drought periods. Infrastructural planning for new urban developments will integrate wastewater and stormwater management into the overall water strategy. Embracing household and community-level water capture, storage, treatment, and monitoring technologies will be pivotal, fostering the emergence of a burgeoning water service sector. Furthermore, it underscores the mainstreaming of advanced sanitation and wastewater solutions, promoting alternatives like low- or zero-water toilets and smart waste management practices that reduce reliance on extensive potable water usage for waste disposal. Encouraging industries to transition toward a circular water reuse paradigm and investing in suitable technologies to enhance water independence and climate resilience are also key facets of this thrust (NACI, 2019).

4.2 WA2: Embedding the Water Sector in the Fourth Industrial Revolution

The Fourth Industrial Revolution (4IR) represents the fusion of physical, digital, and biological realms through technological advancements. Cyber-physical systems emerge as a central theme within these technologies, holding promise for transformative shifts in managing water and sanitation resources and services. Key areas deserving attention encompass (a) integrating suitable ICTs into water resource management and service delivery, (b) elevating the data-driven nature of decision-making processes, (c) leveraging cyber-physical systems to encourage appropriate behaviors, (d) fostering innovation in water sector monitoring and data aggregation, and (e) enhancing skills to navigate the data-intensive landscape of the water sector. Establishing trust and cultivating relationships will be essential for implementing these changes effectively (NACI, 2019).

This thrust aims to yield ICT innovations (such as diagnostics, robotics, artificial intelligence, and distributed sensor networks) interconnected with water resources management and service delivery through robust smart networks, facilitating timely preventive maintenance. Water sector data, augmented by big data analytics and fortified by robust cybersecurity measures, will become more accessible, up-to-date, and informative for decision-making processes. Hydrological monitoring will be revolutionized through the adoption of smart, cost-effective technologies. Additionally, there's a need to reconsider revitalizing South Africa's hydrological monitoring capabilities as a central hub (NACI, 2019).

Exploring partnerships within the Southern African Development Community (SADC) and across Africa for regional solution deployment and collaboration is paramount. Furthermore, there is merit in evaluating a heightened role for the private sector in bolstering service delivery, potentially bolstered by increased availability of venture capital. Such endeavors will prompt municipalities to reconsider their service delivery approaches and responsibilities, potentially challenging existing business models of organizations holding current monopolies over certain datasets (NACI, 2019).

4.3 WA3: Off-Grid and Decentralized Water, Wastewater, and Sanitation Solutions

One pivotal shift that could propel the water sector toward the opportunities presented by a circular economy involves transitioning to off-grid and decentralized water, wastewater, and sanitation solutions. These solutions hold potential for deployment across various scales and markets: (a) a public market catering to underserved areas and municipalities open to exploring alternatives to conventional large-scale infrastructure planning and (b) a private market supporting solutions tailored for individual households, privately owned housing estates, private companies, and industrial stakeholders with specific water-quality requirements and impacts. This strategic focus underpins the water security thrust (NACI, 2019).

This thrust is geared toward furnishing a range of off-grid and/or decentralized solutions tailored to meet the demands of both private and public markets (including solutions for underserved areas and low-cost options). This approach fosters more direct and circular reuse of water and wastewater on-site. Exploration of innovative business models, funding mechanisms, and partnership structures will be integral to the implementation of off-grid and/or decentralized solutions, fostering the emergence of a robust ecosystem of businesses and associated manufacturing capabilities equipped to handle the marketing and operation of these solutions. This approach also facilitates the establishment of quicker nexus synergies with food and agricultural issues through the deployment of more mobile, decentralized, and rapidly deployable water solutions. There is potential to scale up the adoption of this approach across larger swathes of South Africa, underpinned by a suite of policy and regulatory incentives (NACI, 2019).

This thrust presents opportunities for energy generation from off-grid water systems, thereby broadening the spectrum of options for independent power production. It also catalyzes the creation of new high-value products, bolstering the chemical industry in South Africa and contributing to a low-carbon, circular economy. Critical to the success of this endeavor is funding for research and development (R&D) and early-stage testing to cultivate a diverse pipeline of solutions. This necessitates improved coordination of existing R&D investments within the water research, development, and innovation (RDI) system, alongside the establishment of a well-resourced new R&D program focused on water and sanitation. Additionally, careful analysis of water sector financing and pricing mechanisms will be essential to attract more investment into the sector and foster greater public-private partnerships (NACI, 2019).

5 Conclusions and Key Policy Considerations

South Africa has experienced consistent growth in scientific research outputs within the water security domain, particularly since 2016. Ranking 23rd globally with 6600 publications, South Africa stands out for producing the highest number of publications per GDP and per capita. Despite increasing access to technology, data, and associated decision support, the water sector has not fully capitalized on these resources. This is evident in inadequately maintained hydrological monitoring systems, difficulties in data storage and sharing, reliance on outdated data for decision-making, and challenges in managing complexity as urbanization progresses and resource integration becomes more critical (NACI, 2019).

Some specific policy considerations are as follows:

- "Synergize and harmonize the National Water RDI Roadmap, the National Water and Sanitation Master Plan and the water components of the Industrial Policy Action Plan.
- Ensure improved coordination between all institutions engaged in water research, innovation and solution deployment.
- Support the establishment of co-learning platforms to bring together water researchers, planners, funders, implementers and users.
- Develop a framework that states clearly which parties are involved in or relevant to the water sector and what exactly their roles are" (NACI, 2019, p. 127).

Acknowledgments The chapter was prepared within the framework of the Basic Research Program at the HSE University.

Reference

NACI. (2019). *South Africa foresight exercise for science, technology and innovation 2030*, Synthesis Report, National Advisory Council on Innovation, Department of Science and Innovation, Republic of South Africa (Available at: https://www.naci.org.za/wp-content/uploads/2020/07/South-African-Foresight-Exercise-For-Science-Technology-and-Innovation-2019.pdf, last visited on: June 3, 2024).

Part IV
Policy Implications

The Future of Foresight Exercises for Science, Technology, and Innovation in Africa

Mlungisi Cele

Abstract This chapter delves into the insights gained from the South Africa Foresight Exercises for Science, Technology and Innovation (SAForSTI) 2030 to guide future implementations of science, technology, and innovation (STI) foresight exercises across Africa. It underscores the pivotal role of STI in addressing societal challenges and fostering inclusive and sustainable socio-economic development. The chapter is structured into three sections, first situating recommendations in global, continental, and domestic contexts, followed by drawing lessons from SAForSTI, and finally, presenting considerations for future foresight exercises in South Africa and the broader African continent. Drawing lessons from SAForSTI, the chapter underscores the immediate impact of the foresight exercise on the STI Decadal Plan, demonstrating the value of foresight in shaping national priorities. Foresight, characterised as a systematic, participatory, and future-intelligence-gathering process, empowers decision-makers to actively influence the unfolding future and construct narratives tailored to the context of developing countries.

1 Introduction

This chapter seeks to draw lessons from the South Africa Foresight Exercises for Science, Technology and Innovation 2030 (SAForSTI) to inform the future implementation of science, technology, and innovation foresight exercises in Africa. The chapter reinforces the view that STI can play a critical role in addressing societal challenges and contributing towards inclusive and sustainable socio-economic development.

Therefore, the chapter consists of three sections:

1. Situating the recommendations within the global, continental and domestic context

M. Cele (✉)
National Advisory Council on Innovation, Pretoria, South Africa
e-mail: mlungisi.cele@dst.gov.za

O. Saritas et al. (eds.), *21st Century Foresight*, Science, Technology and Innovation Studies, https://doi.org/10.1007/978-3-031-69557-5_17

2. Drawing on lessons learned from SAForSTI
3. Considerations for future implementation of foresight exercises in South Africa and the rest of the African continent

2 Situating the Recommendations Within the Global, Regional, and Continental Context

2.1 Global Context

The world is changing rapidly and fundamentally. The drivers of global change are socio-economic (e.g. demographic shifts, urbanisation, rising inequality, and youth unemployment), geopolitical (e.g. the rise of China and India as economic powers), scientific and technological (e.g. blurring of lines between the physical and digital spheres through use of information and communication technologies and the fourth industrial revolution), and environmental (e.g. climate change and its severe consequences for the world, particularly in most vulnerable people). These global megatrends and drivers render even the near future uncertain (Department of Science and Innovation [DSI], 2019) and worsen the global principal contradiction that is characterised by unbalance, inadequate, and uneven development and growing demands for a better life for all. People feel excluded, unequal, or less important than others and that they have no future—in part because their children "do not have equal opportunities to reach their full potential" (Gurria, 2018). Too many people live precarious lives, at the edge of falling through the cracks of unemployment and poverty and at the "mercy of big impersonal forces—globalisation, technological change, large corporations and financial institutions" (Gurria, 2018).

The lingering effects of the global Covid-19 pandemic are deepening the economic, ecological, and social reproduction crises. The pandemic continues to affect progress towards the achievement of the UN Sustainable Development Goals (SDGs). Difficulties in transforming energy systems and meeting globally agreed upon carbon reduction targets mean that temperatures are likely to increase more than 2 °C above pre-industrial levels. Developing countries and emerging economies will be most exposed to water shortages, food insecurity, premature de-industrialisation, and failures in health and social care systems. These challenges make the need for robust and resilient STI capacities and capabilities even more critical. "In an interlinked but inequitable world, domestic systems of innovation are crucial for transforming science and technology into socially useful products and practices", said the Deputy Minister of Higher Education, Science and Innovation at the launch of the 2021 STI Indicators report.[1]

[1] Speech at the launch of the STI Indicators report. Available at https://www.gov.za/speeches/deputy-minister-buti-manamela-2021-science-technology-and-innovation-indicators-webinar-26

Many of these global societal challenges can be addressed through international cooperation. The Covid-19 pandemic crisis underscored the importance of international cooperation in STI. Cross-border cooperation among researchers is needed to accelerate research and avoid duplication of efforts. Cooperation in technology is needed between public research and business enterprises to accelerate the development of diagnostics, therapeutics, and vaccines (Organisation for Economic Co-operation and Development [OECD], 2021).

OECD standards such as the 2021 Recommendation on International Co-operation in Science and Technology and the 1995 Recommendation of the Council for Facilitating International Technology Co-operation with and among businesses can play an important role in ensuring such scientific cooperation is efficient and effective by removing barriers to international cooperation and promoting cooperation that is mutually beneficial to partners and the global scientific community through the development of international guidelines (OECD, 2021).

The 2021 recommendation covers an expanded scope of principles for international cooperation in science and technology to reiterate some elements of the existing principles in the 1988 Recommendation of the Council Concerning a General Framework of Principles for International Co-operation in Science and Technology. These principles reinforce the importance of basic research as both a goal and an enabler of international cooperation and the exchange and mobility of research personnel. It also aims to articulate some of the implicit actions that would facilitate international cooperation in science and technology. This includes data sharing, publication of results and transparency, promoting access to international research infrastructures, and removing gender-specific barriers to mobility (e.g. unequal pay and social benefits).

The mobility of scientists, in particular, is a continuing challenge. Although mobility benefits researchers and scientific communities through the circulation of ideas and talent, it also generates some risks, notably that of long-term brain drain, even among OECD member countries and not only developing countries (OECD, 2021). International cooperation is a long-term endeavour, since it builds on consensus both at a scientific and societal level, within the bounds of national legal and regulatory systems. In times of peace and stability, timing is not a problem. But it can be a challenge at times of crises as, in many cases, the mechanisms for international scientific cooperation do not avail themselves to rapid responses in terms of redirecting research funds, redeployment and mobility of staff, or the sharing of research data, especially data generated at the local data level and/or by citizens (OECD, 2021).

One of the findings from the OECD Committee for Scientific and Technological Policy (CSTP) project on international cooperation in science and technology has been that the mechanisms for funding collaboration that were developed in the second half of the twentieth century have largely been focused on advanced scientific cooperation for scientific aims, rather than on funding cooperation on global challenges. That is not to say that the subjects of societal or grand challenges have been entirely absent from international research cooperation. Rather, global challenges have been viewed through the lens of scientific challenges and national challenges or priorities (e.g. national security or industrial policy). Today, there is a clear understanding by OECD member countries that some global challenges can only be addressed through international STI cooperation (OECD, 2021).

Long-standing trends and recent disruptions have created a new operating environment for STI policy. The three most recent and salient disruptions—climate change, the Covid-19 pandemic and growing geopolitical tensions—have had far-reaching, cascading effects, including they have seen concepts such as "technology sovereignty" and "open strategic autonomy" strongly emerge as frames for STI policy. Coupled with a growing emphasis on "shared values" in technology development and research, these developments could lead to a "decoupling" of STI activities at a time when global challenges, notably climate change, require global solutions underpinned by international STI cooperation. A challenge for multilateralism will be to "square the circle" of growing strategic competition and the need for growing international cooperation to address global problems (OECD 2023).

2.2 *Regional and Continental Context*

Following centuries of colonialism, imperialism, apartheid, conflict, and underdevelopment, there is a beacon of hope in Africa, given the emerging recognition of the role of STI in the continent's socio-economic development. The African Union (AU) articulated an inspirational, futuristic, and ambitious Agenda 2063, which serves as a master plan for transforming Africa into the global powerhouse of the future. Agenda 2063 is a concrete manifestation of the drive for unity, self-determination, freedom, progress, and collective prosperity pursued under pan-Africanism and the African renaissance. Agenda 2063 calls for the diversification of sources of growth and sustenance of Africa's current economic performance and, in the long run, lifting large sections of our population out of poverty (2015). The strategy further fosters social transformation and economic competitiveness through human transmission to an innovation-led, knowledge-based economy. It fosters social transformation and economic competitiveness through human resource development, innovation, value addition, industrialisation, and entrepreneurship. It also identifies strategic partnerships and collaboration at the bilateral and multilateral levels as essential for jointly solving global challenges (AU, 2015). Inasmuch as the tool is efficient, its benefits are only realisable in the right environment, particularly with the right implementation (AU, 2015).

In June 2014, the 23rd Ordinary Summit of the AU adopted the Science, Technology and Innovation Strategy for Africa 2024 (STISA-2024). The strategy is part of Agenda 2063, which is underpinned by STI as multifunctional tools and enablers for achieving continental development goals.

STISA-2024 places STI at the centre of Africa's socio-economic development and growth. Its mission is to "accelerate Africa's transition to an innovation-led, knowledge-based economy. It aims to achieve this by: (a) improving STI readiness in Africa in terms of infrastructure, professional and technical competence, and entrepreneurial capacity; and (b) implementing specific policies and programmes in STI that address societal needs in a holistic and sustainable way" (AU, 2014).

STISA-2024 will contribute to the achievements of the AU, due to the cross-cutting nature of STI. It is designed to meet the knowledge, technology, and innovation demands of various AU economic and social sector development frameworks (AU, 2014). STISA-2024 has a leading role to play in increasing efficiency and eliminating duplication of effort in the design and implementation of national, regional, and all policies on STI (AU, 2014).

2.3 *Domestic Context*

South Africa's reintroduction into the global community after the end of apartheid held significant implications for locating its national system of innovation (NSI) vis-à-vis other national systems. That was manifest in two areas. Firstly, the country's economic relations with industrialised economies in the West strengthened considerably with the end of sanctions. Secondly, and probably more significantly, there was the opening up of South-South political and economic relations that had hitherto been proscribed during the apartheid era. The two main fronts of those new sets of South-South relationships were—with respect to the continent especially sub-Saharan Africa—South Africa's relations with Brazil, Russia, India, and China (and later its membership of BRICS) and partnerships with other emerging economies such as Malaysia, Argentina, Chile, and Iran. Those two fronts are strongly related in the sense that the placement of the South African system of innovation within BRICS and by proxy at the global level is strongly dependent on its role in the process of economic integration in sub-Saharan Africa and more specifically within the southern African region (Scerri, 2016).

Aligned to the South African National Development Plan, the 2019 White Paper for STI sets the long-term policy direction for the South African government to ensure a growing role for STI in a more prosperous and inclusive society—accelerating inclusive and sustainable socio-economic development and enhancing competitiveness, quality of life, and well-being. The white paper and draft strategic framework for STI diplomacy in Africa highlight the importance of international cooperation in ensuring that STI addresses regional and continental societal challenges and driving the developmental goals. The white paper states that Southern African Development Community (SADC) and AU strategies highlight the potential role of STI in achieving SADC goals and the broader objectives of pan-Africanism. South Africa's linkages with the rest of Africa present opportunities to build and consolidate national, regional, and continental systems of innovation. The country's cooperation strategies in Africa should therefore prioritise efforts to strengthen its own partner countries' STI systems. The aim should be to promote and facilitate cross-border research networks, shared technology innovation platforms, mutual learning, an integrated African STI agenda that encourages development and competitiveness, as well as funding partnerships (2019:20).

The white paper notes that South Africa's strategic science diplomacy focus has already recognised STI as an important means of driving the developmental agendas of the AU and SADC. For instance, major mineral beneficiation projects, as highlighted in the SADC Industrialisation Strategy and Roadmap (2015–2063), will depend on cross-border cooperation in energy and transport infrastructure and input supply and on the development of globally competitive industry sectors (2019).

The DSI's draft South Africa's Science, Technology and Innovation Diplomacy in Africa 2020–2030 (draft framework) argues that the geographical proximity and cultural, political, social, and economic linkages that South Africa shares with the rest of Africa are an advantage to consolidate South Africa's position as the research, development, and innovation (RDI) gateway into Africa. South Africa possesses mostly unparalleled scientific, technological, and innovative capability in select areas; hence opportunities to pursue high-return collaboration with continental and extra-continental partners remain high (DSI, 2021a, b).

The goal of the draft framework is to integrate South Africa's STI system into the African socio-economic system and foster competitive knowledge-based companies from the country that will contribute to capable STI states in Africa. Specifically, this will be done by pursuing the following outcome-oriented goals:

- Capable knowledge production and management.
- Capability in technology development and innovation management.
- Knowledge and innovation utilisation for sustainable development.
- Technology innovation and utilisation for industrial competitiveness.
- Enhanced RDI commercialisation and intra-Africa trade in medium high-technology products.

The primary purpose of these goals is to ensure strategic alignment and integration across the NSI to increase synergies and economies of scale through cross-border collaboration. Furthermore, these goals will ensure that South Africa's engagements in Africa are aligned with the AU and SADC's STI agenda and that of bilateral partners, to better leverage synergies and avoid duplication. To that end, the goals in Agenda 2063, STISA-2024, SADC Regional Indicative Strategic Development Plan 2020–2030, the SADC Industrialisation Strategy and Roadmap (2015–2063), and globally the SDGs will also serve as important multilateral policy frameworks guiding and complementing South Africa's STI cooperation in Africa (DSI, 2021a, b).

The draft framework will also focus on supporting cross-border initiatives to strengthen indigenous knowledge (IK) system policies, including strengthening the recording, protection, use, and commercialisation of IK, to the benefit of the knowledge holders and the country. To that end, a programme on IK biotechnology innovation will also be established to serve as a basis for partnership with African countries in order to exploit Africa's IK competitiveness and accelerate the IK bio-innovation industry in Africa.

3 Drawing on the Lessons Learned from SAForSTI

It is significant that SAForSTI found immediate traction and influence over the Decadal Plan for STI. From the beginning, the National Advisory Council on Innovation (NACI) worked to persuade the political and senior officials of the DSI to embrace foresight exercises to identify the new priorities for the Decadal Plan for STI. The DSI agreed to commence the development of the Decadal Plan after the completion of the SAForSTI. DSI officials were involved in various stages of the SAForSTI, including training that took place before the commencement of SAForSTI. The intention was to equip as many relevant people as possible with understanding of the foresight exercise process itself, and its purpose, methodologies, tools, and value.

The director-general of the DSI led and guided the process of analysing and determining the utility of the SAForSTI results. The DSI embraced SAForSTI domains as the basis of the STI Decadal Plan priorities (see Fig. 1). To deepen the understanding of the potential impact of SAForSTI domains on socio-economic development, the DSI, NACI, and HSE embarked on follow-up study. A high-level panel of experts participated in workshops and offered independent advice on the recommendations. One of the recommendations was for the development of a

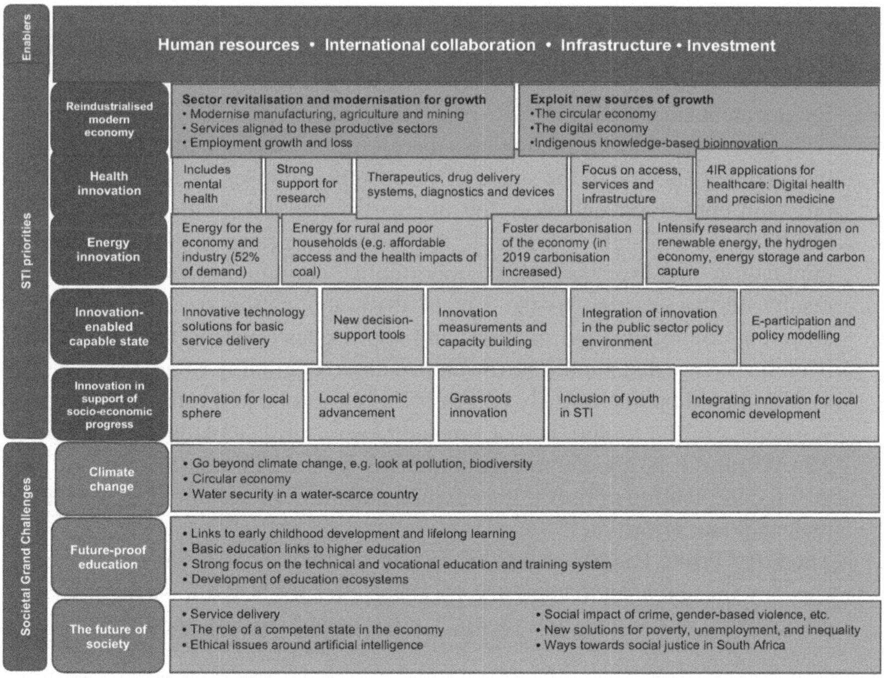

Fig. 1 Societal grand challenges, STI priorities and enablers

vaccine manufacturing capability that has since commenced. If successful, this initiative will benefit South Africa as well as the rest of the continent.

The DSI was heavily invested and engaged in the foresight exercises, which enabled the uptake of SAForSTI results into the Decadal Plan. The involvement of the high-level panel of experts improved the quality and robustness of the outcomes. It also gave policymakers the necessary confidence to use SAForSTI in setting priorities. The collaboration of local and international experts gave SAForSTI an added advantage, while the use of big data gave SAForSTI a unique feature.

4 Considerations for Future Implementation of Foresight Exercises in South Africa and the Rest of the Continent

3.1 SAForSTI highlighted the importance of context in informing and framing the object of the foresight exercise process. South Africa's context includes the societal challenges that the country faces, namely, inequality, poverty, unemployment, food security, health, education, water security, climate change, etc.

3.2 As seen at several stages of the SAForSTI project, engagement with policymakers in the foresight exercises is essential. The high-level support and participation of decision-makers should be secured early in the process. This will require continuous engagement and communication between the foresight technical team and policymakers (at director-general level). Such engagement gives decision-makers the opportunity to identify the potential use or uptake of emerging results or areas of learning where necessary.

3.3 Engagement should also include continuous communication and engagement with other important stakeholders; therefore a communication and engagement strategy is vital to the success of foresight exercises. Better and continuous communication enhances the interest of the stakeholders in using the results of the foresight exercise. It also reduces the potential for unnecessary duplication and overlap. Most importantly, communication connects the results of the exercise to the site of their application or implementation. Different communication methods and approaches ought to form part of the overall strategy that should be developed at the beginning of an exercise. The results should be made easily accessible to the public and other stakeholders. Awareness of the importance of foresight exercises as a planning tool should also be raised. A good initiative under the SAForSTI was a webpage[2] that enabled the sharing of useful information and allowed stakeholders to vote for the candidate priorities.

3.4 The 2019 White Paper assigns this responsibility to NACI. Given the limited capacity that exists in the country, it is necessary for NACI to develop its core competency in the area and recruit human resources with the skills that are

[2] https://www.naci.org.za/STIForesight2018/index.php/i-want-to-participate

needed or may be needed in the future. NACI should identify potential partners domestically and internationally. South Africa should prioritise establishing a foresight exercise hub consisting of NACI as an anchor and other institutions (including universities) as partners. In this regard, the Co-operation Framework on Innovation Systems between *Finland* and South Africa, launched in 20,026, *can serve as a reference.*

3.5 There should be a targeted capacity-building programme for the effective implementation of foresight exercises. Such a programme ought to prioritise participants with zero or little experience of foresight exercises. However, the programme should also aim to attract the experienced senior policymakers who are likely to play a key role in the uptake of the foresight exercise results. The programme should provide the necessary background and technical and analytical material and tools, as well as clarifying expectations and roles.

3.6 There should be adequate resourcing for the implementation of foresight exercises. The wider the scope of an exercise, the more resources will be required for its effective implementation. The SAForSTI exercise demonstrated the value of accessing and collaborating with experts wherever they are. However, collaborations require funding, and, in the future, such funding should be incorporated into the project plan from the beginning. Funds also must be set aside to ensure fair and equitable participation, for example, covering travel, subsistence, and/or accommodation costs for lower-income participants. The use of virtual communication tools could help to reduce the cost of face-to-face workshops, while achieving the same objectives. However, we need to understand more about the impact of virtual workshops and about the establishment of long-term networks, which are often a secondary objective of foresight exercises.

3.7 There should be adequate infrastructure and data, including big data, to support the foresight exercises. The deployment of big data analytics played a crucial role in the SAForSTI 2030 exercise. It is important to develop local big data capabilities, including human resources. Existing STI data sources (for scientometrics) need to be mobilised to provide technical and analytical support for the foresight exercise process.

3.8 Partnerships and networks should be formed. The SAForSTI exercise benefited from domestic and international partnerships, which added value. The public and private sectors could establish joint or complementary foresight exercise processes. It is also possible to use bilateral or multilateral agreements and partnerships for cross-border or multiple country foresight exercises. Such collaborative efforts could promote access to expertise and resources that would otherwise be impossible. Therefore, foresight exercises should be a part of international relations and cooperation.

3.9 Monitoring and evaluation should be an integral part of the foresight exercise process. It is important to subject various elements of the foresight exercise to a rigorous monitoring and evaluation process, which should be developed at the same time as the foresight exercise process is designed. The intention would be to learn the necessary lessons, with the aim of informing and strengthening the implementation and future design and development of foresight exercises.

3.10 Learning should be a permanent feature of the foresight exercise process. As the world and the nature of societal problems changes, so should the tools and methodologies of foresight exercises. This type of initiative could be supported through the South African Research Chairs Initiative or other support initiatives. The focus should be on foresighting and other STI future planning tools (e.g. technology road mapping, scenario planning, and vision creation).

Foresight has been defined as "a systematic, participatory, future-intelligence-gathering and medium to long-term vision-building process aimed at enabling present-day decisions and mobilising joint action" (Miles & Keenan, 2003). So, typically, foresight exercises should consider multiple futures, which may include possible, probable, plausible, and desirable or preferred futures. The process should be participatory and action-oriented (Miles et al., 2016). Foresight is a process that seeks to create shared "visions" of the future; appraisals that stakeholders are willing to endorse by the actions they choose to take today. Foresight empowers decision-makers and policy planners to use new ways of thinking and talking about the unfolding future and implementing strategic plans that are compatible with such a future. The premise of foresight is that the future is still in the making and can be actively influenced or shaped, rather than already decided by past thinking and actions and to be merely awaited and accepted.

Foresight allows the governments of developing countries to construct their own narratives of a desired future, instead of relying on futures foreseen by highly developed countries, which are not necessarily relevant to the contexts of developing countries. Foresight enables public service organisations to frame future policy environments better and present decision-makers with more and better choices for inclusive growth and social justice. Participatory foresight frees countries from over-reliance on foreign or local technical experts and invites citizens to contribute to discussions and decisions concerning their future.

References

AU. (2014). *AU science, technology and innovation strategy for Africa 2024.*

AU. (2015). *Agenda 2063 framework document.*

DSI. (2019). *White paper on science, technology and innovation.* Department of Science and Innovation.

DSI. (2021a). *Strategic framework for South Africa's science technology and innovation diplomacy in Africa (2020–2030).* Unpublished.

DSI. (2021b). *South Africa 2030 science technology and innovation foresight follow-up data analysis to support the choice of priorities and quantify the possible effects of the priorities for the STI 2023 decadal plan.* Pretoria: Department of Science and Innovation.

DSI. (2022). *Science, technology and innovation decadal plan 2022–2032.* Department of Science and Innovation.

Gurria, A. (2018). *Ten years on: Lessons from a crisis.* OECD Observer No 315 Q3 2018.

Manamela, B. K. (2021). *Speech at the launch of the STI Indicators report*. Available at https://www.gov.za/speeches/deputy-minister-buti-manamela-2021-science-technology-and-innovation-indicators-webinar-26

Miles, I., & Keenan, M. (2003). *Practical guide to regional foresight in the UK*. European Commission.

Miles, I., Saritas, O., & Sokolov, A. (2016). *Foresight for science, technology and innovation*. Springer Nature.

OECD. (1988). *Recommendation of the council concerning a general framework of principles for international co-operation in science and technology*. OECD Publishing.

OECD. (1995). *Recommendation of the council for facilitating international technology co-operation with and among businesses*. OECD Publishing.

OECD. (2021). *Recommendation of the council on international co-operation in science and technology*. OECD Publishing.

OECD Science Technology and Innovation Outlook (2023). Science technology and innovation Policy in Times of Global Crises. www.oecd-library.org//science-and-technology/oecd-sciencetechnology-and-innovation-outlook

SADC. (2015). *SADC industrialisation strategy and roadmap 2015–2063*.

SADC. (2020). *Regional indicative strategic development plan 2020–2030*.

Scerri, M. (2016). *The emergence of systems of innovation in South(ern) Africa: Long histories and contemporary debates*. Mapungubwe Institute for Strategic Reflection.

The manufacturer's authorised representative in the EU is Springer Nature Customer Service Centre GmbH, Europaplatz 3, 69115 Heidelberg, Germany. If you have any concerns regarding our products, please contact ProductSafety@springernature.com

Printed and bound by CPI Group (UK) Ltd, Croydon, CR0 4YY

01/10/2026

02225925-0019